# Contemporary Climatology

A. Henderson-Sellers
and P. J. Robinson

Longman
Scientific &
Technical

Copublished in the United States with
John Wiley & Sons, Inc., New York

**Longman Scientific & Technical,**
Longman Group UK Limited,
Longman House, Burnt Mill, Harlow,
Essex CM20 2JE, England
*and Associated Companies throughout the world.*

*Copublished in the United States with*
*John Wiley & Sons, Inc., 605 Third Avenue, New York 10158*

First published 1986
Reprinted 1987, 1989, 1991

**British Library Cataloguing in Publication Data**
Henderson-Sellers, Ann
    Contemporary climatology.
    1. Climatology
    I. Title    II. Robinson, P.J. (Peter John) *1944–*
    551.6      QC981

ISBN  0-582-30057-6

**Library of Congress Cataloging-in-Publication Data**
Henderson-Sellers, A.
    Contemporary climatology.

    Bibliography; p.
    Includes index.
    1. Climatology.      I. Robinson, P.J. (Peter John).
    1944–      II. Title.
    QC981.H48    1986      551.6      85–11293
    ISBN  0-470-20664-0 (USA only)

Produced by Longman Group (FE) Limited
Printed in Hong Kong.

*For*
Richard, Michael, Philip, Stephen, Nicholas and Laurence

# Contents

Preface   ix

Acknowledgements   x

List of Symbols   xv

**Chapter 1   The Scope and Controls of the Climate   1**
1.1   The science of climatology   5
1.2   The development of climatology   7
1.3   Climatic elements   12
   1.3.1   Measured elements   13
   1.3.2   Derived elements   18
   1.3.3   Proxy elements   18
1.4   The climate system   18
1.5   Climatological space and time scales   21

**Chapter 2   The Radiation Budget and Energy Systems of the Earth   27**
2.1   The nature of radiation   31
2.2   Radiation from the Sun   36
2.3   Interaction of radiation with the atmosphere   42
2.4   Solar radiation at the Earth's surface   46
   2.4.1   The availability of solar energy – an application of climate information   53
2.5   Longwave radiation   59
   2.5.1   The temperature structure of the atmosphere   62

2.6    Monitoring radiation from space    62
2.7    The global radiation budget    66
2.8    Surface radiation budgets'    73
2.9    Energy and temperature    75
2.10   Surface energy budgets    76
2.11   Temperatures at the Earth's surface    79
2.12   Applications of temperature
       information    83

**Chapter 3   The Hydrological Cycle**    **91**
3.1    Evaporation    93
3.2    Moisture in the atmosphere    98
3.3    Clouds and cloud-forming processes    103
3.4    Hydrostatic stability    114
3.5    Clouds and climate    120
3.6    Precipitation formation    127
3.7    Precipitation    132
3.8    Global precipitation distribution    139
3.9    Local scale precipitation    142
   3.9.1   Thunderstorms    142
   3.9.2   Applications of precipitation information    146
3.10   The water balance    148

**Chapter 4   The General Circulation and
             Global Climate**    **151**
4.1    The function of the general circulation    154
4.2    Atmospheric pressure    159
4.3    Air movement around a rotating planet    164
4.4    Barotropic and baroclinic conditions    170
4.5    The general circulation of the
       atmosphere    175
4.6    The mid-latitude baroclinic zone    181
4.7    Large-scale effects of the surface
       boundary    185

4.7.1  Oceans                                                          185
4.7.2  Cryosphere                                                      192
4.7.3  Continents                                                      195


**Chapter 5   Regional Climates**                                      **201**

5.1      Climate classification                                        204
5.1.1    Development of thunderstorm regions
         for the United States: an example of
         climatic classification                                       205
5.1.2    Methods of classification of climate                          208
5.2      The empiric approach                                          209
5.2.1    The Köppen classification                                     210
5.3      Tropical climates – the Hadley cell                           214
5.3.1    Hurricanes                                                    226
5.4      Tropical climates – the monsoon regime                        232
5.5      Mid-latitude weather                                          237
5.6      Mid-latitude climate regions                                  248
5.7      Polar climates                                                257
5.8      Changes in regional climates                                  262


**Chapter 6   Local Climates**                                         **269**

6.1      Factors controlling local climates                           272
6.2      The importance of surface type                               278
6.3      The urban climate                                            284
6.3.1    Airflow around obstacles and its
         implications for city planning                               292
6.4      The influence of topography                                  296
6.4.1    Local winds                                                  305
6.5      The influence of larger-scale atmospheric
         features                                                     310
6.6      Defining and measuring local climates                        315
6.7      Inadvertent climate modification                             317
6.7.1    Air pollution                                                318

6.8    Deliberate climate modification                        326
6.9    The human response to climate                          329

**Chapter 7    The Future – Climate Change,**
             **Climate Models, Climate Impacts    341**
7.1    Mechanisms of climatic change                          344
  7.1.1  External causes of climatic change                   344
  7.1.2  Internal causes of climatic change                   349
7.2    Global-scale climate models                            354
  7.2.1  Energy balance climate models                        363
  7.2.2  One-dimensional radiative–convective
         climate models                                       366
  7.2.3  Two-dimensional climate models                       368
  7.2.4  General circulation climate models                   369
  7.2.5  Climatic feedback effects                            373
7.3    Local climate models                                   377
  7.3.1  Agricultural production and climate                  378
  7.3.2  Water resources                                      383
7.4    Changes and cycles in the climate
       system                                                 386
  7.4.1  Geological record of climate                         387
  7.4.2  Shorter time scale climatic changes                  391
  7.4.3  Historical climatic change                           394
7.5    Climatic changes and their impact on
       Man                                                    397
7.6    Future climates – the probable impact
       of carbon dioxide                                      404
7.7    Epilogue                                               415

       Suggested Further Reading                              421

       Glossary                                               423

       Appendix SI units                                      431

       Index                                                  434

# Preface

The study of climate has always been challenging because it draws upon many disciplines. We set out to write a book for undergraduates about the Earth's climate which reflects the various disciplines involved, provides the basic factual information, suggests ways in which this information can be used and indicates where the challenges, and excitement, lie. We have tried to emphasise the importance of climatic information without making its acquisition and understanding seem too daunting. To assist those who are less well acquainted with meteorological and climatological terminology there is a glossary at the end of the book. Information about Système Internationale (SI) units is given in an appendix.

People who study or need to be able to understand the climate come from a wide range of backgrounds with a large variety of motives. Indeed, one of the major difficulties which has beset the study of climate has been the naming of the people who do it! Speaking of climatology in 1978 in an address reported in the *Bulletin of the American Meteorological Society* (**60**, 1171–1174), Professor Kenneth Hare said, '. . . you hardly heard the word professionally in the 1940s. It was a layman's word. Climatologists were the halt and the lame . . . in the British service you actually had to be medically disabled in order to get into the climatological division . . . It was clearly not the age of climate. Now it is. It's the respectable thing to do . . . This is obviously the decade in which climate is coming into its own.' We hope that this text will encourage our readers in their quest for climatological excellence.

A.H-S. & P.J.R.
Departments of Geography
Universities of Liverpool
and North Carolina
(8 November 1984)

# Acknowledgements

There are a great many people who deserve thanks for helping with this book. Not least are our two institutions and our departments and colleagues and Professor Stan Gregory of Sheffield University who started the whole venture. We have both used preliminary versions of the text as lecture notes in undergraduate courses and we thank our ex-students who wittingly and unwittingly helped us. Postgraduates and colleagues have also helped by reading, encouraging and correcting. In particular we should thank Dr Kendal McGuffie for his untiring efforts as unpaid editor, Frances Drake for drafting the glossary and Dr J. Graham Cogley for reading and re-reading all the preliminary efforts. We are also very grateful to Mrs Sandra Mather who did such a marvellous job with the illustrations.

Special thanks are due to Mr L. Dent, Principal Meteorological Officer, Manchester Airport and Mr Kiff at the Meteorological Office Training College at Shinfield, both of whom allowed us to photograph their operational meteorological equipment. The staff at the National Weather Service Forecast Office, Raleigh-Durham Airport and at the National Climatic Data Center, Asheville have unfailingly answered numerous questions with courtesy and wit.

Fortunately we, respectively, married spouses who can add up and integrate and spell and punctuate. For these and all the many efforts Brian and Shirley have made to help us complete this book we are indeed grateful.

We are grateful to the following for permission to reproduce copyright material:

the author, L H Allen and Academic Press for fig 4.25 from figs 46, 48 pp 187, 190 *Descriptive Meteorology* by Willett & Saunders (1959); American Association for the Advancement of Science for figs 7.4(b), 7.5, 7.33 from figs 5, 2, 7 in the article 'Climatic impact of increasing atmospheric carbon dioxide' by J Hansen et al pp 973, 960, 965 *Science* Vol 213 (August 1981) Copyright 1981 by the AAAS, 7.18 from the article 'The Surface of the ice age earth' by T C McIntyre (CLIMAP) p 1132 *Science* Vol 191 (19/3/76) Copyright 1976 by the AAAS; American Chemical Society for fig 6.36 by G E Likens from pp 29–44 *Chemical Engineering News* Vol 54 (22/11/76) Copyright 1976 American Chemical

Society; American Meteorological Society for figs 2.11, 2.12, 6.17 from figs 1, 2, 9 in the article 'A. Heat, Radiant and Sensible' by D M Gates pp 1–26 *Meteorological Monographs* Vol 6 No 28 (July 1965), 2.18, 2.19 from figs 1, 2 in the article 'Solar energy climatology of North Carolina' by P J Robinson & W E Easterling pp 1732, 1734 *Journal of Applied Meteorology* Vol 21 No 11 (Nov 1982), 2.22 from fig 12 in the article 'Thermal equilibrium of atmosphere with convective adjustment' by S Manabe & R F Stickler p 378 *Journal of Atmospheric Sciences* Vol 21 (July 1964), 4.26 from figs 3, 8 in the article 'On the observed annual cycle in the ocean atmosphere heat balance over the Northern Hemisphere' by A H Oort & T H Vonder Haar pp 721–800 *Journal of Physical Oceanography* Vol 6 (Nov 1976), 5.13 from fig 10 in the article 'Precipitation patterns in West Africa' by R P Motha et al p 1571 *Monthly Weather Review* Vol 108 (Oct 1980), 6.9(a) from fig 2 in the article 'Notes and correspondence' 'Deep soil temperature and urban effects in Paris' by Dettwiller p 80 *Journal of Applied Meteorology* Vol 9 (Feb 1970), 6.9(b) from fig 7 in the article 'Further observations of the urban heat island in a small city' by R J Kopec p 604 *Bulletin of the AMS* Vol 51 7 (July 1970), 6.24 from fig 3 in the article 'Dynamical model simulation of the morning boundary layer development in deep mountain valleys' by D C Bader & T B McKee p 347 *Journal of Climate and Applied Meteorology* Vol 22 No 3 (March 1983), 7.11 from fig 3 in the article 'Efficient three dimensional global models for climate studies: models I and II' by J Hansen et al p 611 *Monthly Weather Review* Vol 111 No 4 (April 1983), 7.20(c) from fig 2 in the article 'Variations in surface air temperatures: part I Northern Hemisphere 1881–1980' by Jones, Wigley & Kelly *Monthly Weather Review* Vol 110 No 2 (Feb 1982); Edward Arnold (Publishers) Ltd for figs 3.5, 7.14, 7.15 from figs 3.5, 6.1, 6.8, 6.9 pp 62, 183, 196 *Causes of Climate* by J G Lockwood (1979); Edward Arnold (Publishers) Ltd and the Copyright Agency of the USSR for fig 2.16 from fig 4.2 p 84 *Causes of Climate* by J G Lockwood (1979); Association of American Geographers for fig 3.27 from fig 1 in the article 'The role of climate in the distribution of vegetation' by J R Mather & G A Yoshioka p 33 *Annals, Association of American Geographers* Vol 58 (1968); the author, E C Barrett and Academic Press for fig 3.21 from fig 1.2 in *The Use of Satellite Data in Rainfall Monitoring* by E C Barrett & D W Martin (1981); Cambridge University Press for fig 7.31 from fig 13.6 in the article 'Global influences of mankind on the climate' by W W Kellogg p 220 *Climatic Change* ed J Gribbin (1978); Central Intelligence Agency for fig 4.3(a)(c) from figs 1, 2 in the article 'Meteorology and climatology of the seasonal ice zone' by R G Barry pp 133–150 *Cold Regions Science and Technology* Vol 2 (1980); the author, M Changery for fig 5.1 from fig 2 in the paper 'US Thunderstorms' presented at the Third Conference on Applied Climatology, AMS, Hot Springs, AR (Oct 1983); Crop Science of America for fig 2.13 from fig 1

*Acknowledgements*

in the article 'Variation in the response of photosynthesis to light' by J D
Hesketh & D N Moss p 108 *Crop Science* Vol 3 (1963); Wm Dawson &
Sons Ltd for figs 7.22(a) (b), 7.23(a) (b), 7.24 from figs 20, 21, 25, 26, 34
Springs, AR (Oct 1983); Crop Science Society of America for fig 2.13
from fig 1 in the article 'Variation in the response of photosynthesis to
light' by J D Hesketh & D N Moss p 108 *Crop Science* Vol 3 (1963); Wm
Dawson & Sons Ltd for figs 7.22(a)(b), 7.23(a)(b), 7.24 from figs 20, 21,
25, 26, 34 pp 85, 86, 101, 102, 121 *Climatic Change Agriculture and Settlement*
by M L Parry (1978), 7.22(c), tables 6.10, 7.5, 7.6 from fig 2.5,
tables 4.1, 4.8, 4.7 pp 38, 87, 106, 105 *Applied Chemistry* by J E Hobbs
(1980); Elsevier Science Publishers BV for figs 2.9 from figs 3.3, 3.9
pp 55, 61 *Solar Radiation* by N Robinson (1966), 6.31 from fig 21 in the
article 'The climates of North America' p 224 *World Survey of Climatology*
Vol 11 ed Bryson & Hare (1973); Europaischer Wetterbericht for
figs 4.5, 5.26, 5.28 *European Meteorological Bulletin*; W F Freeman &
Company for fig 3.13 from fig 4.8 p 111 *Understanding our Atmosphere* by M
Neiburger et al W H Freeman & Co Copyright (c) 1982; Dr W M Gray
for fig 5.16 from fig 1.18 p 36 *Atmospheric Science* by J M Wallace & P V
Hobbs (pub Academic Press 1977); the author, J Hansen and the
American Association for the Advancement of Science for table 7.2 from
table 1 in the article 'Climatic impact of increasing atmospheric carbon
dioxide' by J Hansen et al p 959 *Science* Vol 213 (August 1981)
Copyright 1981 by the AAAS; Harvard University Press for fig 6.21
from fig 217 p 399 *The Climate Near the Ground* (trans of 4th German edn)
by R Geiger (1965); the Controller of Her Majesty's Stationery Office for
fig 3.15 (Crown Copyright), fig 3.7 from p 38 *Geophysical Memoirs,
Meteorological Office 102*, table 3.3 from pp 87–88 *A Course in Elementary
Meteorology* by D E Pedgley (1962), table 6.4 from chapter 5 'Comfort
and safety conditions' pp 40–42 *Wind Environment Around Buildings* by
A D Penwarden & A F E Wise (DOE B R E Crown Copyright);
Hutchinson Publishing Group Ltd for table 5.2 from p 60 *The Restless
Atmosphere* by F K Hare (1961); Macmillan, London and Basingstoke for
fig 7.2 from fig 42 *Ice Ages: Solving the Mystery* by J & K P Imbrie (1979);
McGraw-Hill Book Company for figs 3.19, 4.14, table 2.2 from figs 5.1,
9.14, table 3.3 pp 93, 161, 63 *Introduction to Meteorology* by Petterssen
(1969), 3.24 from fig 10.5 *Introduction to the Atmosphere* by H Riehl (1965),
4.33 from fig 7.1 p 239 *An Introduction to Climate* by G T Trewartha (4th
edn 1968); **Charles E Merrill Publishing Co for fig 5.2 from Elements
of Meteorology by A Miller & J C Thompson. Columbus, Ohio: Charles
E Merrill Publishing Co. Copyright 1970, 1975 by Bell & Howell Co;**
Methuen & Co Ltd and the World Meteorological Organisation for
table 6.3 from table 8.4 *Boundary Layer Climates* by T R Oke (1978);
Minister of Supply and Services Canada for fig 6.15 from the article
'The use of lysimetric methods to measure evapotranspiration' by W L
Pelton pp 106–122 *Proceedings of Hydrology Symposium* No 2

Cat R32–361/2; Dr Y Mintz and Prentice-Hall Inc for fig 4.19 from fig 4.5 *Atmospheres* by R M Goody & J C G Walker; Munksgaard International Publishers Ltd for fig 3.18 by J M Wallace & P V Hobbs from pp 258–259 *Tellus* 10 (1958); New York Academy of Science for fig 7.4(a) from fig 2 in the article 'Climate impact of increasing atmospheric carbon dioxide' by J Hansen et al pp 575–586 *Annals of the New York Academy of Sciences* Vol 338 (1980); the author, J E Oliver and John Wiley & Sons Inc for table 6.6 from table 5.2 p 150 *Climate and Man's Environment* by J E Oliver (1973); the author, Dr A H Oort and Academic Press Inc for figs 1.2, 4.20, 4.21 from figs 2, 7, 39 from 355-489 *Theory of Climate* by A H Oort & J P Peixoto Vol 25 (1983); the author, Dr A H Oort for figs 2.32, 4.4 4.34 from figs A7, A12 pp 123, 128 in the article 'Global atmospheric circulation statistics' by A H Oort *NOAA Prof Paper 14;* Oxford University Press for table 6.2 from table 16.1 p 108 *Applied Climatology* by J F Griffiths (2nd edn 1976) (c) OUP 1966, 1976; Pergamon Press Ltd for fig 6.2 from fig 14 p 37 *Agricultural Physics* by C W Rose (1966); Prentice-Hall Inc for figs 2.6 from fig 1.7 p 17 *Atmospheres* by R J Goody & J C G Walker, 2.33, 5.8, 5.9, table 5.1 from figs 2.19, 5.17, 4.22 table 6.6 pp 34, 126, 99, 156 *General Climatology* by H J Critchfield (4th edn 1983); D Reidel Publishing Company for figs 3.4 from fig 1.1 p 6 *Evaporation into the Atmosphere: Theory History and Application* by W H Brutsaert (1982), 7.26, 7.28 from figs IC, 6 pp 664, 671 *Climatic Variations and Variability* ed J Smagorinsky & A Berger, table 7.9 from table 1 in the article 'Scenarios of cold and warm periods of the past' by H Flohn p 691 *Climatic Variations and Variability: facts and theories* ed A Berger (1981); the author, H Riehl for fig 5.4 from p 392 *Tropical Meteorology* (1954); Royal Meteorological Society for figs 2.15 from fig 1 in the article 'Estimation of insolation for West Africa' by J A Davis p 361 *Quarterly Journal of the Royal Meteorological Society* Vol 91, 389 (1965), 6.38, 6.39 from figs 2, 3 in the article 'Mortality in the June–July 1976 hot spell' by D G Tout pp 223, 225 and fig 6.41, table 6.12 from fig 2, table 2 in the article 'Some agricultural effects of the drought of 1975–76 in the United Kingdom' by M G Roy et al pp 67, 73 *Weather* Vol 33 (1978), 7.6 from fig 1 in the article 'Modelling climate and the nature of climate models: a review' by K P Shine & A Henderson-Sellers pp 81–94 *Climatology* Vol 3 No 1 (1983), 7.13, table 7.4 from fig 7, table 4 in the article 'Climatic variation and the growth of crops' by J L Monteith pp 769, 771 *Quarterly Journal of the Royal Meteorological Society* Vol 107 (1981); the author, W D Sellers and the University of Chicago Press for figs 3.28, 4.3, 6.1, 6.18, 6.34, 6.35, table 6.1 from figs 26, 34, 41, 43, 46, 44, table 20 pp 84, 115, 149, 179, 194, 150 *Physical Climatology* (1965); the author, Professor G R Rumney for fig 6.27 from fig 13.5 by J E Oliver p 248 *Climatology and the World's Climates* (pub Macmillan 1968); Springer-Verlag for table 1.1 from table 1 in the article 'The Atosphere' by M Schidlowski *The Handbook of Environmental Chemistry* ed

Acknowledgements

O Hutzinger Vol 1 Part A (1980); the author, Dr A N Strahler for fig 2.8 from *Introduction to Physical Geography* (1965) Copyright (c) 1971 by A N Strahler; the author, Dr A E Strong for fig 4.29 from the article 'Improved ocean surface temperatures from space comparisons with drifting buoys' by A E Strong & E P McLean pp 138–142 *Bulletin of American Meteorological Society* Vol 65 (1984); the author, G Tanner for figs 5.10–5.12, 5.19, 5.31–5.33, 5.35–5.38 from *A Collection of Selected Climographs*; University of Colorado Solar Output Workshop for fig 7.3 from p 57 *Solar Output and Its Variations* ed O R White, Boulder Co: Colorado Associated University Press (c) 1977; the authors, J M Wallace & P V Hobbs and Academic Press for figs 3.26, 4.11, 4.13, 4.15, 4.22, 5.15, 5.20 from figs 5.21, 8.12, 8.16, 9.11, 9.10, 5.30, 9.8 from pp 241, 385, 433, 432, 253, 425 *Atmospheric Science* (1977); the authors, J M Wallace & P V Hobbs, Academic Press and the World Meteorological Organisation for fig 4.32 from fig 7.20 p 349 *Atmospheric Science* (1977); the author, R L Wardlaw for fig 6.16 from *The Aerodynamics of Wind Flows about Buildings in the Pedestrian Wind Environment – A Summary of Research and Practical Experience* by R L Wardlaw & C D Williams; George Weidenfeld & Nicholson Ltd for table 4.1 from table 7 *Climate and Weather* by H Flohn (1969); the author, Dr T M L Wigley and Macmillan Journals Ltd for fig 7.32 from figs 1, 2 in the article 'Scenario for a warm high-$CO_2$ world' by Wigley et al pp 17–21 *Nature* Vol 283 Copyright (c) 1980 Macmillan Journals Ltd; John Wiley & Sons Inc for figs 1.5, 6.10 from figs 3.24, 17.4 pp 633–686, 396 *Weather and Climate Modification* ed W N Hess (1974), 4.30, 4.31(b) from figs 4.3, 4.9 pp 66, 75 *Glacial and Quaternary Geology* by Flint (1971); World Meteorological Organization for figs 1.3 from 'The physical basis of climate and climate modelling' in *GARP Publ. Ser. 16* (1975), 6.12 from fig 2 in 'Urban climatology and its relevance to urban design' (After Davenport 1965) *WMO Technical Note No 149* (1976), 7.20(a)(b), 7.25, table 7.8 from figs 2, 3, 4a, table 3 pp 10, 7, 26 '(a) Report of JSC/CASE Meeting of Experts' in *WMO Projects on Research & Monitoring of Atmospheric $CO_2$* NO WCP–29.

We are unable to trace the copyright holders in figs 6.6 from fig 38 p 111 and 6.14 from fig 32 p 100 *Descriptive Meteorology* by Munn (1966), 6.25 from R S Scorer (1959), table 6.5 by C Thurow p 10 *Planning Advisory Service Report* No 376, Chicao, Illinois, and would appreciate any information that would enable us to do so.

# List of Symbols

All symbols used to represent constants and variables are defined at their first occurrence in the text. A limited number are used in other textual locations separate from this definition and these are collected here for easy reference.

*Roman*

|  |  |  |
|---|---|---|
|  | $A$ | albedo |
|  | $c_\mathrm{p}$ | specific heat at constant pressure |
|  | $C$ | specific heat |
|  | $C^*$ | conductive capacity |
|  | $E$ | energy |
|  | $E^*$ | radiant energy |
|  | $g$ | acceleration |
|  | $G$ | heat flux into the ground |
|  | $H$ | sensible heat flux |
|  | $K^*$ | thermal diffusivity |
|  | $K$ | thermal conductivity |
| *or* | $K$ | solar radiation ($K\downarrow$ = downward, $K\uparrow$ = upward) |
|  | $L$ | longwave (terrestrial) radiation ($L\downarrow$ = downward, $L\uparrow$ = upward) |
| *or* | $L$ | latent heat of vaporisation of water |
|  | $LE$ | latent heat flux (so defined because it equals the product of ($L$ and rate of evaporation) |
|  | $p$ | pressure |
|  | $P$ | precipitation |
|  | PET | potential evapotranspiration |
|  | PWV | precipitable water vapour |
|  | $Q^*$ | net radiative flux at the surface |
|  | $R\downarrow$ & $R\uparrow$ | net incoming and outgoing planetary radiation |
|  | $S_\mathrm{F}$ | solar (flux) constant (= 1370 W m$^{-2}$) |
|  | $S$ | instantaneous top-of-the-atmosphere solar flux (= $S_\mathrm{F}/4$) |

*List of symbols*

| | |
|---|---|
| $t$ | time |
| $T$ | temperature |
| $T_d$ | dew point temperature |
| $V_g$ | geostrophic wind |
| $z$ | height (in the atmosphere) |
| $Z$ | solar zenith angle |

*Greek*

| | |
|---|---|
| $\gamma$ | environmental lapse rate |
| $\Gamma_d$ | dry adiabatic lapse rate (DALR) |
| $\Gamma_s$ | saturated adiabatic lapse rate (SALR) |
| $\triangle$ | indicates a small change in the associated variable (e.g. $\triangle T$ = small change in temperature) |
| $\epsilon$ | emissivity |
| $\phi$ | potential temperature |
| $\lambda$ | wavelength (when a subscript indicates occurrence at a specific wavelength) |
| $\rho$ | density |
| $\sigma$ | Stefan–Boltzmann's constant ($= 5.67 \times 10^{-8}$ W m$^{-2}$ K$^{-4}$) |
| *or* $\sigma$ | standard deviation |
| $\tau$ | optical thickness (of atmosphere or cloud) |
| $\theta$ | latitude |
| $\Omega$ | angular rotation rate of the Earth |

# Chapter 1
# The Scope and Controls of the Climate

1.1     The science of climatology

1.2     The development of climatology

1.3     Climatic elements

   1.3.1     Measured elements

   1.3.2     Derived elements

   1.3.3     Proxy elements

1.4     The climate system

1.5     Climatological space and time scales

# Chapter 1
# The Scope and Controls of the Climate

The envelope of air that surrounds the Earth affects us in many ways as we go about our day-to-day activities. Sometimes we respond to it almost unconsciously, as when we choose the type of clothes we will wear. At other times a conscious decision is needed: do we carry an umbrella today? On a longer time scale, our houses reflect the influence of climate, since if winters are likely to be cold, we install a heating system. To alleviate hot summers, industrial societies install air conditioning, while non-industrial societies select building sites and designs that allow natural cooling. We predict future conditions when we decide what to plant in our gardens or fields, or when we schedule the time and place of our vacation. Institutions, as well as individuals, are influenced by the atmosphere. An electricity generating company must ensure that it has enough capacity to meet the demand on the coldest, or hottest day. A water supply authority must plan ahead to ensure that it has enough storage to supply the needs during a long drought. A construction firm must determine the strongest winds likely to be encountered to ensure that its buildings do not collapse.

Such problems, local and practical, and almost as old as civilisation itself, are climatological problems, requiring 'predictions' of future conditions. Contemporary climatology seeks to find ways to answer such questions. The scope of climatology over the last few years has increased immensely and there are now an array of approaches available to the climatologist seeking answers. The traditional role of climatology was to synthesise the many years of observations of the 'elements' that constitute climate and to analyse them to gain insight into the processes controlling the present climate. Although the observations were mainly of surface conditions, and the insights thus could be only partial explanations of climate, this work still provides a storehouse of information. The climatologist can use these past records, together with understanding of climatic processes, to answer many practical problems associated with conditions over the next few years.

This traditional approach, with its partial view of climate, has been transformed by the advent of satellite observations. They have allowed us to view the climate of the Earth both as a global entity and in a three-dimensional way. This has forced an explicit realisation that there is a *climate system* in which the climate of a particular place is constantly changing and dependent not only on the climate of all other places on Earth, but also on the changes that are taking place in the oceans, within the Earth's snow and ice cover, and on the land itself.

This realisation has also stimulated advances in our understanding of the climate system. These advances are reflected in the development of climatic models. These models, usually couched in terms of mathematical equations expressing the physical laws governing atmospheric behaviour, are beginning to allow us to predict long-term changes in climate and to understand the possible causes of the changes. At the same time that these developments are taking place, there has been an increasing public awareness of climate. Climatic disasters, such as the Sahel drought of the early 1970s (which is still continuing), and the growing concern that Man may alter the climate by adding carbon dioxide to the atmosphere as the result of fossil fuel combustion, have stimulated this awareness. This in turn has encouraged the climatologist to advance our understanding of atmospheric processes and our ability to predict future conditions.

Thus the purpose of contemporary climatology is clearly to 'predict' future climatic conditions. These predictions may involve conditions a few years ahead in a specific locality, where the approach would be to use the historical climate records to generate the required information. At the other end of the scale, the predictions may be needed for a time far in the future, covering a major portion of the globe. Here a climate modelling approach could provide the answer. The present state of development of climatology is such that we are far from providing 'definitive' answers in either of these conditions. Nevertheless, it is clear that significant progress has been made recently and that more can be done. It is the aim of this book to indicate the present state of our knowledge, point out where further work is possible and suggest areas where much basic research is needed.

In order to understand this modern climatological viewpoint, this chapter provides a general introduction to the subject of climatology, looking first at the modern viewpoint in more detail. Thereafter a brief review of the development of the field, emphasising the constraints placed on our understanding by observational and theoretical barriers, will be presented. The final two sections then consider observation and theory in more detail.

## 1.1 The science of climatology

The atmosphere is a body of matter which is constantly in motion. The scales of the motion can range from the molecular, which we sense as heat, to the global, creating the large-scale 'prevailing' wind systems of the Earth. These motions, on all scales, themselves lead to modifications in the structure and composition of the atmosphere, most notably in the cycling of water and water vapour which leads to cloud formation and precipitation. All of these motions and their effects are part of climatology and will be considered in detail subsequently. However, as an 'organising framework' for the whole of climatology it is advantageous to use the concept of the energetics of the atmosphere.

The source of energy for all atmospheric motions is the Sun. Energy from the Sun passes through the atmosphere to the Earth's surface. During its passage a little energy is absorbed and leads to atmospheric heating, but most of the energy is absorbed at the surface. This in turn warms and heats the overlying atmosphere, so that the Earth's surface becomes the main source of heating for the atmosphere. The amount of heating depends greatly on the type of surface. In addition, it varies spatially and temporally. The unequal distribution of heat leads directly to the horizontal motions we know as winds, and to the vertical motions which create clouds and precipitation. Eventually the energy that has been received from the Sun and has taken part in the various activities within the atmosphere is returned to space. Hence the climate as we know it can be viewed as a series of energy transformations and exchanges within and between the atmosphere and the underlying surface. These exchanges and transformations act in such a way as to distribute energy over the globe and to maintain an energy balance by returning as much energy to space as is received from the Sun.

All of the processes associated with these energy flows obey the laws of physics. Thus in order to understand how the atmosphere operates it is necessary to understand the relevant physical laws and principles and apply them in an appropriate way. Since these laws are usually couched in mathematical terms, a basic understanding of mathematics is also required. However, most of these concepts can be explained and understood without detailed reference to the mathematical and physical derivation. Throughout this book climatological concepts are developed in a physically realistic way and the meanings of the results interpreted.

Our understanding of atmospheric processes is expanding as we apply the physical laws in an increasingly realistic way. Advances are the result of both better observations and improved physical insight.

5

In particular, our understanding is reflected in the development of increasingly sophisticated mathematical models which simulate atmospheric processes, give us insight into them and allow us, for the first time, to make reasonable estimates of future conditions.

Although the use of the applicable physical laws remains fundamental for understanding atmospheric phenomena, it is becoming increasingly clear that chemical effects are very important in some areas. Certainly the way in which energy interacts with the atmosphere depends on its chemical composition. At present the atmosphere is dominated by nitrogen and oxygen (Table 1.1). During the evolution of the Earth and its atmosphere, however, that composition has been changing. For example, the second most abundant gas, oxygen, has been produced by photosynthetic green plants and was not present in the primaeval atmosphere. Recently, interest has been focused on the minor constituents, such as $CO_2$, $SO_2$, $NO_2$ and $O_3$, whose abundance have begun to change since the industrial revolution. Their effects on the climate could be considerable. An understanding of these effects, however, demands an understanding of the physics and the chemistry of any interaction they may have with the energy streams.

Climate depends greatly on the conditions at the surface of the Earth and any changes in the composition of the surface must lead to climatic changes. Such changes go on continuously, as the result of changes in the ocean surface caused by currents or overturning, as the result of seasonal changes in ice and snow extent and as the result of vegetation changes. All of these changes are themselves influenced by the climatic conditions. Hence a full understanding of climate also requires appreciation of some aspects of oceanography, glaciology and biology.

Although the climatologist is frequently concerned with suggesting possible conditions some considerable time in the future, the only practical way to compare the results of any prediction schemes with real information is to compare them to past climates. Past climates, however, can only be revealed through a cooperative effort among people in many disciplines. Archaeologists, historians, anthropologists, geologists and glaciologists, among others, are providing their skills for the benefit of the climatologists. Fortunately this exchange has been a two-way process and climatological insights have benefited other disciplines.

Concepts from many disciplines must therefore be incorporated if the climatologist is to understand the causes of climate and its fluctuations more fully. If this understanding is to be utilised in a practical way to provide some form of prediction of future climate to assist in the decision making process of individuals, institutions, or society, a further interdisciplinary approach is required. Only the decision maker knows fully the possibilities and options for a particular problem solution, while the climatologist has only the climatological expertise.

**Table 1.1 The composition of the atmosphere**

| Constituent | Chemical formula | Abundance by volume |
|---|---|---|
| Nitrogen | $N_2$ | 78.08% |
| Oxygen | $O_2$ | 20.95% |
| Argon | Ar | 0.93% |
| Water vapour | $H_2O$ | variable (%–ppmv[a]) |
| Carbon dioxide | $CO_2$ | 340 ppmv |
| Neon | Ne | 18 ppmv |
| Helium | He | 5 ppmv |
| Krypton | Kr | 1 ppmv |
| Xenon | Xe | 0.08 ppmv |
| Methane | $CH_4$ | 2 ppmv |
| Hydrogen | $H_2$ | 0.5 ppmv |
| Nitrous oxide | $N_2O$ | 0.3 ppmv |
| Carbon monoxide | CO | 0.05–0.2 ppmv |
| Ozone | $O_3$ | variable (0.02–10 ppmv) |
| Ammonia | $NH_3$ | 4 ppbv[a] |
| Nitrogen dioxide | $NO_2$ | 1 ppbv |
| Sulphur dioxide | $SO_2$ | 1 ppbv |
| Hydrogen sulphide | $H_2S$ | 0.05 ppbv |

[a] ppmv and ppbv are parts per million and parts per billion by volume.

So a dialogue between the two is necessary. Several examples of the results of such a dialogue are given in this book.

Fortunately, for the climatologist, practising or aspiring, detailed knowledge of all of these disciplines is not necessary. All that is required is expertise in climatology as a physical science and an awareness of the assistance that can be provided to and from other fields of study, together with a willingness to discuss problems with other experts.

## 1.2 The development of climatology

The rise of climatology as a science is closely linked with the increase in our ability to observe the atmosphere. As with many sciences, new observations often provide the basic information needed to give us new insights into how the atmosphere works, while often new theories of such workings demand that we obtain new measurements to test them. Certainly any understanding of how the world's climate system works, how it varies from time to time and place to place, and any use that can be made of resources provided by climate, depends on observing the climate at many places over a long period of time.

### Descriptive climatology

The earliest climatological observations were simply visual or otherwise 'sensed' observations about nature made without instruments or the benefit of sophisticated techniques. The march of the seasons and the annual flooding of the Nile in ancient Egypt are obvious examples. These observations at specific places had become sufficiently organised by the time of the ancient Greek civilisation for the Greeks to be able to divide the world into three zones: torrid, temperate and frigid. Although explanations for the observed phenomena abounded there was little of what we would regard as scientific enquiry into their nature and causes.

The emphasis on description, rather than explanation, of the atmosphere continued for centuries. The development of the barometer and thermometer, together with the maintenance of records of wind direction and rainfall amount, added a quantitative dimension to our knowledge. By the late nineteenth and early twentieth centuries it was possible to describe the climate of much of the Earth's land surface, and some ocean areas, in reasonable detail. These descriptions relied heavily on the available observations, mainly of precipitation and temperature. Probably it is no coincidence that these were the main elements. Not only were instruments available to measure them, but also they were extremely useful and important, particularly in agriculture.

As the number of observations grew, one great problem was encountered by people trying to describe the climate: there were so many numbers that a simple tabulation was impractical and some form of summation was needed. Monthly values provided a convenient method. However, since monthly values vary from year to year, further averaging over several years became necessary. The result was the development of the concept of the 'climatic normal': an average over at least 30 years. It was felt that this long period was sufficient to smooth out the small-scale, year-to-year fluctuations and thus provide a true measure of the climate. Today, to most people, the climate is described by the monthly normals of average temperature and total rainfall, a concept that is frequently used even where it is misleading or of very limited value.

The introduction of the climatic normal provided a means of summarising data for a single station. Summary of data spatially was provided by the introduction of the concept of the climatic region. It was found that stations could be grouped together because they had similar normal values, or a similar monthly pattern of normals. Several schemes for climatic regions were proposed. Although defined almost entirely by analysis and comparison of available data, these climate classifications were often developed with a view to their eventual use. Most divided the climate in ways which had some significance for plant growth, and so the climatic regions reflected vegetation regions. Consequently they could be used, for example, in assessing the suitability of a plant, currently growing in a particular region with a particular climate, for introduction into a new region. Although the division into regions in this way paid no account to the causes of spatial variation, it was possible to infer causes in some cases. Indeed, it is still sometimes necessary to provide this kind of descriptive approach, using different climatic elements, not only to yield practical information, but also to get some insight into causes of variability.

## *Meteorological advances*

At the same time that these descriptions of climate were being generated, an entirely different approach was being taken by workers in the then young field of meteorology. With the advent of rapid telegraphic communications it became possible to collect observations from a variety of places together quickly at a single point and analyse them with a view to forecasting future weather. The prime interest initially was in the short range forecasting of storm tracks. Nevertheless, the questions needing to be answered to solve this problem led to a search for understanding of the physical laws governing the atmosphere. As understanding increased, so did the demand for observations. Pressure, wind speed and direction, visibility, cloud type and amount and hourly temperatures were all needed and therefore observed. Later, as aviation meteorology gained in importance, the need for upper air observations grew. In response to this, the radiosonde was developed. Still later, after the Second World War, methods of radar observations of clouds and precipitation were developed. All these new observations and observational techniques served to advance our theoretical understanding of atmospheric processes and improve our ability to forecast the weather.

During this time of great advance in meteorology, our understanding of climate was progressing slowly. The traditional observations continued to be made and some theses to explain global climate were attempted. The climatological data themselves often proved to be of great benefit; for example, during the Second World War, when operations in unfamiliar territory with equipment which could be sensitive to atmospheric conditions demanded both weather forecasts and longer estimates of the probability of occurrence of particular conditions. After the wartime demonstration of the potential importance of applying

9

climatological information to a variety of problems, several advances in applications were made. A prime example is the development of the concept of the climatic water balance and its use for agriculture introduced by Thornthwaite and co-workers in the USA and by Penman in England. The former group relied heavily on the traditional date of monthly temperature normals, while the latter introduced a method that required more complex observations, thus stimulating the use of such observations for agricultural purposes.

## Questioning the 'constant' nature of the climate

Many developments in both meteorology and climatology until the middle of the 1950s were based on the assumption that the climate was constant. The weather varied with time and was responsible for the minor variations in climate from year to year, but in the long term the normals did not change. Although this was disputed by many climatologists, who could easily cite the Ice Ages several thousand years ago and the Little Ice Age a few hundred years ago as evidence of change, the concept of a static climate was a useful concept to retain while a basic theory of climatic processes was being developed.

It is now clear that climate is never constant. Indeed, it is the departures from supposedly 'normal' conditions that often provide great insight into climatic processes, as well as having the greatest human impact. There have been several such departures in recent history which have captured the attention of politicians and the media because they resulted in food shortages and human suffering. The 30 year drought period at the beginning of this century, which culminated in the 'Dust Bowl' disaster in the USA, reduced the output of the five main corn growing states of the USA by more than 15%. A similar sensitivity to climate is demonstrated by the grain harvests of the Soviet Union. Despite efforts to increase overall output by increasing agricultural acreage and improving technology, a climatic deterioration caused production to fall well below the projected targets in many years after the middle of the century, so that the USSR became a net importer of grain in the early 1970s. Even the world's money markets now feel the effect of climatic variations as the price of gold is affected by large-scale selling by the USSR to provide funds for purchase of grain in years of poor harvests.

The effect upon third world countries of climatic variations can be devastating. In contrast with the developed nations, there exist few food reserves and no possibility of purchase in the world markets. Instead many developing countries have had to rely more and more upon aid from their developed world neighbours. The human impact of the onset of drought in the Sahel region on the southern edge of the Sahara Desert in the early 1970s was traumatic. During the period 1968–72 the rainfall upon this desert margin was around 50% of the

1931–60 average. The length of the growing season decreased dramatically and available water resources were reduced almost to vanishing point; water table levels dropped and the surface area of Lake Tchad was reduced by 65%. By the mid-1980s, over a decade of sustained drought had produced a climatic catastrophe and a human tragedy of enormous proportions across the whole of Africa in the area immediately south of the Sahara and in the Horn of Africa.

A similar climatically induced drama was played out off the Peruvian coast in 1972. In this region the anchovy fishery industry provides both a staple food supply and is an important component of the export trade. An oceanic fluctuation known as El Niño, which results in a sudden rise in water temperature, destroyed the fish and resulted in a drop in anchovy catch from around 13 million tonnes to less than 2 million. The El Niño phenomenon seems to have even wider implications since an extraordinarily large El Niño event occurring in November 1982 has been associated with large-scale flooding on the western seaboard of North America and catastrophic droughts in Australia. Even areas where careful analysis of climatic conditions was made prior to the introduction of 'green revolution crops' have been plagued with problems. Analyses were usually based on the 'normal' period 1925–55. This arbitrary period turned out to be highly anomalous (see Fig. 7.19, Ch. 7), resulting in the development of crops which were sadly unsuited to the climatic conditions of the 1970s and 1980s. Climate-related crop failures have included coffee production loss in Brazil, wheat loss in Australia and catastrophic losses of staples such as sorghum and rice in many parts of Africa and southeast Asia.

These highly visible examples of climatic fluctuations ensured that by the 1960–70s the concept of a static climate had become untenable. Mounting evidence, amongst which figured archaeological, historical and glaciological data, indicated climatic change to be the norm. So climate was seen to be anything but static. A new problem therefore automatically arose: how can one predict climatic changes? Along with this new problem came new observations: those from satellites.

### Satellite input to climatology
The coming of information from satellites provided a new dimension for climatology. In the past all our information was surface based, for a fixed location on the surface and at a particular instant in time. Satellites allow global coverage almost instantly and yield a three-dimensional picture. In addition they measure the energy fluxes entering and leaving the atmosphere, information not readily obtained from other sources. This new information has led to increases in our understanding of atmospheric processes, stimulated climate model development and allowed us to approach prediction of climatic processes and climatic changes.

Satellites, of course, generate a tremendous amount of data, but fortunately the last few decades have seen the rapid development of computer technology, which enables us to handle these data, along with data from more traditional sources, in an efficient fashion.

So at the present time climatology has three sources of data: (a) the traditional surface-based observations, point specific but with a long period of record; (b) upper air data, again approximately point specific, and with a reasonably long record length and (c) satellite data, with wide areal coverage of the globe but not too long a record. Climatology must use all these types in order to fulfil its mission. Much of our basic knowledge, and much of the material in subsequent chapters, has come from surface and upper air observations. The advent of satellite data is allowing us to refine, and in some cases revolutionise, our understanding. Satellite information so far has been used mainly in studies of the workings of the atmosphere on a global scale. Investigations on a more local scale, and in most cases application of climate information, have not yet been significantly affected by these new data. However, this picture is already changing and one can anticipate additional use, as satellite information increasingly complements the more traditional modes of observation.

## 1.3 Climatic elements

The preceding paragraphs indicate that there are a variety of 'elements' which constitute climate. For many years climatology focussed on two of these: temperature and precipitation. They are probably the two things about the atmosphere that we notice most readily and which have the greatest impact on our everyday life. Certainly they are very important in describing the atmosphere and have a far reaching impact, but they are by no means the only climatic elements. Wind speed and direction, cloud type and amount, sunshine duration, atmospheric humidity, air pressure and visibility are obvious additions to the list of elements that we notice every day. Other elements may be equally, or more, important in particular situations. For example, soil moisture, soil temperatures and evaporation are vital in agriculture, pollutant concentration and the acidity of precipitation are of concern for human health, whilst radiant energy fluxes are of great interest to the climatologist seeking to understand atmospheric processes.

Climate therefore consists of a 'mixture' of these elements. Indeed, our individual definition of climate probably reflects the relative importance that we place on each. A person responsible for forecasting electricity demand in an area may think of climate simply in terms of temperature, since this element plays the greatest role in changing demand from day to day. A farmer, on the other hand, may have a more complex definition. Precipitation may be the main source of his water supply, which may be depleted by evaporation. Temperatures will often determine the length of the growing season, as well as affecting the rate of evaporation. Cloudy conditions may prevent his plants from maturing, while a high wind or a hail storm at an inopportune time may destroy his crop.

As climatologists, we must be concerned with all of the elements. Not all of them, of course, will be equally important in all contexts. Furthermore, we know relatively little about some of them. This is particularly the case for those elements for which we have few observations. Some, such as the global extent of lying snow, can only be observed by satellite so that we have a relatively short period of record. Others, such as atmospheric turbulence, require delicate instruments, so that measurements are mainly restricted to specialised research sites. Still others, the acidity of precipitation for example, have not been recognised as important until recently, so that few long-term observations exist. On the other hand, elements such as temperature and precipitation are regularly measured at tens of thousands of sites worldwide. In general we know and understand more about the elements that are commonly measured than about those for which we have only a small number of observations.

Although we shall deal with each element in various sections of this book, it is useful at the outset to consider the elements and techniques for observing them in a general way. The elements of interest to climatology can conveniently be divided into three types: measured elements; derived elements; and proxy elements.

### 1.3.1 Measured elements

The most common and familiar type of climatic elements are those that are measured directly. The instrument used may give either *contact measurements*, where the instrument is in direct contact with the entity being measured, or *remote sensing* measurements, where the instrument measures radiation from the entity, which is then converted into an observation of the required element. With minor exceptions, observations from the surface and in the upper air are contact measurements, whilst those from satellites are remote measurements.

Surface-based observations are point specific and refer only to the time of observation. Usually there is a network of stations designed so that the results from one site can be compared easily with those from another. Hence observations are taken using standardised instruments with standardised exposures and using an agreed procedure. One example would be the common rain gauge. Rainfall is collected in what is essentially a bucket with a standard orifice, size, height above ground, shape and distance from upstanding obstacles (Fig. 3.22, Ch. 3). This gauge is emptied at the same time (or times) each day and the precipitation that has been collected is measured in a standard fashion. Many standards have been established by international agreement. Individual national meteorological services implement these and ensure that individual sites comply with them. Data obtained from these sites are therefore of known quality and accuracy. Many 'unofficial' sites are also in operation, which may or may not adhere to standard procedures. Any data from these stations must be carefully checked for accuracy before they are used.

The temperature measurement that corresponds to the standardised precipitation measurement is that measured by a thermometer in a shelter about 1 m above the ground (Fig. 1.1(a)). The resultant observation is variously called 'surface temperature', 'air temperature' or 'shelter temperature'. The height of the sensor is chosen so that it provides a value for the layer where life goes on, which is representative of a relatively large surface extent. Placing the thermometer at a relatively high level leads to an integration of the effects of the underlying surface types since, right at the surface, temperatures can change very rapidly over short distances. Thus the results would depend on the exact location of the sensor and comparison between sites would be difficult. Sheltering the instrument serves primarily to prevent direct absorption of solar radiation, but also shelters it from the effects of energy flowing up from the ground. The result is therefore more nearly a measurement of the temperature of the air at shelter level, and as such represents conditions over a wide area. Other types of exposure, higher or lower, sheltered or not, are possible and are used for specialised applications.

As the consideration of temperature indicates, the size of the area that is actually represented by a particular instrument measuring any climatic element is not clearly defined. The definition in part depends on the element and on the character of the area. Temperatures in mountainous terrain, or in an urban setting, may change considerably a few metres away from the instrument, while on a flat open plain conditions may not change much for tens of kilometres. Precipitation variability over an area will depend on the terrain and the type of cloud producing the precipitation. In almost all conditions wind speeds, which are measured using an anemometer like that shown in Fig. 1.1(b), are particularly susceptible to local irregularities (as described in sects 6.3 and 6.4, Ch. 6).

The definition of area sensed also depends in part on the use to which the information may be put. A detailed investigation of the influence of climate on plant diseases may require many measurements in a few square metres or less. At the other extreme, a single station may be regarded as representative of an area of thousands of square kilometres when used to characterise conditions throughout the distribution area of an electric power supply industry. However, in the more

Fig. 1.1(a). A Stevenson screen houses meteorological instruments, particularly ther-mometers and thermohydrographs, so that they are well ventilated but protected from direct radiation. The screen is a double-louvred white-painted wooden box erected in an open position so that the thermometer bulbs are 1.25 m above the ground. (Two screens are shown in the photograph.)

Fig. 1.1(b). Anemometers must be installed at a number of heights in the lower atmosphere if the effect of surface friction on wind speed is to be measured. In complex terrain, such as an urban centre, a large array of anemometers may be required as wind speeds can vary considerably over short distances.

15

general, common and traditional use, a surface-based measurement is used to characterise the conditions in some ill defined area, while a series of standardised stations is used to determine differences between locations.

Surface-based measurements have been taken for many years and they provide a great deal of information about changes over time. The results must be used with caution, however. Few stations have survived for many years without a change in location. Each change, even a minor one of a few metres, exposes the instruments to different conditions. Frequently it is necessary to look at the surface character-istics in order to determine what effects have been caused by instrument re-location. Even without such changes, the surroundings may change, whether through the removal of a tree, introduction of legislation limiting air pollution, or the encroachment of urbanisation. These will again influence the instrumental record. Therefore, although a long record at a place may provide valuable information about climatic change, the record itself must be examined closely before any conclusions are drawn.

Observations over the oceans are, not surprisingly, less common than those over land (Fig. 1.2). Some countries maintain fixed location

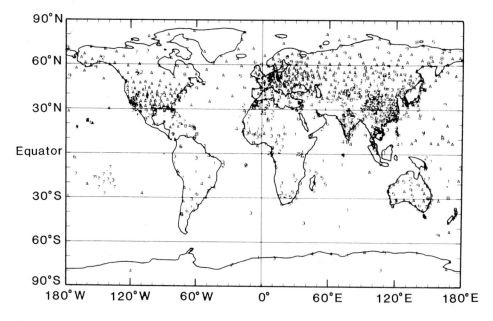

Fig. 1.2. Distribution of observational data available from rawinsonde ascents showing the number of years of data available from each location in the period 1963–73. The scale ranges from 1 to 10 years (where 1 to 9 are figures and A represents 10). Data over land areas are from meteorological stations whilst oceanic data are drawn predominantly from ship reports.

weather ships, which take standard observations. Increasingly ocean buoys are being used to supplement our knowledge of ocean climate conditions. These, as far as possible in the often hostile environment, take observations in a standardised way and transmit them to land-based facilities, often *via* a satellite link.

The most common method of obtaining upper air data is by using a radiosonde. This is a package suspended below a balloon, which senses temperature, humidity and pressure as it ascends, the pressure observation being used to determine altitude. Twice daily ascents at noon and midnight GMT (i.e. at 12Z and 00Z) are made at several hundred stations scattered around the world. Some stations use a more sophisticated version, called a rawinsonde, which additionally measures wind speed and direction (Fig. 1.2).

Specially instrumented aircraft also measure conditions in the free atmosphere, but this is not done on a routine basis. Their use is restricted to experimental investigations of particular atmospheric features. A similar role, usually for the higher reaches of the atmosphere, is played by instrumented rockets.

Most of these measurements are 'contact' measurements since the instrument is actually in contact with the things being measured. They are thus also point measurements. Accuracy is determined by the inherent accuracy of the instrument, the care taken in maintaining, exposing and reading it and the care taken in transforming the reading into a form that can be used by a climatologist.

### Satellite observations

A whole new set of accuracy problems arises when satellite observations are considered. A satellite is a platform for remote sensing instruments. The instrument, which may be measuring surface or atmospheric temperatures, or water vapour amounts in various atmospheric layers, or simply making a digitised 'image' of clouds and the surface, actually senses the radiation emitted or reflected by the object under investigation. Hence there are problems of making sure that it is looking at the right thing, known as 'registration'. Furthermore, problems are associated with radiation transfer through the atmosphere, which may interfere with the measurement. These are not insoluble problems, but great care has to be taken to ensure that the accuracy remains sufficiently high for the measurements to be useful. The result, after correction, is a radiance which is an integrated value over the whole field of view of the radiometer, generally at least a few $km^2$ in area. Thus it differs fundamentally from surface observations. The great advantage of such measurements, which make this extra effort worthwhile, is that satellites provide access to remote areas with no alternative data source and that they give continuous global coverage.

### 1.3.2 Derived elements

In addition to the elements that are actually measured, there are elements which are derived from them. Several of these will be discussed in greater detail later in the book. Examples include heating or growing season degree days, the summation for a given time period of temperatures above a certain threshold, which is used in calculating heating demand (Ch. 2); the probability that snowfall will disable a city (Ch. 3); and the wind chill factor, which gives a quantitative estimate of how cold a human being feels (Ch. 6). These, while being derived from direct observations, play an important role in describing the climate for particular applications. Evaporation is a special case for a derived element. Direct measurement of evaporation is possible, but as an accurate result is costly to produce and requires great care, direct measurements are restricted to a few locations. Evaporation, however, is very important in many aspects of water management and so techniques have been devised to derive estimates by calculation from other, more commonly and easily measured elements.

### 1.3.3 Proxy elements

Finally there are elements which can be called 'proxy' elements. These are usually non-atmospheric indicators of atmospheric conditions, commonly used to infer past climatic conditions in the time before instrumental records. Tree rings, pollen, lake varves and dates of grape harvests from diaries have all been used as proxy variables and are used in Chapter 7. Probably other such proxy elements will be discovered and utilised in the future.

## 1.4 The climate system

The key to the present understanding of climate and climatology is the way in which climate is viewed. There is a *climate system*, linking not only the many parts of the atmosphere but also the various types of underlying media, to produce an integrated whole (Fig. 1.3). The systematic viewpoint is not entirely new, but the emphasis on the effects of the Earth's surface adds a new dimension. It has been found impossible to understand the flows and cycles of energy and matter in the atmosphere without considering the material which underlies it.

Any consideration of energy exchanges, which we have already indicated provide the starting point and general framework for the study of climatology, must incorporate surface effects. The atmosphere is heated from below as a response to surface absorption of solar energy, but different surfaces react in different ways to the receipt of this energy. Ice and snow reflect much of it, land surfaces heat rather rapidly, while the oceans store the energy without experiencing a

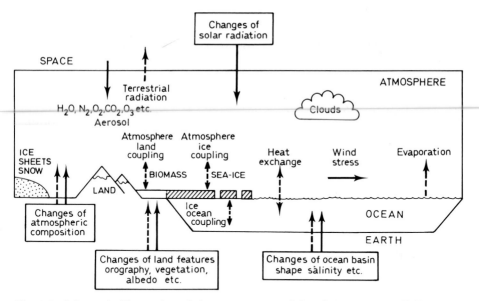

Fig. 1.3. Schematic illustration of the components of the climate system. Full arrows are examples of external processes and dashed arrows are examples of internal processes.

significant temperature rise. This stored energy in the oceans may be moved about, being redistributed by ocean currents or taken to great depth only to be released at the surface after being stored for many years. Thus the various types of surface have various *response times*, so that there is no rapid, automatic transfer of energy from the surface into the atmosphere. The consequences of these differences is only now being realised, as we begin to view the whole of the climate system.

In a similar way, we must be concerned with the movement of matter within the atmosphere and with exchanges between the atmosphere and the surface. Climatologically, water in its various forms is of most immediate concern, as it moves from the surface into the atmosphere, is transported by the atmosphere and eventually returned to the surface. Many details of the water cycle are well known and the systematic approach has been used for many years. However, other materials are beginning to receive attention, most notably carbon dioxide. A major problem in predicting any future climate that might arise from an increase in atmospheric concentration concerns the actual atmospheric concentrations that are likely to occur. This concentration depends greatly on how much of the excess $CO_2$ is taken up in surface processes. Various surfaces have different response times, which at present are poorly known. Hence, again, there is a great need to consider the whole of the climate system.

### Energy cascades and transformations

The vital role of the surface even when considering atmospheric motions in isolation is indicated by the *energy cascade* (Fig. 1.4). This cascade shows, in summary form, the ways in which energy is transformed to produce atmospheric motions, and thus the climate. The starting point is 'strong surface heating', which is created because the vast majority of the energy input from the Sun is absorbed at the surface forming the fundamental energy source for all atmospheric motions. Spatial variations in this heating lead directly to both large-scale horizontal temperature gradients and to local convective instability. Any air mass, heated from below (input), tends to rise, thus increasing

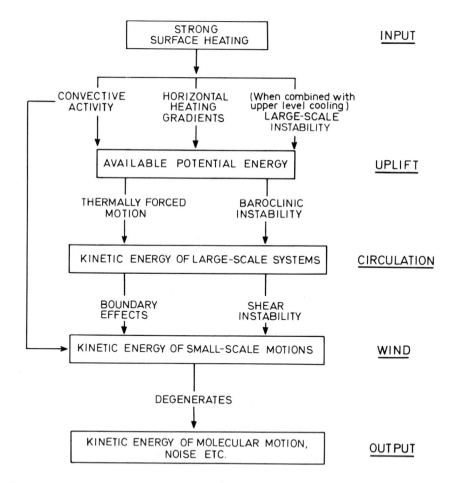

Fig. 1.4. Atmospheric transformations of energy. Almost all energy input to the atmosphere is the result of surface heating following absorption of solar radiation. Large-scale motions degenerate into smaller scale movement and finally into kinetic energy of molecular motion.

the available potential energy (uplift) (as described in Ch. 3, sect. 3.4). Either this potential energy is released in convective activity or, through horizontal energy gradients, in large-scale horizontal motions and synoptic scale weather patterns (circulation) (Ch. 4, sects 4.4 and 4.5). As the resultant winds pass over the Earth's surface small-scale irregularities on the surface give rise to shear instabilities[†] and other boundary effects (wind) (sects 4.3, 6.3 and 6.4). The final fate of either energy path is dissipation into random molecular motions (output). All these energy transformations which constitute the energy cascade are degenerative, leading to lower level forms of energy such as noise and heat. Certainly some of the methods of final dissipation are interesting to consider. For instance, the blowing over of a dustbin creates movement and sound, while an electrical storm produces light and sound. On a somewhat larger scale energy is imparted directly to the oceans through induction of waves, while it has recently been demonstrated that fluctuations in the length of the day are a result of the angular momentum interchange between the atmosphere and the solid Earth. Wind and water power schemes 'tap' the atmospheric kinetic energy as do yachts and gliders and even commerical jet aircraft. Indeed climatology could be defined as the study of local to global scale fluctuations in this energy cascade over time periods long enough to 'smooth out' fluctuations due to the weather.

## 1.5 Climatological space and time scales

As indicated above, there are various scales of activity in the atmosphere, each associated with a particular observing and, we may add, analytical technique. In fact, since the atmosphere is continually in motion, scales of interest could vary from an individual plant or house to the whole globe. To attempt to describe and understand all the activity on all the scales would be a mammoth undertaking. Fortunately it is possible to simplify the situation somewhat. Traditionally meteorologists have divided the whole range into a manageable number of categories: micro, local, meso and macro (Fig. 1.5). This division is based on a number of factors, notably the observational techniques needed to describe the phenomena on that particular scale and the analytical techniques required to obtain an understanding of those phenomena. A major point to note is that short time scale phenomena tend to occur on a small space scale, while long-term phenomena affect large regions. Of course, long-term phenomena are made up of lots of short-term features and regional phenomena are a collection of local effects. So, although the divisions emphasise differing

---

[†] Many of the terms introduced here and throughout the book are explained in the Glossary (p. 423).

ways of looking at things, there are very close linkages between the scales, and our divisions are by no means watertight.

We do not mean to suggest that it is best to think of the climate as a summation of small-scale phenomena finally producing large features. Indeed we shall first consider the large-scale forcing mechanisms of climate as expressed in the energy and water budgets. From these we shall deduce the way that the global atmospheric and oceanic circulations must operate. Thereafter we will consider how these global conditions in turn affect the regional and local scale climate. Thus large-scale features which are fairly consistent in time are considered first. From these we deduce the surface climate with which Man is primarily concerned.

All of these features indicated in Fig. 1.5 have been studied in considerable detail. They have all been considered aspects of meteorology, although the 'core' of that science has traditionally been associated with the scales marked as 'synoptic', which deals with regions the size of nations in Europe or states in the USA and covers a time scale of a few days. The techniques used in this area are many, varied

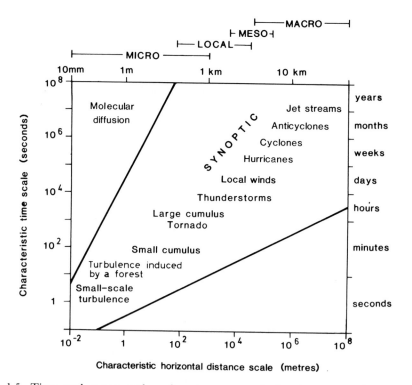

Fig. 1.5. Time and space scales of various atmospheric phenomena. There is an approximately linear relationship to be seen between the size of atmospheric features and their time scales.

and frequently complex. The emphasis of meteorology has been on developing an understanding leading to site and time specific short-term forecasts. The climatologist is more interested in an understanding which leads to longer term predictions or probability forecasts that are certainly not very time specific and probably only in a general way site specific. Hence many of the meteorological techniques need not greatly concern us here. Nevertheless, much of the understanding developed in meteorology must be incorporated as an integral part of our understanding of climatology. Indeed, as meteorology and climatology come closer and closer together, more and more interaction is required.

Although as climatologists we must make great use of results from meteorological investigations, the concepts of Fig. 1.5 must be expanded if we are to give a realistic view of climatology. Firstly, we must expand our time scale to encompass conditions from decades to millennia. Secondly, we must also incorporate the *perception* of the elements of the climate. Such human perception must be included if increases in our understanding of climatic processes are to be trans-

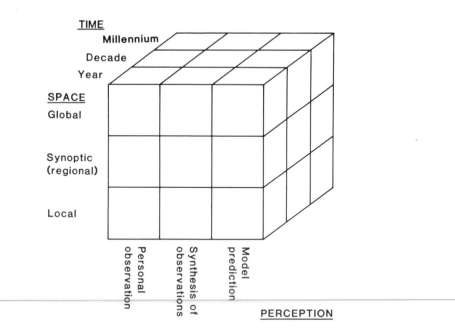

**THE CLIMATE CUBE**

Fig. 1.6. The climate cube. Climate can be viewed as existing in at least three domains: time, space and human perception. The divisions of these domains depicted here are arbitrary but serve to illustrate the importance of interaction between domains.

lated into societal actions which utilise or ameliorate climatic impacts. Hence Fig. 1.6 has been proposed as a comprehensive view of climatology. The 'meteorological' aspects shown in Fig. 1.5 can be thought of as dominating the area near the left front edge of this climate cube. It is also this edge that we know most about. The effort to extend our understanding throughout the rest of the cube is the focus of contemporary climatology.

## Physical climatology

Certain basic processes go on at all scales, or irrespective of scale. These we can think of as the climate controls, and their study is usually called 'Physical Climatology'. This is considered in Chapters 2 and 3. Since the atmosphere is driven by energy, energy considerations provide our starting point. Thus in Chapter 2 we treat atmospheric processes as a series of activities which takes the energy which arrives as sunlight, converts it to other forms of energy, moves it around in the atmosphere and eventually returns it to space. Also in this chapter some of the direct consequences of this energy system, notably the distribution of temperature, are discussed. One of the most important consequences of energy exchanges is the development of a water cycle which includes ice in the polar regions, liquid water in the oceans and in rain, and vapour in the atmosphere. Water in all its forms is not only vital to life but also modifies, and is driven by, the energy system. Thus water is not only part of the climate, but also plays a great role in creating climate. Chapter 3 explores this role.

## Dynamic climatology

As soon as we add consideration of horizontal air motions – wind – to our study of climate, not only do we need to change the name to 'Dynamic Climatology', but we also introduce the effect of scale. We have divided the dynamic climatology section into three scales in order to demonstrate more clearly how the climate controls generate what we now think of as the complete climate. Thus in Chapter 4 we will be concerned with the largest scale: the whole globe or major portions thereof. The time scale will range from seasons to decades, a scale where we can use actual observations to develop our understanding. Following the global scale we have, in Chapter 5, the regional scale – the synoptic scale treated in a climatological way. Chapter 6 is concerned with the smallest scale, from a field to a city and from minutes to a day or so. Each of these scales needs different techniques, and a differing emphasis, within the general framework of our understanding of the physical laws governing atmospheric activity.

The quantitative analysis of the science of climatology is drawn together in Chapter 7, where we describe the methods employed in constructing climate models. Study of the climate on longer time scales, decades to millennia, requires a largely different technique, mainly because we lack observations or measurements of a direct nature. Proxy variables do not necessarily give us the 'right' parameter at the 'right' place at the 'right' time. We have to use what we can get and use the physical understanding developed earlier to fit all the bits and pieces together to find out how, and possibly why, the climate has changed in the past. Historical climatology is also reviewed in Chapter 7. Finally, in section 7.6, we can put everything together to discuss ways of answering the question 'What will the future climate be?' Even this is scale dependent, so that there are several approaches that we shall follow in order to see how we can answer this question, what we need to know to answer it and even what type of information constitutes a useful answer.

# Chapter 2
# The Radiation Budget and Energy Systems of the Earth

2.1    The nature of radiation

2.2    Radiation from the Sun

2.3    Interaction of radiation with the atmosphere

2.4    Solar radiation at the Earth's surface

  2.4.1    The availability of solar energy – an application of climate information

2.5    Longwave radiation

  2.5.1    The temperature structure of the atmosphere

2.6    Monitoring radiation from space

2.7    The global radiation budget

2.8    Surface radiation budgets

2.9    Energy and temperature

2.10   Surface energy budgets

2.11   Temperatures at the Earth's surface

2.12   Applications of temperature information

# Chapter 2
# The Radiation Budget and Energy Systems of the Earth

All aspects of the climate system of the Earth – the winds, rain, clouds and temperature – are the result of energy transfers and transformations within the Earth/atmosphere system. These energy exchanges, which create and drive the climate, are the focus of this chapter. The whole process starts when energy from the Sun arrives at the top of the atmosphere in the form of radiant energy, or radiation. It is transferred down through the atmosphere, interacting with it, so that some energy is reflected back to space, some is absorbed and transformed into heat and some is transmitted to the surface of the Earth. The radiation that penetrates to the surface and is absorbed can heat the surface, evaporate water, melt snow and heat the underlying soil. This transformed energy is eventually transferred to the atmosphere and finally returned to space again in the form of radiation. Variations in the amount of radiant energy received from the Sun and variations in the interaction with the Earth and atmosphere create the temporal and spatial variations in energy exchanges which lead to our climate.

*Energy cascade*
The whole process of energy exchange in the Earth/atmosphere system can be summarised by the 'energy cascade' (Fig. 2.1). This indicates the many and various forms of energy that are possible, the tremendous variations in their amounts and the ability of the system to store energy. Energy exchanges are going on all the time on a variety of time scales. Some processes are rapid enough to be seen. A familiar example is the formation of scattered clouds on a warm sunny afternoon after a cloudless morning. Energy from the Sun is used to heat a water surface throughout the day. Some of this energy is used for evaporation, which is a form of energy transfer. This creates humid air which, again as the result of solar heating, ascends. The vigour of the ascent usually increases throughout the morning until, by afternoon, it is

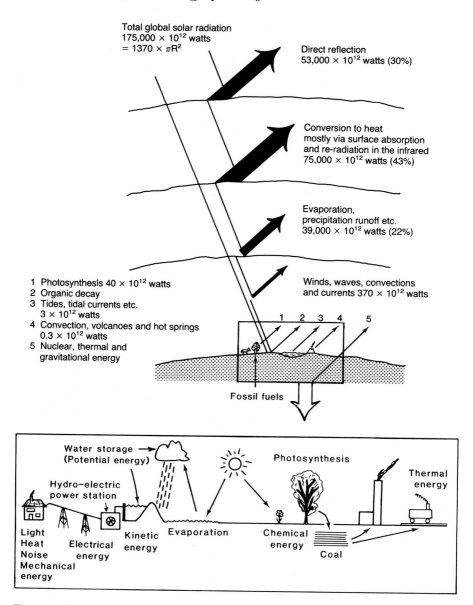

Fig. 2.1. The global energy cascade. Top-of-the-atmosphere irradiance is equal to the solar irradiance at the Earth's orbital distance multiplied by the instantaneous area of the planet illuminated, $\pi R^2$ (see Fig. 7.7 in Ch. 7). The diagram shows schematically how solar energy, arriving at the top of the atmosphere, is absorbed by the surface and how this, when re-emitted, is distributed amongst the various features of the energy cascade. This cartoon complements the energy transfer systems shown in Fig. 1.4. The lower box illustrates schematically some of the features of the energy cascade which most directly affect life on the surface of the Earth.

strong enough to force the water vapour to condense and form clouds. Energy is stored in these clouds as potential energy. This will soon be released if conditions are appropriate for the development of rain. Thus there is short-term energy storage. On a much longer time scale, when we heat our homes using coal or oil, we are experiencing the final stages of an energy exchange process that has taken millions of years to complete. Solar radiant energy was used directly to build plant and animal tissue which was subsequently stored when buried and converted to coal or oil. We now use this fossilised radiant energy to create heat.

*Energy balance of the Earth*

The globally and annually averaged values for the various energy exchanges are given in Fig 2.2. It shows that the surface loses as much energy as it receives. The same is true for the atmosphere and for the whole planet. Thus there is a radiation balance maintained on the global, annual scale considered here. Without this balance very rapid climatic changes would occur. On this global, annual scale, of course, all the horizontal motions in the atmosphere and oceans, which transfer energy, are omitted. The effects of vegetation, topography, day and night, and the seasons are ignored. The numbers thus represent a generalised picture of the energy exchange, a picture we can later refine. In Fig 2.2 radiation has been divided into two categories: the incoming short wavelength solar radiation and the outgoing long wavelength terrestrial radiation emitted by the Earth. This fundamental division between solar and terrestrial radiation arises directly from the nature and properties of electromagnetic radiation which are described in the next section. This division has not only great implications for the Earth's climate system, but is also of practical significance for satellite observations of the Earth and its atmosphere.

# 2.1 The nature of radiation

Radiation is a form of energy that is emitted by all objects having a temperature above absolute zero. It is the only form of energy that can travel through the vacuum of outer space. Thus energy receipt and disposal from the planet Earth must be in the form of radiation.

*Electromagnetic spectrum*

A fundamental characteristic of radiation is the wavelength of propagation. All bodies radiate at a large number of wavelengths. The complete set of possible wavelengths is called the *electromagnetic spectrum* (Fig. 2.3). Climatologically important radiation is within the wavelength range 0.1 $\mu$m to 100 $\mu$m. Within this region the human eye responds to a very small portion, which we call *visible light*. Our sense

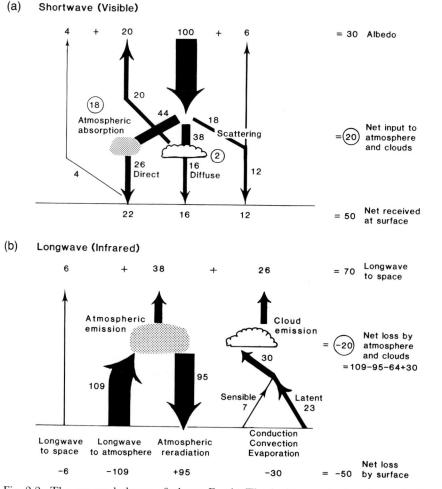

**(a)    Shortwave (Visible)**

4    +    20        100    +    6        = 30  Albedo

20

(18)
Atmospheric
absorption        44        18
                    38    Scattering        =(20) Net input to
                                                  atmosphere
                                                  and clouds

4        26        16    (2)
         Direct    Diffuse        12

22        16        12        = 50  Net received
                                    at surface

**(b)    Longwave (Infrared)**

6    +    38    +    26        = 70  Longwave
                                    to space

Atmospheric
emission                Cloud
                        emission        =(−20) Net loss by
                                                atmosphere
                            30                  and clouds
        95                                  =109−95−64+30
109                Sensible        Latent
                    7            23

                            Conduction
                            Convection
Longwave    Longwave    Atmospheric    Evaporation
to space    to atmosphere  reradiation
−6        −109        +95        −30        = −50  Net loss
                                                   by surface

Fig. 2.2. The energy balance of planet Earth. The incident solar irradiance is shown as 100 units. Thus the reflected, transmitted and absorbed components of both incident solar and emitted terrestrial radiation are percentages of this value. For example, the diagram shows a globally averaged value of surface albedo of 4/26 or 0.15. The short-wave radiation (a) is balanced by the longwave radiation (b) at (i) the top of the atmosphere (the planetary budget); (ii) the atmosphere; and (iii) the surface. Note the major contribution to the planetary albedo (20%) made by clouds which also make a significant contribution to the emitted radiation. Despite the importance of these features the primary site of absorption of shortwave radiation is the surface.

of colour depends very much on the wavelength of the light our eyes receive. Wavelengths around 0.40 $\mu$m give violet light. As wavelength increases we see the colours of the rainbow until at 0.70 $\mu$m we reach red light. Regions adjacent to this visible portion are given names associated with the nearest colour. The region with wavelengths slightly shorter than 0.40 $\mu$m is termed *ultraviolet*, while radiation with

32

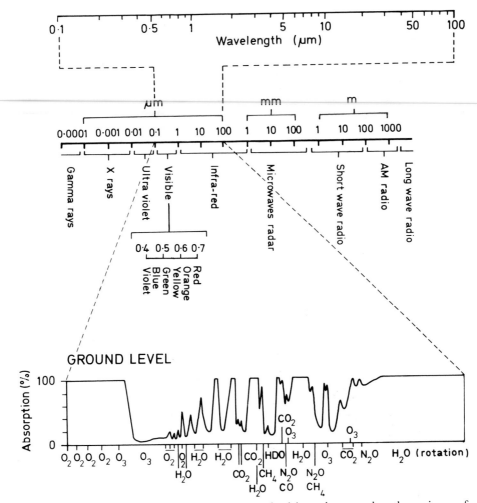

Fig. 2.3. The electromagnetic spectrum, emphasising the wavelength regions of importance for climatology which range from ~0.1 $\mu$m to ~100 $\mu$m. The lower part of the diagram shows atmospheric absorption over this wavelength range. The atmosphere is seen to be transparent (little absorption) in the visible part of the spectrum but exhibits considerable absorption by $O_3$ in the ultraviolet and by $H_2O$ and $CO_2$ and other molecules in the infrared part of the spectrum.

wavelengths longer than ~1.0 $\mu$m (and $\leqslant$ 1 mm) is termed *infrared radiation*.

## Radiation laws

The wavelength of propagation depends on the temperature of the emitting body. The basic law of radiant emission is *Planck's Law*

$$E_\lambda^* = c_1/[\lambda^5 \ (\exp(c_2/\lambda T) - 1)] \qquad\qquad [2.1]$$

33

where $E_\lambda^*$ is the amount of energy (W m$^{-2}$ $\mu$m$^{-1}$) emitted at wavelength $\lambda$ ($\mu$m) by a body at temperature $T$ (K). The two constants $c_1$ and $c_2$ have values of $3.74 \times 10^8$ W $\mu$m$^4$ m$^{-2}$ and $1.44 \times 10^4$ $\mu$m K respectively. This equation is displayed graphically in Fig. 2.4(a, b) for perfectly radiating bodies that have temperatures that correspond approximately to those of the Sun (5800 K) and the Earth (255 K). For a given temperature, $T$, these Planck curves are uniquely defined and have a characteristic shape. The radiating or black body temperature of the Earth is about 33 K less than the mean surface temperature. The higher surface value is due to the greenhouse effect which will be explained more fully in section 2.5.

The wavelength of maximum emission for a body at a particular

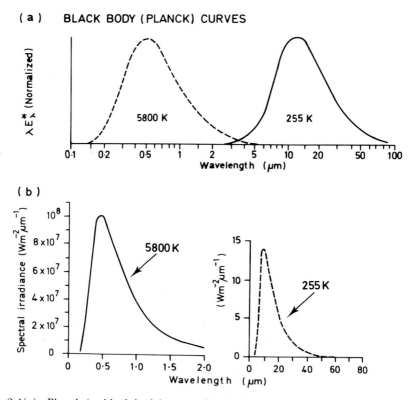

Fig. 2.4(a). Planck (or black body) curves for the Sun (temperature ∼5800 K) and the Earth (temperature ∼255 K). The vertical scale is normalised so that the two curves have the same vertical range. The peak of the solar curve is at ∼0.5 $\mu$m while the peak of the terrestrial curve is at ∼11.4 $\mu$m. The very small spectral overlap (at ∼3.5 $\mu$m) means that there is little chance of confusing the sources of remotely sensed radiation.

Fig. 2.4(b). Quantitative vertical scales (W m$^{-2}$ $\mu$m$^{-1}$) show the differences in total energy ($E^*$) emitted by the Sun and the Earth. The value of $E^*$ is given by the area beneath the Planck curves.

temperature is inversely proportional to the temperature and is given by *Wien's Law*

$$\lambda_{max} = 2897/T \qquad [2.2]$$

This equation is obtained by differentiation of equation [2.1] and indicates maximum emission wavelengths for the Sun and the Earth as 0.50 $\mu$m and 11.4 $\mu$m respectively.

The total energy emitted by a body, the area under each curve in Fig. 2.4(b), can be found by integration of equation [2.1]. This gives the *Stefan–Boltzmann Law*

$$E^* = \sigma T^4 \qquad [2.3]$$

where $\sigma = 5.67 \times 10^{-8}$ W m$^{-2}$ K$^{-4}$ is the Stefan–Boltzmann constant. Thus the amount of energy emitted increases with temperature, as indicated by Fig. 2.4(b).

These three laws apply directly to bodies that are theoretically perfect radiators. Such bodies are called *black bodies*. The degree to which a real body approaches a black body is given by the emissivity of the body

$$\epsilon_\lambda = E_\lambda/E_\lambda^* \qquad [2.4]$$

where, for the wavelength $\lambda$, $\epsilon_\lambda$ is the emissivity and $E_\lambda$ and $E_\lambda^*$ are the emissions of the real and black bodies respectively. The value of the emissivity is dependent on wavelength. In general, solids and liquids radiate across a continuous spectral interval with a more or less constant emissivity, usually between 0.9 and 1.0. Gases, on the other hand, radiate only at specific wavelengths, and so have variable emissivities, a characteristic which complicates calculations of how much energy they emit at a given temperature but which can be exploited in satellite-based measurements of atmospheric properties.

Although, so far, we have been concerned with emission, there is a very simple relationship between emissivity ($\epsilon_\lambda$) and absorptance ($\alpha_\lambda$)

$$\epsilon_\lambda = \alpha_\lambda \qquad [2.5]$$

This is known as *Kirchhoff's Law* and indicates that, at the same wavelength, good emitters are equally good absorbers. Note that the equation refers to emissivity and absorptance, not actual amounts of absorption or emission. An earthbound body with a high emissivity at solar wavelengths must absorb a great amount of solar energy but may emit very little because its temperature is such that Planck's law indicates that very little energy can be emitted at that wavelength.

Energy that is not absorbed when it impinges on a body must be either reflected or transmitted. The proportion of each that will occur in a given situation depends on the wavelength of the incident radiation and the properties of the body.

The basic radiation laws emphasise that when considering radiation in the climate system it is advantageous to use two distinct radiation regimes: the shortwave (solar) radiation originating in the Sun and the longwave (terrestrial) radiation emitted by the Earth and its atmosphere. Although the two overlap slightly, they are sufficiently distinct to be treated completely separately. This separation, indicated in Fig. 2.4(a), is very convenient since the two regimes interact with the climate system in different ways, which we explore in the rest of this chapter.

## 2.2   Radiation from the Sun

Nuclear reactions within the core of the Sun produce the radiant energy that is emitted, mostly from the photosphere (Fig. 2.5(a)). This radiation is a good approximation to that of a black body. Thus the emitted radiation essentially follows the Planck curve for a temperature of 5800 K. Only in the high energy region, at wavelengths shorter than about 0.3 $\mu$m, does emission fall significantly below the theoretical level (Fig. 2.5(b)). The wavelength of peak emission is observed to be at 0.474 $\mu$m. This suggests a Sun with a blue-green colour. The observed

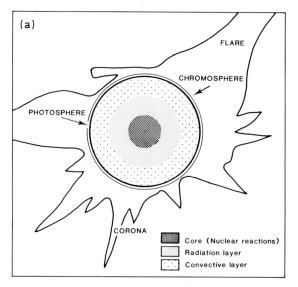

Fig. 2.5(a). A cross-sectional diagram showing the regions of the Sun. The radiant energy passes from the core to the photosphere and finally through space to the planets.

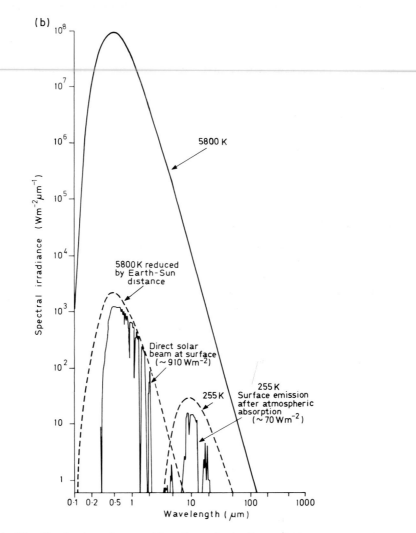

Fig. 2.5(b). The distribution of spectral irradiance for bodies radiating at temperatures of 255 K and 5800 K. The solar irradiance incident at the top of the atmosphere (solid 5800 K curve) is reduced by the Earth–Sun distance (dashed curve). Note that only at wavelengths shorter than ~0.3 $\mu$m and longer than ~2.0 $\mu$m do the curves of top-of-the-atmosphere irradiance and irradiance received at the surface differ significantly. This is due to the absorption, particularly in the stratosphere, of high energy radiation and near infrared absorption in the troposphere respectively. The effect of absorption by tropospheric gases depletes the terrestrial radiation (dashed 255 K curve) to a greater extent resulting in the highly wavelength-dependent irradiance (solid curve) finally emitted from the top of the atmosphere.

37

yellow Sun is a result of the relative sensitivity of the human eye, interaction of radiation with the Earth's atmosphere and the relative intensities of emission because of the shape of the Planck curve. As a result of this shape, 99% of the energy is emitted in the wavelength region 0.15 $\mu$m to 4.0 $\mu$m. The distribution of emitted energy is such that 9% is in the ultraviolet, 45% in the visible and 46% in the infrared (Fig. 2.5(b)).

### Received solar energy

The top of the Earth's atmosphere receives a fraction approximately equal to $4.5 \times 10^{-10}$ of the energy output of the photosphere. This represents the energy available to drive the climate system. The amount is usually expressed in terms of the *solar constant*: the amount of energy passing in unit time through a unit surface perpendicular to the Sun's rays at the outer edge of the atmosphere at the mean distance between the Earth and the Sun. Although numerous observations of the solar constant have been made, its value is not precisely known. The current best estimate is 1370 W m$^{-2}$.

The solar constant is known to vary on a number of time scales. Although the importance of these for climatic change will be considered in Chapter 7 (sect. 7.1), it is useful here to review briefly the main time scales and causes of variations. The longest of these is the variation caused by the evolution of the Sun itself. Over the lifetime of the solar system the temperature and hence (from equation [2.3]) the emitted energy from the Sun is believed to have increased by between 20% and 40%. This increase has not only changed the solar constant, but also the spectral distribution of the energy.

On a much shorter time scale, variations in energy output from the Sun result from changes in activity within the Sun itself. These are associated with sunspots, which result from convective activity which constantly mixes the upper layers of the photosphere. Sunspots appear as dark regions having a radius of about 4000 km. They generally occur between 5° and 35° of the solar equator. The number of spots varies in an approximately cyclic way, a complete cycle taking about 22 years. The magnetic activity within the active sunspots leads to solar flares, enhanced ultraviolet and X-ray emissions. These variations influence the value of the solar constant and are one main reason for our uncertainty concerning its exact value.

In addition to the processes acting within the Sun, the energy received at the Earth also varies as a result of the astronomical relationship between the Sun and the Earth. This relationship, which is well known and can be exactly specified, accounts for the systematic daily and seasonal changes in the solar energy received at the top of the atmosphere, and exerts great influences on some longer term variations.

### Earth's orbit around the Sun

The main characteristics of the orbital geometry of the Earth are shown in Fig. 2.6. The Earth revolves around the Sun in an elliptical orbit with a period of one year. The *eccentricity* of the orbit, a measure of the variation in the Earth–Sun distance, is such that at present the Earth is closest to the Sun (perihelion) in the Northern Hemisphere winter. The eccentricity of the Earth's orbit is very small and thus its orbital path is similar to that of the planet represented in the lower part of Fig. 2.6. Although the eccentricity changes very slowly with time, the present eccentricity has a relatively minor influence on the variation in radiation receipt. Much more significant for seasonal variations is the inclination of the Earth's axis. The axis, the line through the centre of the Earth joining the two poles, is tilted by 23.5° from the perpendicular to the plane of the ecliptic, the plane of the Earth's orbit around the Sun. Since the axis retains the same orientation with respect to the galaxy, the effect is to create seasonal variations in the amount and intensity of radiation received at a given spot at the top of the atmosphere and, eventually, at the surface of the Earth.

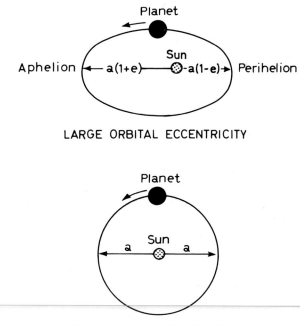

Fig. 2.6. Planet–Sun geometry illustrating the orbital configuration of a planet having a large orbital eccentricity (upper) and that of a planet like the Earth having an almost zero orbital eccentricity (lower). The point in the orbit at which the planet is furthest from the Sun is called aphelion, while the closest point is known as perihelion.

NORTHERN HEMISPHERE SUMMER
( Aphelion )

NORTHERN HEMISPHERE WINTER
(Perihelion )

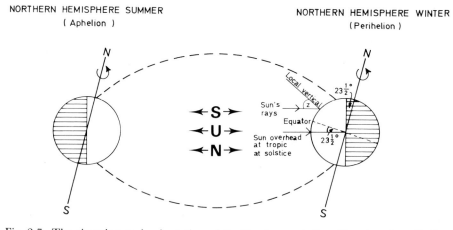

Fig. 2.7. The situation and orientation of the Earth at aphelion (June) and perihelion (December). The solar zenith angle, Z, is shown.

## Daily and seasonal variations in solar radiation

The consequences of the *axial tilt* can be seen by considering Fig. 2.7. Both the duration of daylight and the height of the Sun in the sky vary with time. To a good approximation we can say that the length of daylight is proportional to the fraction of the latitude circle that is unshaded. At the extreme positions the noontime Sun is overhead at either the Tropic of Cancer, giving the summer solstice, or Tropic of Capricorn, creating the winter solstice. At the latter, in areas north of the Arctic circle there is 24-hour night whilst south of the Antarctic circle there is 24-hour daylight. The equinoxes, at which there is 12 hours daylight everywhere on the Earth, occur when the Sun is overhead at the equator on 22 March and 22 September.

The height of the Sun in the sky is usually given in terms of the solar *zenith angle*. This is the angular distance between the Sun's rays and the local vertical, and is given by

$$\cos Z = \sin \theta \sin \delta + \cos \theta \cos \delta \cos h \qquad [2.6]$$

where $\theta$ is the latitude of the place, $\delta$ is the solar declination and $h$ is the hour angle, which is zero at local noon and increases in magnitude by $\pi/12$, (15°), for every hour before or after noon. The 'half day length', $H$, can be calculated from equation [2.6] by noting that at sunrise and sunset $\cos Z = 0$ ($Z = \pi/2$, (90°)). Thus

$$\cos H = - \tan \theta \tan \delta \qquad [2.7]$$

The amount of incoming energy on a horizontal surface at the top of the atmosphere $I_0$ is related to $Z$ by

$$I_0 = S_F \, (d/d')^2 \cos Z \qquad [2.8]$$

where $S_F$ is the solar constant and $d/d'$ is the correction factor for the variable Earth–Sun distance, $d$ being the mean Earth–Sun distance and $d'$ its instantaneous value. Combining equations [2.6] and [2.8] and integrating from $-H$ to $+H$, the daily total radiation at the top of the atmosphere is

$$I_{DT} = \frac{86\,400}{\pi} S_F (d/d')^2 \, (H \sin \theta \sin \delta + \cos \theta \cos \delta \sin H) \qquad [2.9]$$

where $H$ is expressed in radians.

The resultant variation in the incident solar energy, or insolation, at the top of the atmosphere as a function of latitude and month is shown in Fig. 2.8. It can be seen that while the insolation near the equator is never as great as the maxima achieved at the poles, it is consistently high throughout the year. Polar insolation, on the other hand, is zero during the polar night but peaks during the polar day when the Sun is permanently above the horizon. This spatial and temporal variation in radiation that enters the atmosphere is vital for driving the climate system. In the next section we follow the radiation as it enters the atmosphere and interacts with it.

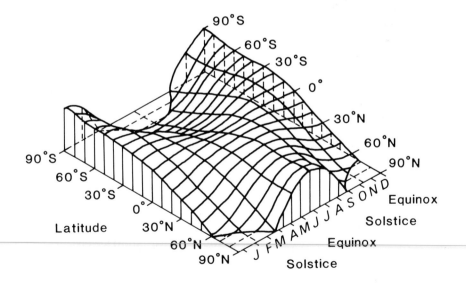

Fig. 2.8. The variation of insolation (at the top of the atmosphere) as a function of latitude and month for the whole globe. The very large amounts of radiation received at the poles in summer are the result of the 24 hours of daylight at that time.

## 2.3 Interaction of radiation with the atmosphere

Once solar energy impinges upon the top of the Earth's atmosphere the global energy cascade (Fig. 2.1) begins. This radiant energy can be returned directly to space, stored within the Earth/atmosphere system, or absorbed and transformed. As solar radiation starts to pass down from the top of the atmosphere, scattering and absorption of the radiation commences. Although absorption leads to heating, which in turn gives longwave radiation emission and, later, absorption by atmospheric gases, we shall be concerned only with absorption, scattering and transmission of shortwave radiation in the present section. Absorption and re-emission of longwave radiation will be mentioned in places but the main discussion of longwave radiation is postponed until section 2.5.

### *Scattering of solar radiation*

Scattering of radiation occurs whenever a photon impinges on an obstacle in the atmosphere without being absorbed. The only effect of scattering is to change the direction of travel of the photon (Fig. 2.9). The change can be in any and all directions, but it is usually convenient to think at first in terms of only two directions: upwards and downwards relative to the Earth. Upward scattered radiation, unless it is rescattered downwards and thus takes part in multiple scattering, is lost to space and can take no further part in energy processes. Downward scattered radiation, on the other hand, is still in the system and susceptible to further interactions.

Scattering particles such as atmospheric gas molecules which are small compared to the wavelength of the incident radiation produce *Rayleigh scattering* (Fig. 2.10(a)). The amount of scattering can be shown to be inversely proportional to the fourth power of the wavelength. Thus for visible radiation the scattering of blue light (wavelength $\sim 0.4\ \mu$m) is $\sim 10$ times greater than for red light (wavelength $\sim 0.7\ \mu$m) (Fig. 2.9(a)). This type of scattering is characteristic of an atmosphere composed mainly of the normal atmospheric gases without contaminants. The major, and obvious, consequence of Rayleigh scattering is the blue colour of the daytime sky. The effect of the very efficient scattering of blue light can be viewed on any evening at around twilight. The very long path length of the radiation through the atmosphere at that time causes most of the shorter visible wavelengths to be scattered many times and hence dispersed from the direct solar beam, leaving only the red component unaffected, creating a much redder Sun than during the rest of the day. The reddish sky often associated with this is a result of scattering of red wavelengths slightly out of the direct beam.

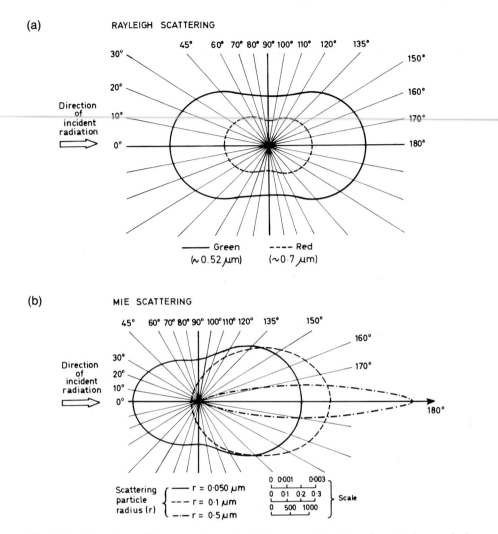

Fig. 2.9. Polar plots of the intensity of radiation as a function of scattering angle in the case of (a) Rayleigh scattering for a particle size $r \simeq 0.025$ $\mu$m for two wavelengths (upper) and (b) Mie scattering for green light (i.e. wavelength $\sim 0.52$ $\mu$m) for three particle sizes (lower). Note that the intensity of the scattered beam increases very rapidly (see inset scale) as the particle size increases.

Whenever clouds, pollution particles, or water droplets are present in the atmosphere *Mie scattering* occurs (Fig. 2.10(b)). This is effective whenever the scattering particles are of a similar size to the wavelength of the incident radiation. In this case scattering is almost entirely in the forward direction (Fig. 2.9(b)). The amount and direction of scattering is a function of both the particle size and the wavelength of the radiation. However, there is a tendency to scatter light of all wave-

43

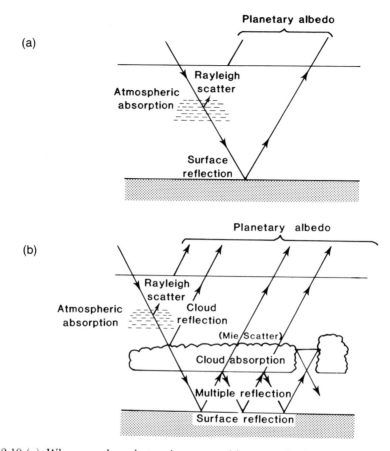

Fig. 2.10.(a) When a solar photon interacts with atmospheric gases the scattering process is well described by Rayleigh theory. (b) However, Mie theory is required to describe the more complex interaction which takes place between incident solar photons and water droplets and particulate material (especially in clouds).

lengths, the effect of which is most readily seen when the atmosphere is turbid (i.e. laden with pollution) and creates a light blue/grey sky, since no wavelengths are preferentially scattered. As most cloud droplets absorb radiation as well as scattering it, the effect is less obvious in cloudy conditions.

Although some photons are only scattered once as they pass through the atmosphere, multiple scattering is more common. Scattering within clouds is always multiple scattering. The effect can also be seen in conditions of a *whiteout* occurring in high latitudes. Here multiple scattering between a bright snow surface and the base of a low cloud layer makes it extremely difficult to see which is which, or where the horizon is located, and to distinguish features of the surface.

When radiation impinges on a much larger body, such as the Earth's surface, there is a complete change in direction for the radiation which is not absorbed and it is conventional to regard this special case of scattering as 'reflection'. Reflection and scattering, whether single or multiple, lead to both upward and downward radiation streams. The ratio of the two radiation streams is the *albedo*

$$A = K{\uparrow}/K{\downarrow} \qquad\qquad\qquad [2.10]$$

where $K$ is solar radiation and the arrows indicate the direction of travel. The albedo at any level of the Earth/atmosphere system can be determined from measurements of the two radiation fluxes. However, two levels are of paramount importance: the Earth's surface and the top of the atmosphere. At the top of the atmosphere satellite measurements enable us to calculate the *planetary albedo* for the whole globe or large regions thereof. The values are vital for an understanding of large-scale climate processes. Figure 2.2 indicates that the planetary albedo is about 0.3. At the surface of the Earth the *surface albedo* is normally calculated from direct measurements for a single type of surface, since the value is highly dependent on the nature of the surface. Variations between surface types have significant consequences for local climates. Often *cloud albedo* is considered separately, since it is important not only in calculations of radiation transfer, but also for its influence on the local climate.

### Absorption of solar radiation

*Absorption* of shortwave radiation by atmospheric gases is relatively small and is usually considerably less important than scattering. As can be seen from the absorption curve shown in the lower part of Fig. 2.3, the atmosphere is almost completely transparent near the peak of the solar Planck curve. Ultraviolet radiation is strongly absorbed by ozone in the lower stratosphere. This results in a shielding of the surface from ultraviolet radiation and is also responsible for the temperature inversion at the tropopause (Fig. 2.21). There are a few narrow regions of the spectrum, called 'bands', where absorption by oxygen, ozone and water vapour occurs. In total, Fig. 2.2 indicates that this gaseous absorption removes only about 18% of the energy arriving at the top of the atmosphere from the beam with absorption by clouds removing another 2%. This is much less than the amount of energy that is absorbed at the surface ($\sim$50%). Since absorption leads directly to heating, this fact has a great consequence for the energetics of the atmosphere.

Although scattering and absorption are separate processes and have different consequences, they occur almost simultaneously and it is often very convenient to consider as a single unit the depletion of solar radiation by the atmosphere. This is best expressed in terms of the

45

*optical thickness*, sometimes referred to as the *optical depth*, of the atmosphere. This is defined for an atmospheric layer of interest in terms of the ratio of the radiation emerging at the bottom to that incident at the top, $I_b$ and $I_t$ respectively, such that

$$I_t/I_b = \exp(-\tau/\cos Z) \qquad [2.11]$$

where $\tau$ is the optical thickness of the layer in question. The whole atmosphere is said to be *optically thin* if $\tau \lesssim 0.2$–$0.5$. From equation [2.11] it can be seen that the atmosphere tends to become *optically thick* as the ends of the day are approached and $Z$ tends to $\pi/2$. The single largest component of the atmospheric optical thickness, both on average and on a given day, is that due to clouds. In general, high thin clouds are optically thin, a readily verifiable concept since it is frequently possible to see the Sun or the Moon through them. However, all other cloud types are generally optically thick and responsible for a great amount of depletion of incoming solar radiation.

The net effects of the interaction between shortwave radiation and the atmosphere are summarised in Fig. 2.2. Of the incoming energy at the top of the atmosphere 26% is backscattered by the air and clouds and 20% is absorbed. Thus 54% reaches the surface of the Earth. We now, therefore, consider this solar energy reaching the Earth's surface.

# 2.4 Solar radiation at the Earth's surface

The final interaction as the solar radiation passes through the atmosphere is with the surface of the Earth itself. Radiation incident on an opaque surface is either absorbed or reflected. The proportion of the incident energy that is reflected, the *surface albedo*, is not highly wavelength dependent and can be treated as a single value for a given surface type (Table 2.1). Most natural land surfaces have albedos between 0.10 and 0.25, with Man-made surfaces having albedos slightly higher. Snow, with a high albedo, is a notable exception. Water is also an exception both in its generally low value and in that the value varies strongly with the zenith angle.

## *Surface albedo*

A high albedo indicates that much of the incident energy is reflected rather than absorbed. For example, the high albedo of snow means that it reflects much of the incident energy, instead of absorbing it and being warmed. Thus a clean snow surface can survive on a sunny day. Dirty snow includes material of a lower albedo within it and will absorb more energy and melt much faster. Similarly, albedo differences between vegetation and nearby Man-made surfaces are one of the

46

## Table 2.1 Albedos and emissivities of common surface types (annual means)

| Type | Albedo (A) | Emissivity ($\epsilon$) |
|---|---|---|
| Tropical forest | 0.13 | 0.99 |
| Woodland | 0.14 | 0.98 |
| Farmland/natural grassland | 0.20 | 0.95 |
| Semi-desert/stony desert | 0.24 | 0.92 |
| Dry sandy desert/salt pans | 0.37 | 0.89 |
| Water (0°–60°)[a] | <0.08 | 0.96 |
| Water (60°–90°)[a] | <0.10 | 0.96 |
| Sea ice | 0.25–0.60 | 0.90 |
| Snow-covered vegetation | 0.20–0.80 | 0.88 |
| Snow-covered ice | 0.80 | 0.92 |

[a] The albedo of a water surface increases as the solar zenith angle increases. Ocean surface albedos are also increased by the occurrence of white caps on the waves.

causes of the temperature differences between them. Although more than albedo must be considered in analysing these temperature differences, the surface albedo is of prime consideration in the establishment of local climates, as is discussed in Chapter 6.

In general, the albedo of the Earth's surface is around 0.15, mainly because of the predominance of water as a surface type. Consequently the majority of the energy reaching the surface is absorbed. Indeed, Fig. 2.2 indicates that approximately 50% of the solar energy reaching the planet is absorbed at the surface, compared to only 20% absorbed by the atmosphere. The planetary albedo is seen to be 0.3, i.e. 30% of the incident radiation is reflected. The predominance of absorption and heating at the surface indicates that the major source of heating for the lower part of the atmosphere is the surface.

Although it is rarely necessary to subdivide shortwave radiation at the surface into specific spectral regions when considering the energetics of the climate system, for certain other applications such detail is needed. A pertinent example here is the variability of plant responses to the wavelength of incoming radiation (Fig. 2.11). The leaf of the plant *Populus deltoides* (cottonwood tree) is reasonably characteristic of most leaves. There is strong absorption in the ultraviolet and at blue and red wavelengths, where the energy is used for photosynthesis. In the near infrared, absorption is weak and most of the energy is transmitted or reflected. This weak absorption, in a region where the solar energy is not required, prevents overheating of the plant tissue. Variations in the spectral composition of the solar radiation thus influence plant growth. A change from complete cloud-free conditions to complete overcast, for example, mainly decreases incoming radiation

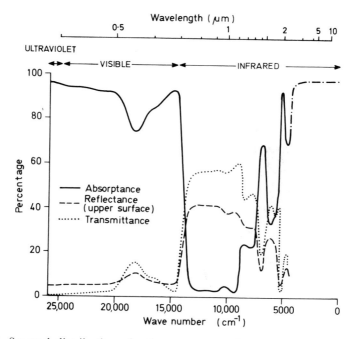

Fig. 2.11. Spectral distribution of reflectance, transmittance and absorptance of the leaves of *Populus deltoides* (cottonwood tree). The sudden increase in reflectance at ~0.7 $\mu$m is easy to see. Note the terms reflectance, absorptance and transmittance are applied when radiation impinges upon a partially transparent substance. In the case of a completely opaque body (e.g. the Earth's surface) the terms reflectivity and absorptivity can be used.

in the wavelengths where transmission occurs (Fig. 2.12). Similar but more persistent and detrimental changes in the spectral composition may be caused by an increase in pollution in the atmosphere or, on a much longer time scale, by changes in the nature of the Sun's radiant output.

### *Photosynthesis controlled by solar radiation*

The amount of photosynthesis, and therefore the amount of growth, in plants depends not only on the spectral composition of the radiation, but also on the amount. In general, other factors such as water availability and carbon dioxide concentration being non-limiting, the amount of photosynthesis increases approximately linearly with incident energy amount up to a certain point (Fig. 2.13). Beyond this point the plant becomes light-saturated and the photosynthetic rate remains constant. This saturation value depends greatly on the species of plant, some shade-loving plants having low values while others rarely receive sufficient energy to reach saturation. At extremely high levels of

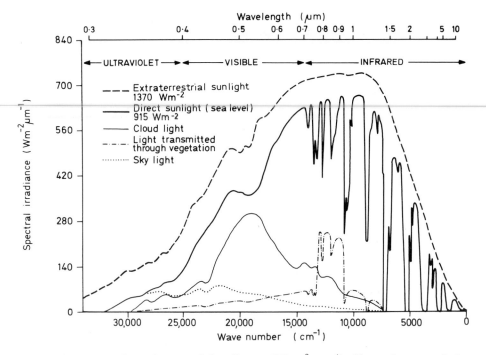

Fig. 2.12. Distribution of spectral irradiance ($W\,m^{-2}\,\mu m^{-1}$) (i) at the top of the atmosphere; (ii) the direct component at sea level; (iii) the component diffused by clouds; (iv) the component typically transmitted through vegetation; and (v) the component diffused by atmospheric scatter alone (i.e. in cloudless conditions.)

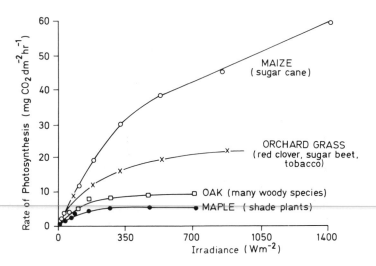

Fig. 2.13. The effect of irradiance upon the photosynthetic rate for four green plants which are typical of the plant groups indicated in parentheses.

irradiance, however, the plant's temperature regulating mechanism may break down and wilting and death occur.

### *Instruments to measure solar radiation*

Knowledge of the amount of shortwave radiation at the Earth's surface is vital, from the viewpoint of both the energetics of the climate system and practical applications, notably in agriculture. Surface-based instruments are commonly of three types:

1. *Pyranometers* (Fig. 2.14(a)), which measure solar radiation at all wavelengths coming from an entire hemisphere. They are usually mounted with the receiver surface horizontal, thus observing the total radiation from the Sun and the sky. A pair of pyranometers, one upward facing, one looking down, are usually used to determine surface albedo. With the use of appropriate filters it is possible to measure radiation in particular spectral bands.
2. *Pyrheliometers* measuring direct solar radiation incident on a collector perpendicular to the Sun's rays. Again, these can observe the total spectrum or certain wavelength intervals.
3. *Diffusographs*, which observe radiation from the sky only. This instrument consists of a normal pyranometer surrounded by a 'shade ring' which obscures the direct beam from the Sun (Fig. 2.14(b)).

Pyrheliometers are usually precision instruments, which must be carefully used and more or less constantly maintained. Consequently

(a)

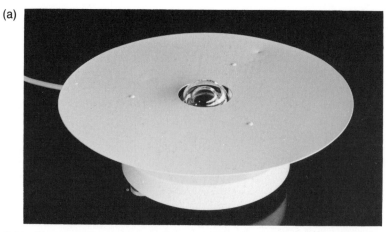

Fig. 2.14. Photograph of (a) pyranometer (Kipp and Zonen solarimeter); (b) the method of mounting a shade ring above such an instrument so that only diffuse solar radiation is monitored. By subtracting the diffuse irradiance measured by one such pyranometer from the total radiation measured by an unshaded pyranometer a value of the direct solar radiation is achieved. (c) A Campbell–Stokes sunshine recorder. On bright days with little or no cloud the Sun's rays, concentrated through the glass sphere, burn the card encircling the recorder.

(b)

(c)

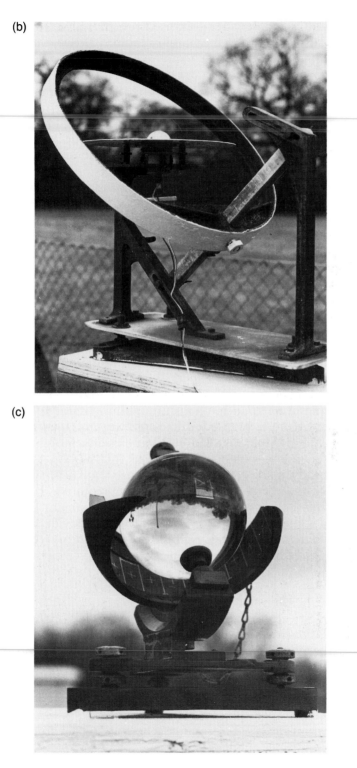

they are usually found only at specialised research sites. However, they are instruments which can measure radiation directly as an energy flux. Pyranometers, on the other hand, are designed to be more robust 'field' instruments and usually operate on a 'differential absorption' principle. Part of the sensor surface is coated with a paint with a high albedo, the rest with a low albedo paint. The temperature difference between the surfaces is usually measured electrically. The relationship between the observed temperature difference and the amount of radiation must then be established by 'calibrating' the instrument against an absolute instrument such as a pyrheliometer.

A simpler instrument that records the number of hours of bright sunshine is the *Campbell–Stokes sunshine recorder* (Fig. 2.14(c)). The burn produced on a graduated card is a very useful, but less exact, measure of the extent and nature of sunshine hours.

### Empirical methods of obtaining solar radiation data

There is a network of pyranometers over the land areas of the Earth making more or less continuous observations. Many have been operated for 25 years or more. They provide detailed information, for a specific location, on the long-term radiation climate and its variability. Attempts have been made to provide similar information for locations without measurements by estimating solar radiation amounts using available sunshine duration, e.g. from a Campbell–Stokes recorder (Fig. 2.14(c)) and/or cloud amount data. The approach adopted is a typical example of *empirical* methods of obtaining data. It is clear that radiation amount must be related in some way to cloud amount and daylight duration. While detailed calculations of atmospheric transfer would be needed to provide the required information for a specific location for a specific day, it is possible to use statistical *regression* techniques if only long-term average values are required. For a station which measures both solar radiation and cloud amount (or sunshine hours), daily values of each are correlated to provide a regression relationship of the form $X = f(Y)$. Usually a linear relationship is assumed, so that the equation becomes

$$X = a + bY, \qquad\qquad [2.12]$$

where $a$ and $b$ are empirical constants. For example, for some Nigerian stations a relationship was found such that $X = 0.19 + 0.60\,Y$, where $Y$ is the daily averaged percentage of possible sunshine and $X$ is the fraction of the extraterrestrial radiation received at the surface. The results illustrated in Fig. 2.15, show that despite the fact that the regression coefficients vary spatially and temporally a good empirical basis exists for extrapolation.

Global maps of radiation receipt can be constructed from the obser-

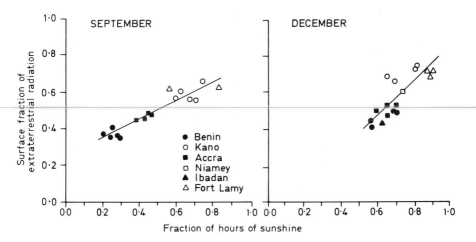

Fig. 2.15. The relationship between solar radiation at the Earth's surface and hours of sunshine for six stations in Nigeria for the months of September and December. The solid line indicates the relationship obtained by statistical regression techniques. The differing slopes between months indicates the temporal variation in the relationship.

vational and empirical data. Although data are sparse, particularly over the oceanic areas, and a good deal of subjectivity is involved in map construction, radiation tends to be 'spatially conservative', not varying rapidly horizontally, so that broad regional patterns emerge (Fig. 2.16). It is clear that the areas of maximum radiation receipt are the desert regions of the Earth, while minima are reached in polar regions.

## 2.4.1 The availability of solar energy – an application of climate information

### Installation of a solar energy collecting system

The overall problem to be considered in this section can be simply stated as: 'Is it worthwhile for me to install a solar energy collecting system as an energy source for my home?' Whether 'worthwhile' is interpreted in an economic or practical sense, the performance of a solar energy collecting system obviously depends on the amount of solar radiation arriving at the collector. Hence, we have a problem with a climatological dimension, which can be treated in some detail to illustrate the basic approach that can be adopted to provide climatological information for practical applications.

Defining exactly what climatological information is required needs care. There are numerous types of systems available for converting solar energy to a form more useful to mankind, all requiring slightly different information. For purposes of illustration, we will consider the information needs for a 'flat plate' solar collector to be used to provide

53

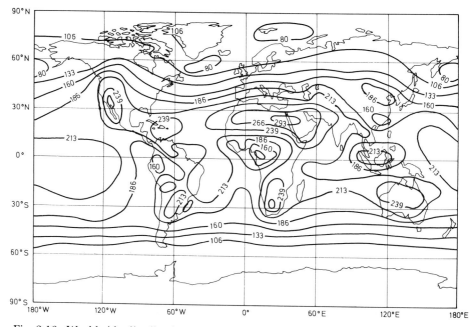

Fig. 2.16. Worldwide distribution of annually averaged global (i.e. direct plus diffuse) surface-received solar radiation (W m$^{-2}$).

space heat and hot water to a house (Fig. 2.17). For this type of system, typical of many, there is a well developed method of relating performance to the climatic parameters. This method requires solar radiation data to establish the energy available and temperature data to determine the energy requirements for space heating. The way in which the climatologist can assist in deciding whether it is worthwhile to install a system, therefore, is by providing answers to the questions: 'How will a system perform, given my own local climate?' and 'Is there plenty of energy for heating when it is cold outside?'.

### Data requirements

The next step in the solution of this problem, assessing the data requirements, follows directly from the statement of the problem. In this case the basic method of calculating performance dictates that we have monthly averages of daily total solar radiation on a horizontal surface and monthly average temperatures. Note that in this context there is no need for more detailed data, since the basic method of assessing performance could not use additional information.

Specifying the data needed may be simple, but assembling these data may be more of a problem. Temperature data are available for many locations. However, solar radiation is routinely measured at only a few places. Again, for purposes of illustration, we will use as an example

(a)

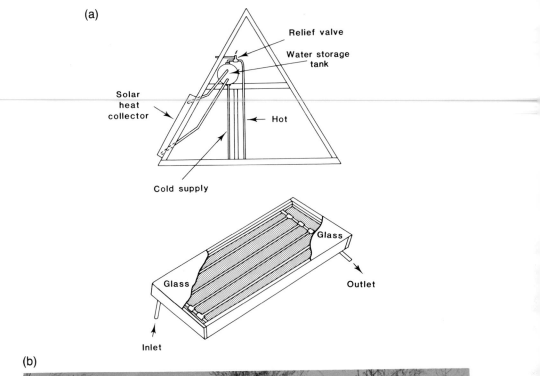

(b)

Fig. 2.17. (a) Layout and installation mode of a solar collector. (b) This method is used to heat this dwelling in Jacksonville, North Carolina. (Courtesy: John J. Busenberg, Astron Technologies Inc.).

the situation in North Carolina, USA. Two stations measure radiation and cloud amount, while four more measure cloud amount alone. It is therefore possible to estimate, using equation [2.12], radiation for these four stations, assuming regression constants appropriate for this region are available. Although six stations is a small number for an area as large as North Carolina, the stations are well scattered through the state and reflect most of the local climate variation, particularly cloud amount, which is likely to influence radiation receipt. Furthermore, since radiation is spatially conservative, it is possible to use these data to construct statewide maps of radiation receipt (Fig. 2.18).

### Application strategy

With the data available, the type of analysis needed must be considered. Solar energy systems are typically constructed with an expected lifetime of around 20 years. Forecasting of day-to-day conditions over this time span is impossible. However, our main concern is with the likely performance throughout the period. We can easily estimate this performance by assuming that the radiation climate during the next 20 years will be essentially the same as that during the past 20 or 30 years. We can then use the past climate record directly. This type of approach, using *historical normal* data, depends on the asssumption of no significant climate change. A check of this assumption for the North Carolina data suggested that there was no significant temporal trend in the radiation data. Indeed, analysis suggested that it would require a change of over 4% in the radiation receipt to change the performance of a collecting system by 1%.

The analysis of the assembled data is usually conceptually quite straightforward if the previous steps have been considered correctly. However, the techniques involved may be complex and considerable data manipulation may be needed.

### Results

The final presentation of the results may not be simple. Of the two results presented here, Fig. 2.18 is easy to interpret. Any system designer could use the map to determine directly the monthly average daily total radiation for a particular place. The second result, Fig. 2.19, is not so easily dealt with. This figure indicates the average radiation receipt to be expected at a given ambient temperature (solid line) and the percentage of the heating season having a particular ambient temperature (dashed line). Careful analysis of the curves, however, allows derivation of a great deal of information for a specific location, say Raleigh, North Carolina, and for a selected building. For example, it can be seen that the coldest days have more radiation than those that are merely fairly cold. However, it is also seen that there are few really cold days during the time a solar space heating system would be in operation. In order to ensure that this information can be related to system performance, two 'system performance lines' ($f = 0.8$ and $f = 0.2$ lines) are added to the figure. These indicate, for a 'typical' system, the '$f$ value', the fraction of the total heating requirement which can be met from solar sources. In this case the warmest days have enough incident radiation to satisfy over 80% of the load. At

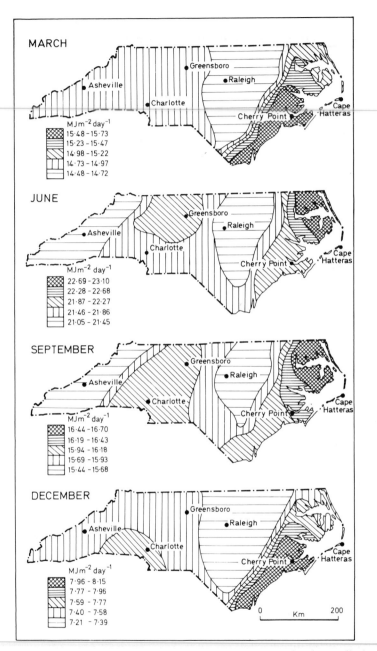

Fig. 2.18. Spatial distribution of monthly averaged daily total solar radiation on a horizontal surface in North Carolina for (a) March, (b) June, (c) September and (d) December. The six named stations are the complete observational network used.

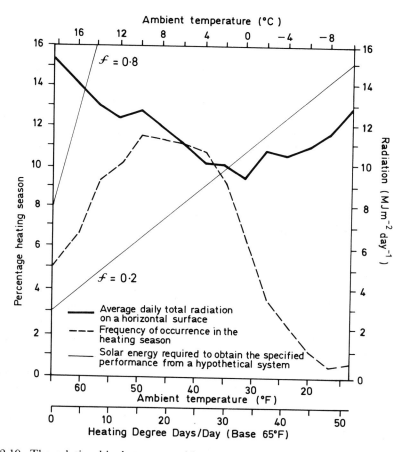

Fig. 2.19. The relationship between ambient temperature and (i) average daily total radiation on a horizontal surface (solid line) and (ii) frequency of occurrence of particular ambient temperatures during the heating season (dashed line). The fine lines indicate the amount of radiation required to obtain a specified fraction, $f$, of the total energy required for a specified building at a selected ambient temperature. The position of these 'performance lines' depends upon the size of the building and the size of the solar energy collection system. Two typical fractions are illustrated here.

temperatures below freezing, however, less than 20% can be met. A similar analysis, using the performance lines for a system of different size, reveals that significant benefits could be gained by installing a bigger, and more costly, system. The actual costs, of course, depend on many non-climatological factors. Nevertheless, this approach allows anyone to relate climatology and performance easily; a vital step in ensuring that the climatological information will be used to assist in the solution of the problem, 'Is it worthwhile for an individual or industry to install a solar energy collecting system?'

# 2.5  Longwave radiation

As indicated in Fig. 2.2, some solar radiation is absorbed directly in the atmosphere but most passes through the atmosphere to be absorbed at the Earth's surface. The absorption of solar radiation leads to heating which, in turn, leads to the emission of longwave radiation. The amount of energy emitted is given by the Stefan–Boltzmann Law (equation [2.3]) modified for the effects of emissivity, $\epsilon$. Thus

$$E^* = \epsilon\,\sigma\,T^4 \qquad\qquad\qquad [2.13]$$

Since the variation of emissivity with wavelength is small for solid objects, but large for gases, it is convenient to treat longwave emission separately for the Earth's surface and for the atmosphere.

### *Surface infrared emissivity*

Although various types of surface on the Earth have different emissivities (Table 2.1), almost all equal or exceed 0.90. Artificial surfaces tend to be slightly below this value, vegetated surfaces just above it, and water considerably above it. In detailed calculations of the energy balance of a particular surface type, or when comparing contrasting surfaces, these differences must be incorporated. Nevertheless for many calculations it is possible to assume a uniform emissivity for land surfaces, and a uniform one for water. Indeed for many purposes it is possible to assume that the Earth's surface acts as a black body for longwave radiation.

Such simple assumptions are not possible for the atmosphere. Values of absorptivity and emissivity vary greatly with wavelength. The value also depends on the amount, temperature and pressure of the emitting gas.

### *Gaseous absorption and emission*

Each gas absorbs radiant energy in a series of very narrow wavelength intervals, called *spectral absorption lines*. Commonly these lines are grouped together forming *absorption bands*. The locations of the bands, and the strength with which they absorb, depend on the molecular structure of the gas. As the amount of gas, its temperature and the total atmospheric pressure increase, these bands are broadened and the amount of absorption increases.

The major absorbing gases in the Earth's atmosphere are water vapour, carbon dioxide and, to a much smaller extent, ozone. Most of the larger absorption bands occur in the infrared portion of the spectrum (i.e. $\lambda \gtrsim 3\ \mu\text{m}$) and within this portion there is some absorption throughout most of the region where significant amounts of energy are emitted (Fig. 2.3). The major exception is the region between 8 and 14 μm, which is the so-called *atmospheric window*.

As a result of these absorption characteristics most of the longwave radiation emitted by the surface of the Earth is absorbed by the atmosphere. Only a small portion is transmitted through the atmospheric window and escapes to space. The portion that is absorbed combines with the absorbed shortwave energy to heat the atmosphere. The heating stimulates the atmosphere to emit radiation. This emission is in all directions, but again we can simplify matters and note that a portion is emitted upwards and eventually is returned to space, while a portion travels downwards and is received at the surface as *incoming longwave radiation*. The complementary term, *outgoing longwave radiation*, can refer either to the energy emitted from the Earth's surface or to the energy leaving the atmosphere for space. The context makes it clear which of these is being discussed.

### Effect of clouds

In a cloudy atmosphere the same basic radiation exchanges take place. However, complications arise because the presence of cloud modifies the wavelength dependence of emissivity. The most important effect is that clouds tend to 'close' the atmospheric window because of their very much greater absorptivity in the infrared region. This effect can readily be seen by comparing temperatures on a cloudy night to those on a cloudless one. In the former the closing of the atmospheric window prevents the significant loss of longwave radiation that occurs in the latter. The result is that the overnight temperature decrease on a cloudy night is much less than on a cloudless one.

### Greenhouse effect

Since the atmosphere is almost transparent to solar wavelengths but absorbs terrestrial radiation strongly, an analogy was drawn long ago between the operation of the atmosphere and that of a greenhouse. The term *greenhouse effect* has passed into the literature to denote the process in the atmosphere whereby solar energy passes almost unimpeded to the surface, creates a warm surface which emits longwave radiation which in turn is absorbed in the atmosphere only to be re-radiated back to the surface. The net effect is to maintain the surface at a higher temperature than would occur if the atmosphere were as transparent to longwave radiation as it is to shortwave energy. We now know that a greenhouse maintains its higher internal temperature largely because the shelter it offers reduces the turbulent transfers of energy away from the surface, rather than because of any radiative considerations. Thus while the concept of the greenhouse effect (Fig. 2.20(a)) remains valid, and vital, for the atmosphere, it might be better to think of the physical

## (a) Inappropriate greenhouse

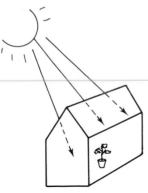
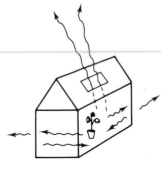

Transparent "glass"
Solar radiation penetrates

Glass absorbs larger wavelength
heat energy, although thermal
radiation escapes through
the atmospheric "window"

## (b) Leaky bucket (better "greenhouse" analogy)

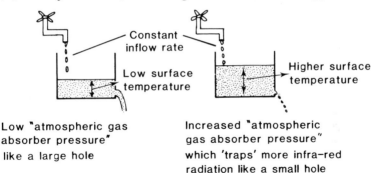

Low "atmospheric gas
absorber pressure"
like a large hole

Increased "atmospheric
gas absorber pressure"
which 'traps' more infra-red
radiation like a small hole

Fig. 2.20. Schematic representation of the (a) greenhouse and (b) leaky bucket analogies which may be used to represent the 'blanketing' effect of the atmosphere upon surface temperature. In the case of the greenhouse analogy, the atmosphere is represented in terms of its transmission (shortwave) and absorption (longwave) properties by the glass of the greenhouse. In the case of the leaky bucket analogy, the surface temperature is represented by the height of the water depth in the dynamic situation in which a greater amount of absorbing gases is represented by a smaller hole while the input from the tap remains constant. The closing of the hole in the bucket could be thought of as analogous to the 'closing' of the infrared 'windows' by increased greenhouse absorbers.

processes in terms of the 'leaky bucket' analogy depicted in Fig. 2.20(b). Here any increase in the amount of a gas with absorption bands in the infrared part of the spectrum is represented by a decrease in the size of the hole in the bucket. The surface temperature, represented by the depth of the water in the bucket, rises as more absorbing gases enter the atmosphere. One of our prime concerns with the current worldwide increase in atmospheric carbon dioxide content is that it may enhance the greenhouse effect, and thereby possibly increase surface temperatures.

As before, Fig. 2.2 summarises the above discussion. It indicates clearly the small proportion of energy that passes through the atmospheric window directly to space and the large amount of longwave radiation exchange between the Earth and the atmosphere. These incoming and outgoing fluxes are of the same order as the incoming shortwave flux. However, they are the result of continuous interactions without the marked diurnal variations characteristic of the solar flux.

### 2.5.1  The temperature structure of the atmosphere

One important consequence of radiation exchanges is the development of a distinct temperature structure in the atmosphere. An averaged temperature profile of the atmosphere (Fig. 2.21) indicates that there are three major 'warm' regions where radiation is absorbed directly: the surface, the stratosphere and the thermosphere. These regions of absorption create a series of distinct layers with distinct temperature gradients within the atmosphere. These layers are termed 'spheres'. The spheres are separated by levels of demarcation called 'pauses' at which the temperature gradient reverses. The cause of the heating at the surface, the surface absorption, has been noted already. Heating in the lower stratosphere is caused by absorption of high energy ultra-violet radiation by ozone. In the thermosphere the heating is also the result of interaction between radiation and atmospheric constituents, through a process known as photo-ionisation.

### *Importance of the troposphere*

Most of the activity which we usually associate with climate occurs in the layer between the lower stratosphere and the surface. This region accounts for approximately 80% of the atmospheric mass and is the region where the atmosphere is characterised by a decrease in temperature with elevation. The dominant mechanism for heating is through surface absorption of solar radiation and re-radiation from the surface. The height of the top of the layer, the tropopause, varies with latitude and season, being generally higher at the equator than the poles and higher in the summer hemisphere than in the winter one (Fig. 2.22). The detailed structure of the troposphere at a given time and place is dictated by the levels at which radiation absorption occurs and is influenced by vertical motions in the atmosphere.

## 2.6  Monitoring radiation from space

Satellites observe the flux of radiation leaving the top of the atmo-

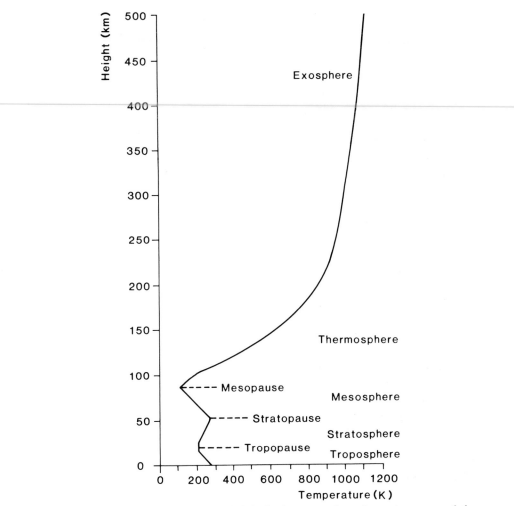

Fig. 2.21. Temperature variation with height in the atmosphere (sometimes termed the 'lapse rate'). Note the three regions of increased temperature where absorption occurs: (i) the surface (visible and near infrared solar radiation); (ii) lower stratosphere (ultraviolet absorption by $O_3$); (iii) thermosphere (high energy absorption causing photo-ionisation).

sphere. Since they can have a variety of orbits, fields of view and wavelength ranges in which they are sensitive, they are capable of providing a wide variety of information. They are complementary to the surface-based observations in that they provide regular, repetitive observations of a large portion of the Earth (Fig. 2.23).

### Polar orbiting satellites

Satellites are generally inserted either into a polar or near polar orbit or into a geostationary orbit. In a *polar orbit* the satellite is usually

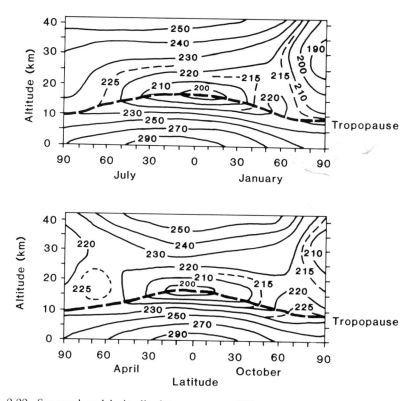

Fig. 2.22. Seasonal and latitudinal temperature (K) structure of the Northern Hemisphere atmosphere for the four mid-season months. Note the variation in height of the tropopause and the lower temperatures in the polar stratosphere in winter.

between 500 and 1500 km above the surface of the Earth and the orbital path crosses the equator at approximately 90°. Characteristic orbital periods are of the order of 90 minutes, each orbit taking the satellite directly over or close to both poles. Usually such satellites are Sun synchronous: as the Earth rotates the new orbit is over an area experiencing the same local time as the area viewed on the previous orbit. Half of each orbit, therefore, is over the dark or night side of the planet. A meteorological polar orbiter typically has a field of view sufficiently large that it takes about 15 orbits to cover the globe, and a specific location is seen approximately twice a day.

### Geostationary satellites

*Geostationary* satellites are very much higher than those of the polar orbiters. They must be approximately 35,400 km above the Earth's surface and in the plane of the equator so that they remain geosynchronous, remaining in the same position relative to the Earth at all times. As a consequence of their great height, they continuously view

**(a)** TYPICAL PATHS OF POLAR ORBITER SATELLITE

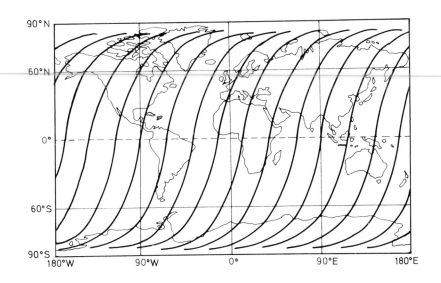

**(b)** COVERAGE BY GEOSTATIONARY SATELLITES

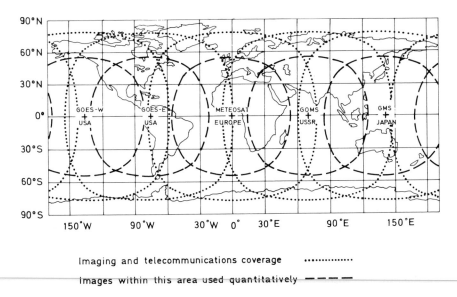

Imaging and telecommunications coverage $\cdots\cdots\cdots$

Images within this area used quantitatively $- - - -$

Fig. 2.23. Areal coverage of satellite surveillance for (a) polar orbiting satellites and (b) geostationary satellites. The upper diagram shows typical swaths; the sensor scans are made perpendicular to the centreline of the swath. Note that each of the swath line curves is a half-orbit path but the projection distorts these great circles into S shapes.

the full disc of the Earth and have much lower resolution than polar orbiting satellites. The curvature of the Earth makes it difficult to provide information for latitudes higher than about 45–50°. Commonly these satellites sample the radiation fluxes over the whole disc once every 30 minutes.

### Satellite sensors

Radiance measurements are generally made in a series of spectral intervals, the number of which varies from satellite to satellite. Two relatively broadband wavelength regions, one within the solar spectrum (approximately 0.3–1.0 $\mu$m) and one within the atmospheric window (approximately 10–12 $\mu$m), are commonly used to identify cloud fields, derive winds and retrieve surface temperatures. In addition they provide the only source of information about planetary albedo and net radiation. Sensors responsive to a narrow spectral interval can provide more specific information. For example, a sensor centred at 6.3 $\mu$m, a water vapour absorption band, measures the amount of absorption in an atmospheric column, which is then related to the column water vapour content. Similarly, several narrow bands in the infrared and microwave regions of the spectrum can be used to obtain vertical temperature profiles which augment radiosonde data.

## 2.7   The global radiation budget

On a time scale of several years there is approximate equality between the amount of solar radiation received from the Sun at the top of the atmosphere and the amount of longwave radiation emitted to space. Hence there is a global radiation balance (Fig. 2.2). Minor variations will occur from year to year; for example, those resulting from the variations in the Sun's output. However, the balance must be maintained if the climate, again on a long time scale, is to remain stable. Certainly the presence of even a small imbalance, if maintained for a considerable time, would lead to climatic variations. Indeed, this may be one of the causes of climatic change.

### Latitudinal gradient in radiative fluxes

However, this energy balance is a global phenomenon. No particular point or region is individually in a state of radiative balance. In fact, it is the local imbalances that drive the climate as we know it. The radiation imbalance at each latitude is readily illustrated by comparing latitudinally averaged values of the solar radiation absorbed and the infrared radiation emitted by the system (Fig. 2.24). The amount of solar energy absorbed is affected by both the total amount incident and the albedo. Thus in high latitudes the prevalence of ice and snow further reduces by reflection the already low level of solar radiation.

Annual (June 1974–February 1978) Zonal average

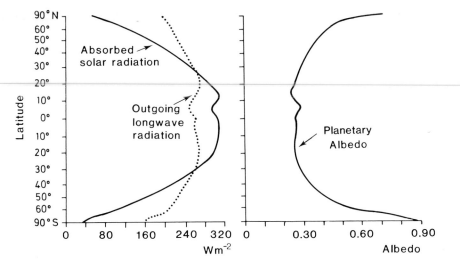

Fig. 2.24. Latitudinally averaged elements of the planetary radiation budget as observed by the scanning radiometers on board NOAA polar orbiting satellites in the period June 1974 to February 1978. Annually averaged values of absorbed solar radiation, outgoing longwave radiation and albedo are shown. (NOAA–NESS, Washington DC).

Similarly the increased albedo just north of the equator, caused by the large amount of cloud common in the area, leads to a decrease in the solar radiation absorbed there.

In contrast to the considerable gradient between equator and poles for the absorbed solar radiation, there is only a slight gradient for the emitted longwave radiation. This suggests that the equator to pole temperature gradient is considerably less than would be expected from considerations of solar radiation alone. In fact the mean annual climatic conditions at every latitude are considerably more hospitable than they would be if each zone were in radiative equilibrium. The atmosphere acts, very efficiently, to redistribute energy. The tropical latitudes are cooled as they export energy to the mid- and high-latitude regions which thus gain energy and are warmed. The redistribution of energy is a direct result of the equator to pole temperature gradient, which is itself a consequence of the latitudinal radiation imbalance. The transport between latitudes is accomplished by horizontal energy transfers using both the atmospheric and the oceanic circulation. These processes operate in such a way that the whole system, driven by the radiation imbalances, tries to achieve equilibrium.

### Seasonal variation in radiative fluxes

Seasonal variations in the radiative fluxes must be added to the annu-

67

ally averaged picture to complete the global radiation budget. Results from polar orbiting satellites indicate that the seasonal global budget is a strong function of albedo variations (Fig. 2.25). The albedo curve reaches a maximum in the Northern Hemisphere winter, underlining the importance of the seasonal snow and ice cover variations in the Northern Hemisphere. The emitted longwave radiation, for similar reasons, has a maximum during the Northern Hemisphere summer. The absorbed solar radiation shows a weaker seasonal variation. The resulting global net radiation is thus at a maximum during the Northern Hemisphere winter.

While these generalised results for seasonal and latitudinal variations

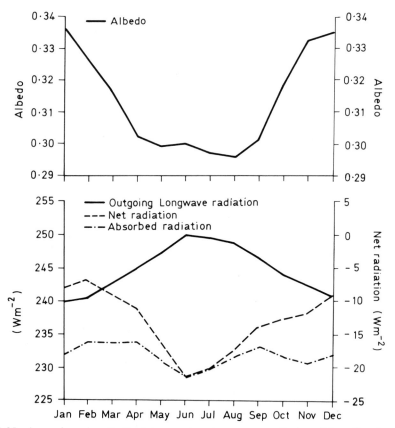

Fig. 2.25. Annual cycle of albedo, outgoing longwave radiation (W m$^{-2}$), absorbed solar radiation (W m$^{-2}$) and net radiation (absorbed − outgoing in W m$^{-2}$) based on the same 45-month satellite data set as used in the construction of Fig. 2.24. The negative value of net radiation throughout the year seems anomalous and may be due to errors in the observations as discussed in the text. (Ohring, G. and Gruber, A. 1983, *Adv. Geophy.*, **25**). (Also Figs. 2.26, 2.27 and 2.28).

are of great importance in understanding the dynamics of climate, it is frequently necessary to consider the actual spatial distribution of the components of the radiation balance. Small changes in one component for one area may, for example, be compensated by changes in the opposite direction at another place along the same latitude circle, thus leaving the zonal average unchanged. The change, however, may have tremendous consequences for the atmospheric dynamics both locally and globally. Site specific values can be obtained from satellite observations but they must be treated with caution. Only a limited number of comparable observations are available for most places, so that temporally averaged conditions cannot be readily determined. Furthermore, satellite technology, while remarkably sophisticated, is still only just past infancy, so that it is sometimes difficult to decide whether observed variations are the result of true climatic fluctuations or factors influencing the satellite sensing system. Nevertheless, the results presented in Figs 2.26, 2.27 and 2.28 represent realistic estimates of the spatial distribution of the radiation components. The maps emphasise the polar areas which play a very significant role in climatic dynamics.

### Geographical variation in radiative fluxes

Regions of high albedo (Fig. 2.26) occur in the polar regions throughout the year. The high reflectivity of the North African desert is also permanent, but other areas of high albedo in the tropics and subtropics, such as in the eastern Pacific and South American regions, vary in position with season. This variation is echoed in the outgoing longwave radiation distribution (Fig. 2.27), with values decreasing poleward throughout the year, but with marked seasonal variations in the area between 30 °N and 30 °S. The most striking characteristic of the net radiation maps (Fig. 2.28) is the seasonal reversal in sign almost everywhere except for the persistent negative values over both poles and the generally positive values near the equator.

These satellite derived results are likely to produce realistic spatial patterns and temporal variations but absolute values may not be so accurate. For example, the annual average global net radiation indicated in Fig. 2.25 is close to $-13$ W m$^{-2}$. It has already been stated, however, that there is a long-term global radiation balance. The discrepancy may be partly a result of the relatively small sampling period used. The consensus from a large number of observations over the last 15 years is that the global outgoing longwave radiation is approximately 250 W m$^{-2}$ and the albedo is 0.310, rather than the values of 245 W m$^{-2}$ and 0.314 shown here. Errors may also have arisen in the retrieval and analysis of the satellite observations themselves, or through inadequate sampling.

Despite these possible problems associated with satellite derived results, satellite data have been available for a sufficient length of time

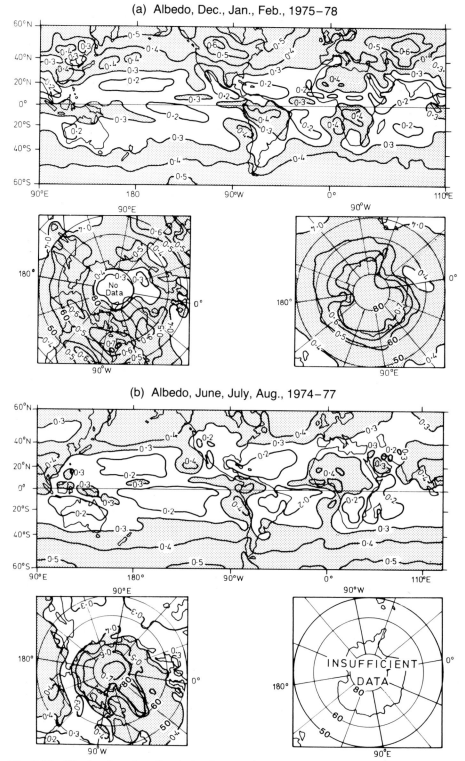

Fig. 2.26. Albedo for the December/January/February season (a) and the June/July/August season (b) displayed in a Mercator projection between 60 °N and 60 °S plus two polar-stereographic projections for the North and South Poles. The data are compiled from the scanning radiometers on board the NOAA polar orbiting satellites for the period June 1974 to February 1978.

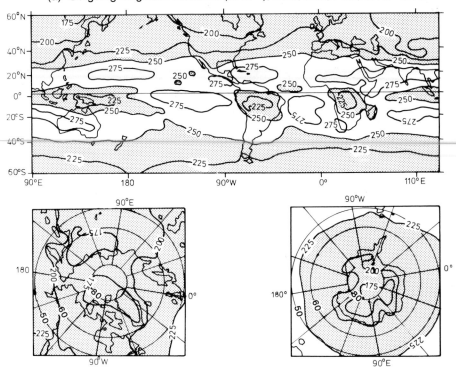

(a) Outgoing longwave radiation (W m$^{-2}$), Dec., Jan., Feb., 1975–1978

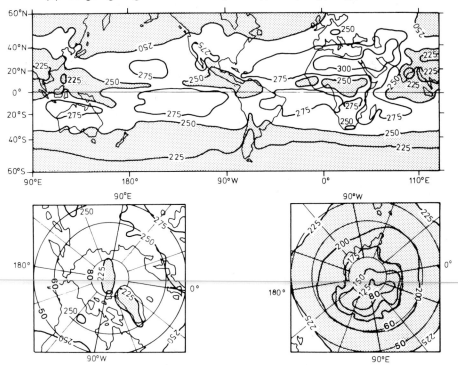

(b) Outgoing longwave radiation (W m$^{-2}$), June, July, Aug., 1974–77

Fig. 2.27. Same as Fig. 2.26 for outgoing longwave radiation.

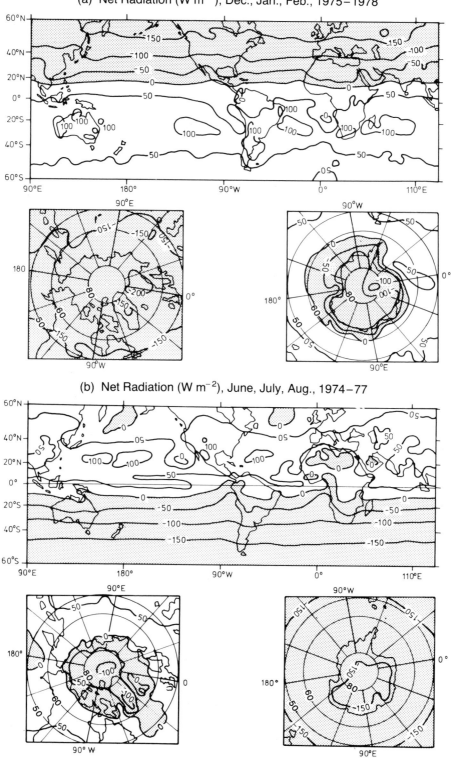

Fig. 2.28. Same as Fig. 2.26 for net radiation.

for self consistent results to be obtained and a global climatology established. While this will certainly be refined as new results are obtained the current results allow us for the first time to present an integrated, global picture of the global radiation budget.

## 2.8 Surface radiation budgets

The surface of the Earth is the location of the most important energy transformation in the global energy cascade: the absorption of solar radiation and the emission of infrared radiation. The surface of the Earth also experiences an energy balance such that on a global annual average as much energy reaches the surface as leaves it. Here, however, it is inappropriate to think solely of radiative fluxes. As Fig. 2.2 indicates, the fluxes of latent and sensible heat, which are not radiative, must be incorporated. Thus there is not a true surface radiation balance. Instead it is useful to consider the total amount of radiation at all wavelengths that is absorbed at the surface. This, the surface *net radiation*, is given by

$$Q^* = K\downarrow - K\uparrow + L\downarrow - L\uparrow = (1-A)K\downarrow + L\downarrow - \epsilon\sigma T_s^4 \quad [2.14]$$

where $Q^*$ is the net radiation, $K$ and $L$ are the shortwave and long-wave fluxes respectively, and the arrows indicate the direction of the radiation streams. The first form of the equation emphasises that the net radiation is the sum of all the fluxes, while the second form emphasises the role of the surface characteristics (particularly the surface albedo, $A$, and the emissivity, $\epsilon$) in determining the amount of radiation absorbed.

### Daily cycle of surface radiation budget

Although the net radiation and its component fluxes vary on a global and seasonal scale in much the same way as we have already discussed for the planetary case, we can here consider variations on a daily basis. In general, the shortwave element, $K$, is the component flux of the net radiation, $Q^*$, that is most variable in magnitude. It varies with latitude, season and time of day, as indicated in equation [2.9]. A typical diurnal variation for a mid-latitude location is shown in Fig. 2.29. Figure 2.29(a) represents cloudless conditions, Fig. 2.29(b) a day upon which cloud was intermittent in the morning and formed a complete overcast in the afternoon. The net solar component simply echoes the incoming component, the magnitude of the variation being damped by the effect of the surface albedo. The incoming longwave radiation is much less variable. The amount, depending on the temperature and humidity of the overlying air, will change through such effects as horizontal movements associated with winds. The overall net radiation,

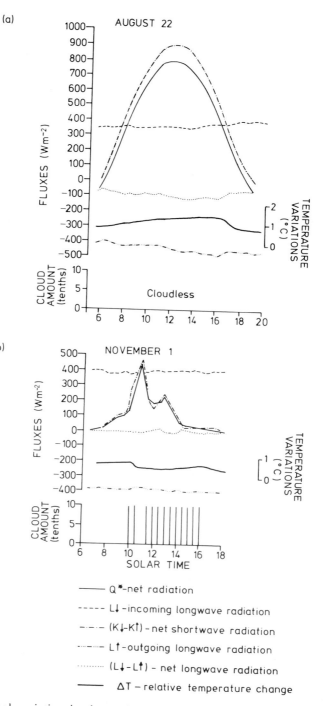

Fig. 2.29. Diurnal variation in the surface radiation budget components over Lake Ontario on a (a) clear and (b) partly cloudy day.

$Q^*$, is seen to follow the net shortwave component quite closely although the correspondence is closer in the cloudless case (Fig. 2.29(a)). Thus no diurnal pattern is completely predictable from radiative considerations alone. If there is relatively little air movement, an increase in $L\downarrow$ is to be expected in the afternoon, as the atmosphere is heated by direct absorption of solar energy and by transfer of heat from the underlying surface. The surface itself will be heated by the absorption of radiation and $L\uparrow$ will vary in response to that heating. The time of maximum $L\uparrow$ (N.B. negative and hence a 'dip' in Fig. 2.29(a)) will be later than the time of maximum $(K\downarrow - K\uparrow)$ because heating will be occurring throughout the period with positive net radiation. During the night $Q^*$ is likely to be negative. Only the longwave components are active and $Q^*$ will then depend on the difference between the radiative temperature of the atmosphere and the surface temperature. Thus there is likely to be greater longwave loss on cloudless nights than when clouds close the atmospheric window.

Although surface temperatures reflect the influence of the net radiation, they are modified by the other energy fluxes affecting the surface. Thus to understand how temperatures are created and how they vary spatially and temporally, we need to consider both the relationship between energy and temperature and the surface energy budget.

## 2.9   Energy and temperature

We have already stated that when a body absorbs energy, its temperature increases. In fact there is a simple relationship between the change in energy of a body and its temperature change

$$\Delta E = \rho C\, \Delta T \qquad\qquad [2.15]$$

where $\Delta E$ is the change in energy and $\Delta T$ the change in temperature, of unit volume in unit time, of a body of density $\rho$ and specific heat $C$. Note that both $\Delta T$ and $\Delta E$ are proportional to rates of change, so that the equation enables us to calculate heating rates directly such that $\Delta E/\Delta t = \rho C\, \Delta T/\Delta t$.

If we consider initially a situation where only radiative energy is involved, we can perform a simple analysis. As the body absorbs solar radiation its temperature will rise in accordance with equation [2.15]. This will lead to an increase in the amount of longwave energy emitted, in accordance with the Stefan–Boltzmann law (equation [2.3]). Neglecting non-radiative energy fluxes, the temperature will increase until the absorption rate is equal to the emission rate. The net change in energy within the body will thus be zero, there will be no further temperature change and the body will be in *radiative equilibrium*. The actual temperature at this point will depend, for a given incoming radiation stream, on the albedo of the body, controlling the amount absorbed, and on its emissivity, controlling the amount emitted for a given temperature. If either the emissivity or the albedo is increased the equilibrium temperature is decreased. The heat capacity of a body is not important in determining this temperature, but is very important in determining the time needed to reach equilibrium.

### Radiative control of temperature

In the free atmosphere radiative exchanges are the major determinant of heating rates most of the time. However, the transfer of other forms of energy from the surface can be very important locally. They provide, for example, the energy needed to drive the circulation in many severe storms. Nevertheless, it is possible to calculate atmospheric heating rates, to a very good approximation, using radiative considerations alone. At the surface of the Earth the calculation of heating rates or equilibrium temperatures is not as simple since the non-radiative energy fluxes are at least as important as the radiative ones.

## 2.10   Surface energy budgets

The temperature at the surface of the Earth is a response to all of the energy fluxes affecting the surface. Thus the energy responsible for temperature changes is given by the *energy budget* equation in which the net radiation, $Q^*$, is

$$Q^* = H + LE + G \qquad\qquad [2.16]$$

where $H$ and $LE$ are respectively the sensible and latent heat fluxes into the air and $G$ is the heat flux into the ground. Sensible energy flows from high- to low-temperature areas predominantly by movement of air warmed by surface contact. Latent energy is associated with the movement of water vapour molecules and exchanges are the result of evaporation and condensation. The flux of heat into the ground is by conduction, i.e. similar to heat flow along a heated rod, although when the underlying surface is water, $G$ can also be by convection.

The energy budget equation (equation [2.16]) is written in a form which suggests that there is a balance between the various fluxes. While this is true for long-term average conditions, most of the time this is not the case. Fluxes vary constantly although there is a strong tendency towards achieving this balance and consequently an equilibrium temperature. However, under the continuously varying atmospheric conditions, the balance is rarely attained. Instead, the imbalance leads to excess energy gains or losses at the surface and thus to temperature changes.

As an example we can again consider the diurnal energy cycle.

Surface temperatures start to rise as soon as the net radiation becomes positive. In most cases the surface becomes warmer than the overlying air and a sensible heat flux upwards is initiated. The net radiation is also likely to provide energy needed for evaporation and latent heat transfer begins. At the same time heat is transferred from the warm surface to lower layers of the underlying medium. This situation could continue throughout the period of positive net radiation. However, changes in the air above the surface may disrupt the simple pattern. A warm airflow may bring air that is warmer than the surface into contact with it and create a sensible heat flux towards the ground. If the underlying medium is water, internal currents may create a similar effect. If the ground dries, there can be no upward latent heat flux. The causes of all of these changes are largely unconnected with the particular spot we are considering, and so can lead to unpredictable changes in the energy reaching or leaving the surface and thus the temperature. However, if we continue our idealised diurnal cycle, once the net radiation becomes negative and radiative cooling dominates, the non-radiative fluxes tend to be directed towards the surface, decreasing the rapidity of cooling. Thus, in general, the non-radiative energy transfers tend to minimise the diurnal temperature changes that would result from radiative exchanges alone.

### Importance of non-radiative fluxes at the surface

The transfer of energy away from the surface by ground heat fluxes can be considered by analogy with heat flowing along a rod. When one end is heated the heat will start to flow from the hotter to the colder region. The rod is thus progressively heated, with a maximum temperature change at the heated end, the change being gradually damped at larger distances from that end. The rate of heat penetration is dependent on the *thermal diffusivity*, $K^*$, of the material (or equivalently dependent upon the *thermal conductivity*, $K = \rho c_p K^*$). At any instant after the start of heating, the depth of penetration, which can be defined as the point where the temperature rise is a small fraction, say 5%, of that at the heated surface, is proportional to $\sqrt{K^*}$. If we have a heating cycle, as with the diurnal cycle, rather than steady heating, temperature waves will spread vertically downwards with their amplitude diminishing as they progress. Eventually a point will be reached where the diurnal cycle will be sufficiently damped to be negligible. Values of $K^*$, together with depths of penetration for various surface types and for the atmosphere, are given in Table 2.2. Although values differ for various land surfaces, the major difference is between solid land, stirred water and stirred air. Penetration rates are more rapid and energy reaches a greater depth in water than in land, whilst penetration in air is greater than both. Heat transfer in solids can only be through molecular interactions, the true *conduction*

**Table 2.2 Thermal properties of air and various surfaces**

| Substance | Heat capacity $\rho C$ (J m⁻³ K⁻¹) | Thermal diffusivity $K^*$ (m² s⁻¹) | Thermal conductivity $K$ (W m⁻¹ K⁻¹) | Conductive capacity $C^*$ (J m⁻² K⁻¹ s⁻¹ᐟ²) | Penetration depth Diurnal (m) | Annual (m) |
|---|---|---|---|---|---|---|
| Ice | $1.89 \times 10^6$ | $1.2 \times 10^{-6}$ | 2.27 | $2.1 \times 10^3$ | 0.6 | 10 |
| Dry sand | $1.26 \times 10^6$ | $1.3 \times 10^{-7}$ | 0.16 | $4.5 \times 10^2$ | 0.2 | 4 |
| Wet soil | $1.68 \times 10^6$ | $1.0 \times 10^{-6}$ | 1.68 | $1.7 \times 10^3$ | 0.5 | 9 |
| Still water | $4.2 \times 10^6$ | $1.5 \times 10^{-7}$ | 0.63 | $1.6 \times 10^3$ | 0.2 | 4 |
| Stirred water' | $4.2 \times 10^6$ | $5.0 \times 10^{-3a}$ | $2.1 \times 10^4$ | $3 \times 10^{5a}$ | $40^a$ | |
| Still air | $1.26 \times 10^3$ | $2.0 \times 10^{-5}$ | $2.5 \times 10^{-2}$ | 5.6 | | |
| Stirred air | $1.26 \times 10^3$ | $10.0^a$ | $1.3 \times 10^4$ | $4 \times 10^{3a}$ | $1500^a$ | Troposphere |

[a] These values are not determined, as are the others, by molecular properties and so cannot be measured precisely as in laboratory experiments. (From Petterssen, 1969)

*process.* However, both air and water can transfer heat through the mass movements associated with stirring: *turbulent* transfer and *convection.*

Since different substances conduct heat away from the surface at different rates, the surface temperature resulting from a particular amount of energy input will also differ. The volume over which the heat is effective is proportional to $\sqrt{K^*}$ and the temperature rise is, similarly, found to be proportional to $\rho C \sqrt{K^*}$. This is often called the *conductive capacity, $C^*$* (Table 2.2). At the interface between two substances the heat will be shared in proportion to their respective conductive capacities. The temperature range at the interface must be the same for both media and will be approximately given by the inverse of the sum of the conductive capacities.

### Contrast between land and ocean

Of primary concern to us at the moment is the different response of land and water to a given energy input. At the surface both are in contact with air. The inverse of the sum of conductive capacities for air and land is about 7; for water and air it is close to 0.14. Consequently the temperature range is about 50 times larger over land than water. It therefore follows that land surfaces heat and cool more rapidly, and have a greater temperature range, than do water surfaces. This result, expressed on the larger space scale in the concept of *continentality*, plays a profound role in establishing the global distribution of temperature.

If we use the values in Table 2.2 with fairly typical values for the radiation fluxes, we find that the annual range of temperature over the oceans is typically only a few degrees, while for land the annual range can be many tens of degrees. These are somewhat larger values than those observed. The differences are largely due to the effects of latent heat fluxes and horizontal motions. Surface cooling by latent heat transfer occurs whenever there is *evaporation.* Horizontal motions moderate the sensible heat flux because of the transfer and mixing of air above the surface. All of these features of the climate system are the direct result of elements of the global energy cascade. Nevertheless, the above discussion allows us to analyse the processes creating the global distribution of surface temperature.

## 2.11   Temperatures at the Earth's surface

The previous consideration of the way energy flows produce temperatures suggest that temperature changes near the surface of the Earth can vary rapidly both horizontally and vertically. In this section we are concerned with the general global distribution of surface temperatures. However, because of these small-scale variations it is first necessary to define surface temperatures and temperature measurements as they are used in this global context.

### Measurement of temperature

The basic instrument for measuring temperature is the *thermometer*. Although many types are available the most common form is the mercury-in-glass thermometer (Fig. 2.30). Mercury in the bulb expands when heated, the expanding liquid being forced along the tube. The temperature is thus indicated by the length of the mercury column in the tube. The instrument must first be calibrated to provide the relationship between length and temperature. Thereafter the length can be expressed directly in temperature units.

**(a)**

**(b)**

Fig. 2.30. (a) Orientation of wet and dry bulb thermometers (vertical) and maximum and minimum thermometers (horizontal) in a Stevenson screen (see Fig. 1.1). (b) The maximum thermometer operates in the same way as a clinical thermometer: a constriction close to the bulb forces the complete column to remain extended showing the maximum temperature until 'shaken down' when reset.

There are numerous variations on this basic design. Two common ones are used in the maximum and minimum thermometers (Figs. 2.30 and 2.31). In the former a constriction is placed in the tube close to the bulb. The expanding mercury is able to force its way past this, but when cooling and contraction takes place the constriction does not allow the mercury to return to the bulb. The mercury is thus left in the tube to record the maximum temperature. A rapid shaking of the tube can restore a continuous thread of mercury across the constriction, thus resetting the instrument for use again. In the minimum thermometer the mercury is usually replaced by alcohol. A small rod called an index and usually barbell shaped, is placed within the alcohol in the tube. As temperatures fall and the liquid contracts this index is drawn towards the bulb by the surface tension of the alcohol surface. The index is left behind when the liquid again expands, the end of the index farthest from the bulb thus indicating the lowest temperature during the period. Tilting the instrument so that the index slides to the liquid meniscus resets it.

### Stevenson's screen
These instruments are generally housed in a shelter of some form (often a Stevenson screen as shown in Fig. 1.1). This serves to shield the instrument from direct radiation, yet provides ventilation so that there is a free flow of air past the thermometers. The instruments thus measure air temperatures above the surface. This type of observation is frequently, and loosely, called 'surface temperature', although it is certainly not the temperature which would be recorded by an instrument in contact with the ground. However, it is this type of measurement that we are considering here. Most nations have established standard conditions for their screen placements so that comparisons between sites are possible. The standard usually requires the screen to be of the order of 1 m above a grass surface.

True surface temperatures can be obtained from satellite observations of upwelling radiation in selected channels in the atmospheric window. The results, after corrections for atmospheric transmission losses and surface emissivity, are true surface temperatures integrated over the field of view of the sensor. Hence they are not strictly comparable with the more traditional surface-based observations. Nevertheless, they are an invaluable adjunct to these surface observations, especially over the oceans where surface observations are sparse.

### Global patterns of mean sea-level temperature
The global distribution of temperature near the Earth's surface is shown in Fig. 2.32 for the summer and winter seasons. In D/J/F the highest temperatures lie in a belt close to the equator over the oceans

81

(a)

(b)

Fig. 2.31. (a) Minimum thermometer, showing the barbell shaped index which is drawn back by the (colourless) alcohol as temperatures decrease but remains in place when the alcohol column again increases in length (here the thermometer is installed with its bulb just touching the blades of short grass. The recorded temperature is termed the 'grass minimum'); (b) an earth thermometer which is installed in a steel tube driven vertically into the ground to standard depths of 30 cm and 100 cm; (c) set of soil thermometers installed so that their bulbs are at depths of 5, 10 and 20 cm approx. (usually 2, 4 and 8 inches).

(c)

and somewhat south of it over the land masses. Maximum temperatures exceed 30 °C over portions of these land areas. Minimum temperatures occur over the polar regions, with the lowest values, below −30 °C, in north central Asia. There are sharp temperature contrasts between land and sea, particularly on the western sides of continents where at a given latitude the Northern Hemisphere land is colder and the Southern Hemisphere land warmer than the adjacent ocean. A similar phenomenon, with the hemispheres reversed, occurs in J/J/A. In this season Antarctica is the coldest region and the warmest areas are those continental areas just north of the equator.

A close comparison of the conditions in the two seasons clearly indicates that seasonal changes in ocean surface temperature are relatively minor, but that mid-latitude continental interiors suffer a much greater range. Figure 2.33 indicates the annual course of mean monthly temperatures for three stations: a continental interior, a coastal location and a tropical situation. These results largely follow from our considerations of the energy cascade and the resultant surface and atmospheric temperatures in previous sections.

## 2.12  Applications of temperature information

The type of surface-based temperature observations we described in the last section is mostly suitable for giving a general indication of

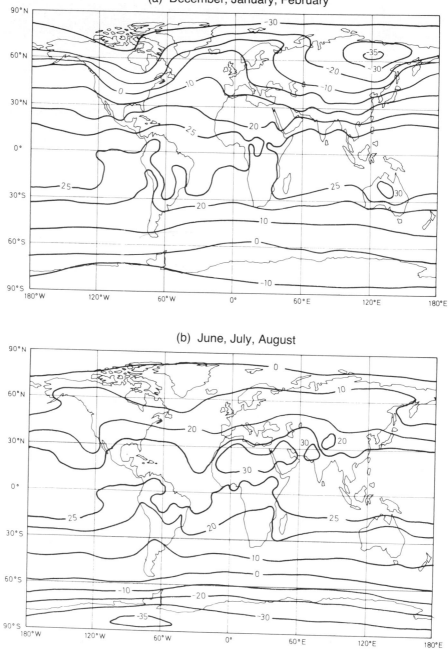

Fig. 2.32. Mean sea-level temperature (°C) averaged for (a) December, January, February (1963–73) and (b) June, July, August (1963–73).

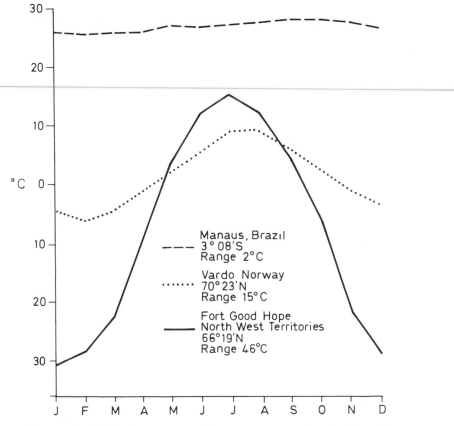

Fig. 2.33. Seasonal variation in mean monthly temperatures at three stations. Manaus, located near the equator in the Amazon rain forest, shows almost no variation in temperature through the year. The effect of the Gulf Stream in reducing the seasonal temperature range in Norway as compared with a station further south in the North West Territories is clearly seen.

conditions in an area, the results then being related to the particular point of interest. As such they have been widely applied in many areas, especially in the agricultural and energy fields.

### Crop growth as a function of temperature
One prime agricultural concern is the length of the growing season. The definition of growing season varies with location. In tropical regions it is frequently associated with the temporal rainfall distribution, while in mid-latitudes it is primarily temperature dependent. Although we concentrate here on the latter, the ideas are equally applicable to the former.

Crop growth is initiated when temperatures in spring rise above a certain threshold value and ceases in autumn when they fall below this value. From the temperature record it is possible to determine for a given place and year the length of time between these events and thus the length of the growing season. Since this length will vary from year to year it is not possible to use data for one year to provide a forecast for another. Long periods of data are required to produce such forecasts. A network of measuring sites is used to prepare a map of the average length of the growing, or in this case, freeze-free, season (Fig. 2.34), enabling the agriculturalist to obtain a general idea of the growing season length at a particular location. Further information can be gained by analysing the long period record to determine the probability with which the growing season will have a particular length (Table 2.3). Using the same method, the probability that the season will start on or after a given date in spring and will end on or before a given date in autumn is calculated. Thus a knowledge of planting and harvesting dates, as well as the time available for crop growth, can be acquired. This information therefore provides a probability forecast for a particular season. The individual farmer can use it to balance his strategy for maximum yields against the likelihood of crop failure for climatic reasons.

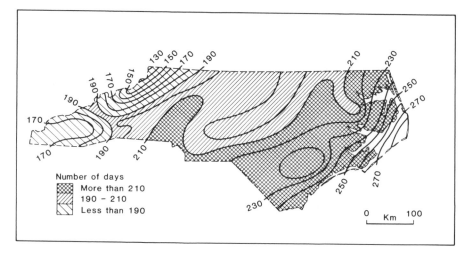

Fig. 2.34. Average length of the freeze-free season in North Carolina (number of days). (Clay, J.W., Orr, D.M. and Stewart, A.W. eds. *North Carolina Atlas*, 1975, University of North Carolina Press, Chapel Hill, NC, 331 pp. ).

The temperature values chosen for Table 2.3 are rather low for most crops. Nevertheless the figure and table are illustrative of the type of information that can be produced for agricultural purposes and are suggestive of applications in other industries, such as the construction industry, where temperatures can dictate work schedules.

## Table 2.3   Freeze date probabilities

Probability of longer than indicated period (days) with temperatures continuously above indicated value

| Probab. Temp (°C) | 0.10 | 0.20 | 0.30 | 0.40 | 0.50 | 0.60 | 0.70. | 0.80 | 0.90 |
|---|---|---|---|---|---|---|---|---|---|
| 0 | 276 | 266 | 258 | 251 | 245 | 239 | 232 | 274 | 213 |
| −2.2 | 318 | 303 | 292 | 283 | 274 | 266 | 256 | 245 | 230 |
| −4.4 | >365 | 329 | 304 | 304 | 295 | 286 | 278 | 268 | 254 |
| −6.7 | >365 | >365 | >365 | >365 | 340 | 330 | 322 | 313 | 303 |
| −8.9 | >365 | >365 | >365 | >365 | >365 | >365 | >365 | >365 | 345 |

Probability of the occurrence of the listed temperatures at a later date in spring (month/day) than the date indicated

| Probab. Temp (°C) | 0.10 | 0.20 | 0.30 | 0.40 | 0.50 | 0.60 | 0.70 | 0.80 | 0.90 |
|---|---|---|---|---|---|---|---|---|---|
| 0 | 4/1 | 3/25 | 3/20 | 3/16 | 3/12 | 3/8 | 3/4 | 2/27 | 2/20 |
| −2.2 | 3/31 | 3/19 | 3/10 | 3/3 | 2/24 | 2/9 | 2/9 | 1/31 | 1/19 |
| −4.4 | 3/17 | 3/5 | 2/25 | 2/18 | 2/11 | 1/27 | 1/27 | 1/18 | 1/3 |
| −6.7 | 2/17 | 2/8 | 2/1 | 1/25 | 1/19 | 1/12 | 1/3 | | |
| −8.9 | 2/1 | 1/23 | 1/14 | | | | | | |

Probability of the occurrence of the listed temperatures at an earlier date in autumn (month/day) than the date indicated

| Probab. Temp (°C) | 0.10 | 0.20 | 0.30 | 0.40 | 0.50 | 0.60 | 0.70 | 0.80 | 0.90 |
|---|---|---|---|---|---|---|---|---|---|
| 0 | 10/25 | 11/1 | 11/5 | 11/9 | 11/13 | 11/16 | 11/20 | 11/25 | 12/1 |
| −2.2 | 11/7 | 11/13 | 11/18 | 11/22 | 11/26 | 11/29 | 12/3 | 12/8 | 12/14 |
| −4.4 | 11/12 | 11/21 | 11/28 | 12/4 | 12/9 | 12/15 | 12/21 | 12/28 | 1/9 |
| −6.7 | 11/30 | 12/10 | 12/17 | 12/24 | 12/30 | 1/6 | 1/16 | | |
| −8.9 | 12/24 | 1/8 | 1/25 | | | | | | |

### *Energy demand as a function of temperature*

The energy industry commonly uses temperature observations in a way which is similar to that described above. Demand for energy, particularly electricity, natural gas and home heating oil, is temperature dependent. This is primarily the result of the increase in building heating needs as the ambient temperature decreases or, in regions where air conditioning is common, an increase in demand as temperatures rise. This demand, of course, is superimposed on the more or less fixed need for energy by machine operations. However, the demand caused by temperature variations is responsible for the transient peaks in demand that dictate a power company's generation or storage requirements.

As a result of many comparisons between temperature and demand a simple linear regression relationship between the two has often been found to be appropriate. This takes the form

$$D = a + b(\text{HDD}) \qquad [2.17]$$

where $D$ is demand and $a$ and $b$ are constants. HDD is the *heating degree day*, defined as the number of degrees by which the average daily temperature falls below a threshold, or base, temperature. (On a daily basis HDD has units of degrees.) HDD can thus readily be calculated from temperature information and values are available for many locations. In North America the threshold value has traditionally been 18.3 °C. If a house is heated to this temperature it will, because of the heating provided by lights, appliances and people, eventually attain a temperature around 22 °C, which was formerly felt to be 'comfortable' for living. Changes in lifestyle, mainly induced by sharply increased energy prices, have lowered the definition of comfortable by 3 or 4 degrees. Europeans, with longstanding high energy prices, have used the lower value for many years.

The accuracy of the demand estimates depends greatly on the nature and location of temperature measuring stations chosen for use in equation [2.17]. It is, of course, highly desirable that the stations have a long record of high quality data. The station must also provide temperatures representative of the population centres where demand is concentrated. This may mean considering not only geographical proximity but also the characteristics of the local site. Thus station selection, if there is a choice of stations available, provides a challenge to the climatologist.

Once the stations are chosen and the regression established, it becomes the job of the short-term weather forecaster to provide the information needed by the company to prepare for the expected demand.

This type of approach to the solution of climate related problems can be generalised as a *regression* approach. The major steps are the development of a regression equation based on historical data, its use in conjunction with short-term weather forecasts to aid in operational decisions and its use in long-range planning. Use is by no means restricted to the energy industry or to temperature information. Crop yield models, for example, are vital in agriculture and require knowledge of several climatic parameters. Hence we will consider them in Chapter 6 after we have discussed the other elements that constitute the climate system.

## Summary

Energy exchanges at the surface of the Earth are the primary driving mechanisms of all elements of the climate system. In this chapter we have reviewed the nature of the incoming solar radiation and the outgoing infrared radiation in terms of the surface radiation budget, the temperature structure of the atmosphere and surface temperatures. The major elements of the energy cascade will be shown in the following chapters to be responsible in addition for the atmospheric moisture and horizontal motion which we recognise as the climate.

Satellite observations have been seen to reveal large-scale features of the Earth's energy budget. These will be drawn upon again in the discussion of the general circulation of the atmosphere in Chapter 4. Focussing upon the surface radiation budget has prompted us to consider other elements of the energy fluxes at the surface. In particular, we have identified the importance of water to the system. The hydrological cycle will be discussed in the following chapter.

# Chapter 3
# The Hydrological Cycle

3.1   Evaporation

3.2   Moisture in the atmosphere

3.3   Clouds and cloud-forming processes

3.4   Hydrostatic stability

3.5   Clouds and climate

3.6   Precipitation formation

3.7   Precipitation

3.8   Global precipitation distribution

3.9   Local scale precipitation

    3.9.1   Thunderstorms

    3.9.2   Applications of precipitation information

3.10  The water balance

# Chapter 3
## The Hydrological Cycle

Clouds and precipitation, along with temperature and wind, are the most striking elements of weather and climate, elements that can change very rapidly with time and space. However, water in all its forms and in all its various activities in the atmosphere plays an important role in sustaining not only the climate but also life itself. In order to understand and predict the actions of water in the climate system it is useful to think of the water as being part of a distinct system, sometimes called the *hydrological cycle* (Fig. 3.1). A complete understanding of this system would, as the figure implies, require excursions into geomorphology, pedology, botany, glaciology, oceanography and, if Man-made structures are included as part of the Earth's surface, civil engineering. Further, the implications of water and water supply would require a knowledge of public health, water supply engineering, agricultural practices and, eventually, appreciation of public policy development and implementation. Whilst such considerations may indicate the broad range of disciplines that are of interest to the climatologist, we must obviously concentrate on those aspects of water that are directly connected with the climate, drawing on the other disciplines only when needed and indicating some areas where climatology can contribute to these other disciplines.

Since we are thinking of water moving in a cycle, there is no one best place to start our discussion of the climatic role of water. However, the previous chapter has emphasised energy exchanges, so it is convenient to start our discussion at the air–earth interface, where these exchanges are vital.

## 3.1 Evaporation

The concept of evaporation was introduced in Chapter 2, but there it was treated as a consequence of energy exchanges. Here we are

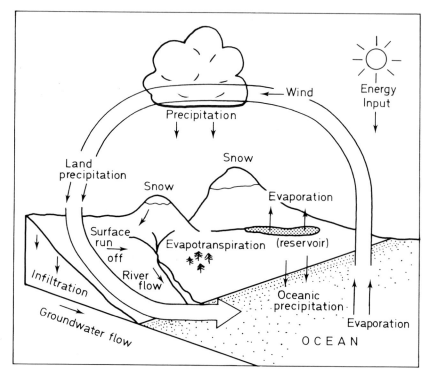

Fig. 3.1. The global hydrological cycle.

concerned with the evaporation as it influences the amount of water at the surface and in the atmosphere. Nevertheless, there is a simple relationship between the energy used for evaporation and the amount of water evaporated, as was demonstrated by our use of the symbol LE for the latent heat flux.

### Evapotranspiration

Water is removed from the surface into the atmosphere by two distinct processes. *Evaporation* occurs when there is a 'free' water surface, be it water in an ocean or between individual soil particles. *Transpiration* occurs when water is removed from the interior of plant leaves through their stomata. Although these two processes are different and distinct, it is possible in most contexts to treat them together, since the forcing mechanisms are similar and the end result is the same. The combined process is called *evapotranspiration*. We shall be dealing with this combined process unless it is specifically noted to the contrary.

The rate of evapotranspiration at any instant, and thus by integration the amount of evapotranspiration, is controlled by four factors: (1) energy availability; (2) the humidity gradient away from the surface; (3) the wind speed immediately above the surface; and (4) water availability.

Viewing the evaporative process on the molecular level, evaporation occurs whenever the flow of individual water molecules away from the water body exceeds the return flow to the water from the atmosphere. Dew is formed if the return flow exceeds the amount leaving the surface. The flow from the surface is largely dependent on the energy available, as discussed in the previous chapter, and on there being a supply of water at the surface. The rate of return flow is highly dependent on the humidity of the air immediately above the surface. As the humidity increases this return flow increases and the net rate of evaporation decreases. Thus for rapid evaporation we need a steep humidity gradient away from the surface, in exactly the same way that a steep temperature gradient is needed for rapid heat flow. In near calm conditions the surface evaporation will lead to a rapid build-up of the moisture content of the air layer immediately adjacent to the surface and a decrease in the evaporation rate. In the presence of wind, with its propensity for turbulent mixing, the low level moist layer will be removed and replaced with less humid air. Usually this replacement is associated with large-scale horizontal advection, which introduces a new, drier air mass to the surface. The net result is that moderate wind speeds tend to maintain steep humidity gradients and high evaporation rates.

High rates also depend on a continuous supply of water at the surface. For an open water surface this is no problem. However, for a land surface, water movement from depth, whether through the soil or through plants, requires a considerable time. There is, for example, an upper limit to the rate at which plants can transpire. Often on a sunny midsummer day plants can wilt in the early afternoon simply because water vapour is being removed from the leaves faster than it can be brought in from the roots. Similarly the surface layers of a soil may become dry although plenty of water remains at depth.

Not all of the energy absorbed at the surface is used to evaporate water, since the other energy fluxes will be maintained. For many land surfaces the soil heat flux is small and can be neglected, so that the major partitioning of energy is between sensible and latent heat. The ratio of the two ($H/LE$), the *Bowen ratio*, is an indication of this partitioning. In general, surfaces act to keep the ratio at a minimum. Indeed a moist surface will increase little in temperature while evaporation is occurring, but there will be a rapid temperature rise once it has dried out and the sensible heat flux takes over as the major energy transfer mechanism.

The action of the four factors controlling evapotranspiration, any of which can be limiting, has led to the development of two concepts of evapotranspiration for practical applications. The first is *potential evapotranspiration* (PET), which is the rate which will occur from a well-watered, actively growing, short green crop completely covering the ground surface. This is essentially identical with the values which would be obtained over a large open water surface. It represents the rate controlled entirely by atmospheric conditions and is the maximum possible in the prevailing meteorological conditions. The second concept, *actual evapotranspiration* (AET), is the amount that is actually lost from the surface given the prevailing atmospheric and ground conditions. Both are important, since PET provides some measure of possible agricultural productivity if, for example, irrigation is initiated, while AET provides information vital for the determination of soil moisture conditions and the local water balance.

## Measurement of evapotranspiration

The measurement of evapotranspiration, potential or actual, is difficult. Many techniques have been devised. The most common method of measuring PET directly is using *evaporation pans* (Fig. 3.2). These are simply containers of a standard size containing water freely exposed to the atmosphere. The water depth is measured at the beginning of the time period of interest and again at the end. The difference, after correction for any precipitation received, is the evaporation. Energy transfer through the sides of the pan, together with turbulence created by the pan itself, make it difficult to relate the results to evaporation from natural open water surfaces. Usually a correction factor, a 'pan coefficient', is employed before the results are used. Rather more sophisticated instruments, *lysimeters* (Fig. 3.3), are available, but these require very careful installation and maintenance if they are to give useful results. They are thus restricted mainly to a few agricultural research establishments. A section of the land surface is removed, a pan placed in the cavity and the land replaced in the pan with as little alteration to its initial structure as possible. Usually the pan is placed

Fig. 3.2. Photograph of evaporation pan. The amount of evaporation is established by measuring the depth changes and compensating for the input of precipitation.

Fig. 3.3. Photograph of a weighing lysimeter being lowered into position. When *in situ*, the lysimeter should be undetectable. The amount of evapotranspiration taking place over a fixed period is calculated by weighing the soil plus biomass at the start and finish of the experiment and carefully monitoring rainfall. The sides and base of the soil container are carefully sealed before emplacement. (Courtesy: John Stewart).

on a weighing mechanism which is used to record the change in weight and thus the amount of evaporation.

As it is difficult to measure evapotranspiration directly, it is usual to estimate it from more commonly measured parameters. Numerous methods are available. Almost all start by estimating PET from atmospheric measurements. Some simple ones, requiring only air temperature as input, can be used for the determination of monthly average values. Others are much more complex and incorporate solar radiation, wind speed, air temperature and humidity measurements, and can be used to estimate daily values. Once PET is determined, some form of book-keeping method is used to track the amount of moisture in the soil and to relate this to the amount that will be available for actual evapotranspiration.

The relative sparseness of observations, together with the difficulty of reconciling various estimation methods, makes it very difficult to present reliable global maps of evaporation. Nevertheless, various attempts have been made and, although the results differ in detail, the general global picture is well established (Fig. 3.4). Maximum rates of actual evaporation occur over the subtropical oceans with a general decrease in amounts poleward. Land values are lower than oceanic ones, the isopleths making a sharp break at the coasts. A rough extrapolation of the oceanic values over the continents gives an idea of the PET of land. This serves to emphasise the considerably lower rates of AET over the land, most marked over the desert regions of the Earth.

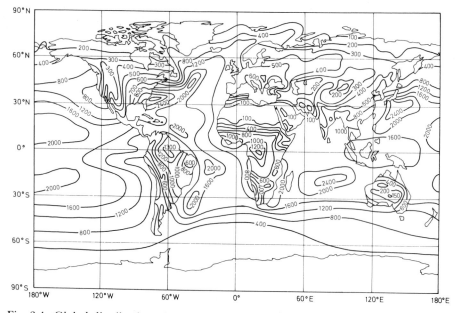

Fig. 3.4. Global distribution of evaporation (mm yr$^{-1}$) based upon a number of recent estimates. Two features are immediately obvious: (i) the discontinuity between adjacent land and ocean regions which might be anticipated as the oceans are an almost infinite moisture source; and (ii) the dependence of oceanic evaporation upon ocean temperatures (especially clear in the North Atlantic).

# 3.2  Moisture in the atmosphere

Evaporated water enters the atmosphere as individual energetic water vapour molecules. There are numerous ways of expressing the resulting moisture content of the atmosphere, each appropriate for particular applications. Prior to discussing these, however, it is necessary to introduce a fundamental concept vital for the understanding of atmospheric processes.

### Saturation of the air

There is an upper limit to the amount of water vapour that the air can hold. This is the point at which air becomes *saturated*. A rigorous definition of saturation, which will be important in later sections, is that it is the maximum water vapour content of air in equilibrium with a plane surface of pure water or of pure ice at the same temperature as the air. Values of saturation differ between the two types of surface. The saturation vapour pressure is smaller over ice than over water at the same temperature because the latent heat required for the solid to vapour transition $(2.834 \times 10^6 \, \mathrm{J\,kg^{-1}})$ is greater than that for the change from liquid to vapour $(2.501 \times 10^6 \, \mathrm{J\,kg^{-1}})$. At the point where saturation occurs the interaction of the large number of high energy water vapour molecules with the lower energy air molecules becomes sufficient to cause the vapour molecules to give up some energy and become liquid droplets (or solid crystals) as they revert to the lower energy state. Latent heat is released in the process.

The saturation value varies with temperature. If we introduce an expression for humidity we can quantify the relationship. Vapour amount can be expressed in terms of the *vapour pressure*: the force per unit area created by the motions of the vapour molecules treated in isolation from all the other gases of the atmosphere. The relationship between the value at saturation, the saturation vapour pressure, $e_s$, and temperature, $T$, can be derived from the second law of thermodynamics. The result is the *Clausius–Clapeyron* equation.

$$\frac{de_s}{dT} = \frac{L}{T} \times \frac{1}{V_2 - V_1} \qquad [3.1]$$

where $L$ is the latent heat of vaporisation and $V_2$ and $V_1$ are the specific volumes (volume occupied by unit mass) of water vapour and liquid water respectively. The value of $V_1$ is usually negligible in comparison with the value of $V_2$. The saturation vapour pressure increases approximately logarithmically with temperature (Fig. 3.5). Since $e_s$ differs between liquid and solid water surfaces, at temperatures below freezing the saturation vapour pressure depends on whether condensation occurs as liquid water droplets or as ice crystals.

### Evaluation of atmospheric moisture content

So far the moisture content has been expressed in terms of the vapour pressure. Alternative means of expression which are used in various places later can be collected and defined here.

The *absolute humidity* is defined as the mass of water vapour per unit volume of air and is expressed in grams per cubic metre.

The *mixing ratio*, strictly the 'water vapour mixing ratio', is the ratio of the mass of water vapour to the mass of dry air in a specified volume. It is expressed in grams per kilogram. The saturation mixing ratio, $w_s$, is frequently used, being directly analogous to and having the same shaped curve as the saturation vapour pressure, $e_s$. Indeed, it can be shown that they can be related in the normal atmospheric situation where $e_s$ is very much less than the total atmospheric pressure $p$. The relationship takes the form

$$w_s \simeq 0.622 \, e_s/p \qquad [3.2]$$

The *relative humidity* with respect to water is the ratio of the actual mixing ratio $w$ to the saturation mixing ratio $w_s$. It is usually expressed

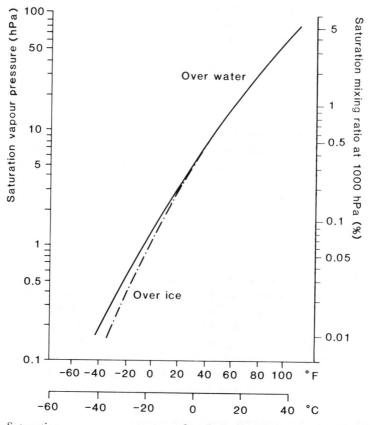

Fig. 3.5. Saturation vapour pressure as a function of temperature. Below 0 °C there are two curves: one for supercooled water; the other for ice.

as a percentage such that

$$RH = (w/w_s) \times 100\% \tag{3.3}$$

The relative humidity of a parcel of air therefore depends upon its temperature since $w_s$ is functionally dependent upon temperature. Consequently the value will alter whenever the temperature changes without a change in moisture content. For instance, the relative humidity may decrease by as much as 50% between morning and noon as temperatures rise.

The *dew point* temperature ($T_d$) is the temperature at which an air parcel would become saturated if it were cooled without a change in pressure or moisture content. Since there is an unique relationship between saturation and temperature, the dew point temperature also has an unique value for any air mass. It must be emphasised that although $T_d$ is a temperature, it is only of interest as a measure of humidity.

### Measurement of atmospheric moisture content

Several types of instruments have been devised to measure humidity. Only those commonly used currently are discussed here. The first is the *dew point hygrometer* (dew cell), which measures $T_d$ directly. A mirrored surface is cooled electrically until dew forms. As soon as this occurs a photoelectric detector senses a change in surface reflectance and switches the cooling circuit to a heating one. The heater remains operative until the dew is evaporated, at which point the cooling cycle is again initiated. The cycling is repeated until a stable temperature is reached, representing the dew point of the air above the mirror surface. Figure 3.6 illustrates another type of hygrometer which senses relative humidity directly.

A less sophisticated instrument is the *psychrometer* (e.g. Fig. 2.30(a)). This consists of two thermometers. One called the *dry bulb*, is unmodified. The bulb of the second, the *wet bulb*, is kept moist by being

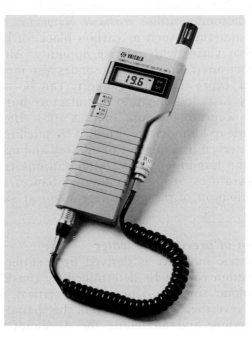

Fig. 3.6. A humidity sensing instrument. The sensor is similar to that used for humidity measurement in radiosondes. It is based on a polymer thin-film capacitor contained in the probe which can be held at a distance from the display. A 1 $\mu$m thick dielectric polymer layer absorbs water molecules through a thin metal electrode. This causes a capacitance change proportional to the relative humidity. This particular instrument permits display of both temperature and the relative humidity. (Courtesy: Vaisala).

101

encased in a wick connected to a water reservoir. Water is evaporated from around the wet bulb. The energy required for this evaporation is extracted from the bulb itself and thus its temperature is lowered. The amount of cooling depends on the evaporation rate, and thus on the humidity of the air. Provided an even flow of water comes from the reservoir, allowing neither flooding nor drying of the bulb, the temperature will become stable. The amount of moisture can then be calculated from the dry bulb or 'air' temperature together with the difference between the dry and wet bulb temperatures ($T_a$ and $T_w$), the latter being known as the wet bulb depression. Note that the wet bulb temperature is not the same as the dew point but that $T_d \leqslant T_w \leqslant T_a$, the equalities holding only when the air is saturated. To obtain reliable estimates of humidity with a psychrometer it is necessary to ensure some airflow over the two thermometers, which are usually held so that they lie parallel to each other a few centimetres apart. This airflow is created either by placing the instrument in a tube and drawing air across it, forming an aspirated psychrometer, or by whirling it through the air manually, as in a sling psychrometer.

Humidity in the free atmosphere is measured by both radiosondes and satellites. Radiosondes utilise *resistance hygrometers*. These use the property of some materials, such as carbon black, of having an electrical resistance that varies with relative humidity. Although relative humidity is thus measured directly, radiosondes also measure temperature, so that other expressions for humidity can be calculated. The instrument is fairly easy and cheap to manufacture but in general the results are rarely accurate to more than ± 10%. Increasing use is therefore being made of humidity determined by satellites. Use is made of the radiances observed in a number of infrared and microwave channels to establish the humidity profile. Simultaneous retrievals of the atmospheric temperatures, using measured radiances in other channels, allow determination of the moisture content at various levels in the atmosphere.

### Global distribution of precipitable water

Although atmospheric moisture is derived from surface evaporation, the spatial distribution cannot be determined by consideration of the source strength alone since horizontal and vertical mixing almost invariably occur. Nevertheless, knowledge of the spatial distribution is vital for understanding and modelling the dynamics of the climate system and, to a lesser degree, understanding the processes of precipitation formation. The atmospheric vapour content is expressed as the *precipitable water vapour* (PWV): the total amount of water vapour in the atmospheric column above a point on the Earth's surface, or in a specific atmospheric layer. The amount between any two pressure levels $p_1$ and $p_2$, is given (in kg per unit area) by

$$\text{PWV} = \frac{1}{g} \int_{p_1}^{p_2} q \, dp \qquad\qquad [3.4]$$

where $p_1$ and $p_2$ are the pressure levels, $g$ is the acceleration due to gravity and $q$ is the specific humidity, which is given as the ratio of the mass of water vapour to the mass of moist air.

The PWV represents, as the name indicates, the amount of water that is available to fall as precipitation. It also indicates the amount of moisture available in the cloud-free atmosphere to interact with the various radiation streams and thus influence atmospheric heating rates.

PWV can be determined directly from moisture profiles obtained by radiosonde ascents or satellites. It can also be estimated from surface observations alone, since most vapour is concentrated in the lowest atmospheric layers, close to its source. The distribution of PWV is such that at any time there is considerable moisture throughout the atmosphere (Fig. 3.7). This is true even for the desert regions. Although this map could be combined with one showing the wind field to indicate the moisture flux over an area, it is clear that the distribution of clouds and precipitation is dependent on much more than simply the availability of moisture. Mechanisms are needed to convert the atmospheric water vapour into liquid or solid form to produce clouds and to cause this moisture to fall as precipitation. These mechanisms are treated in subsequent sections.

## 3.3 Clouds and cloud-forming processes

When air becomes saturated, water vapour condenses and clouds are formed. The type of cloud produced depends greatly on the process by which the air is brought to saturation (Table 3.1). Prior to our discussion of the cloud-forming processes, therefore, it is useful to consider the main types of clouds.

### Classification of cloud types

The first systematic attempt to classify clouds was made by Luke Howard (1772–1864). Although changes have been made subsequently, the names he invented – at a time when it was usual to name scientific phenomena in Latin – have been retained. The infinite variety of individual clouds are classified into four main 'families' (low, middle, high and vertically extended), which are themselves intermixed to form subgroups (Fig. 3.8). *Cumulus* (Cu) represents the family with predominantly vertical development. They can range from the small white 'fluffy' clouds of a summer afternoon (Fig. 3.9(a)) to the towering black threatening cumulonimbus (Cb) – the thundercloud (Fig. 3.9(b)). The term 'nimbus' means rain or snow producing and usually refers to a very well developed member of a particular cloud

103

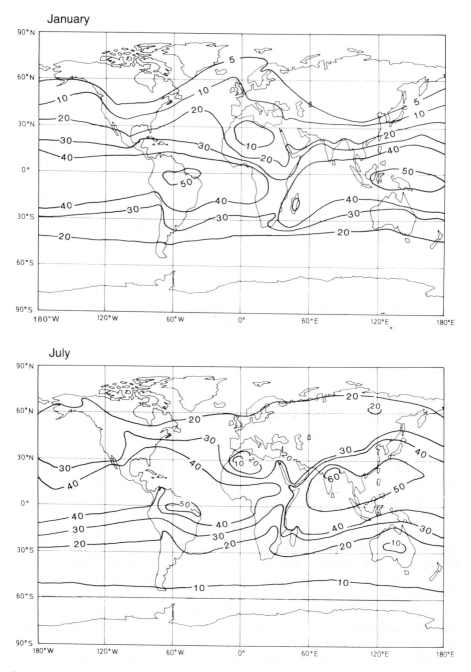

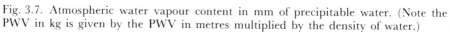

Fig. 3.7. Atmospheric water vapour content in mm of precipitable water. (Note the PWV in kg is given by the PWV in metres multiplied by the density of water.)

## Table 3.1 Cloud types and formation processes

| Process | Common cloud types (abbreviations) |
|---|---|
| A. Without cooling: | |
|     air mass mixing | stratocumulus (Sc) |
| B. Cooling without vertical motions: | |
|     radiative cooling | radiative fog |
|     advective cooling | advective fog |
| C. Cooling with vertical motions: | |
|     orographic uplift | stratus (St), altostratus (As) |
|     frontal uplift | all types |
|     airflow confluence | cirrus (Ci), stratus (St), altostratus (As) |
|     convection | cumulus (Cu) and cumulonimbus (Cb) |

family such as nimbostratus (Ns). Layer clouds with predominantly horizontal development are divided into three families, their distinctive appearance being created because of height and temperature differences. *Cirrus* (Ci) clouds are high and often wispy, being composed of ice crystals (Fig. 3.9(c)). The lowest layered clouds are *stratus* (St), composed of liquid water droplets and giving dull overcast, often drizzly, conditions (Fig. 3.9(d)). Stratocumulus (Sc) clouds (Fig. 3.9(e)) have a distinct cellular structure. The *alto* prefix always appears in conjunction with another family. Altostratus (As) is similar to stratus except that it tends to be less dense and less likely to give precipitation. Such clouds are commonly composed of *supercooled* water droplets with temperatures below 0 °C. Altocumulus (Ac) is a cumulus type cloud, having significant vertical development but with a base sufficiently high to be composed of supercooled water (Fig. 3.9(f)). Finally, it is convenient when considering cloud-forming processes to treat *fog* as a cloud at ground level.

### Cloud formation processes

In our discussion of the processes leading to cloud formation we shall follow the order outlined in Table 3.1, starting with the only process which can create clouds without a cooling of the air. In some circumstances when two air masses with different temperatures and moisture contents converge the two mix together. As a result of the non-linear relationship between saturated vapour pressure and temperature

## (a) CLOUD TYPES

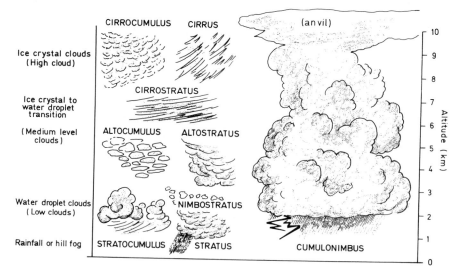

## (b) CLOUD FORMATION

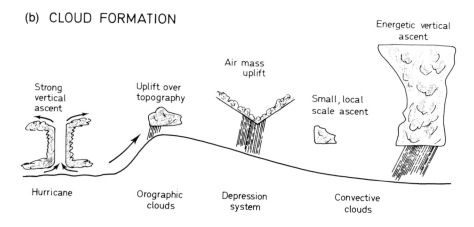

Fig. 3.8. The major cloud types and altitudes (a) and the major cloud formation processes (b).

Fig. 3.9. Photographs illustrating (a) cumulus (Cu); (b) cumulonimbus (Cb); (c) cirrus cloud (Ci); (d) stratus cloud (St); (e) stratocumulus (Sc) viewed from above; (f) altocumulus (Ac) overlying lower level cloud.

(a)

(b)

107

(c)

(d)

St

St

(e)

Sc

**(f)**

(Fig. 3.5), the mixture may be saturated with respect to the new temperature although both initial air masses were unsaturated. The types of cloud created depend on the level of the mixing, but commonly this process can lead to stratocumulus cloud. The process is rarely seen in its pure form, since airstream convergence is likely to lead to widespread vertical motions or frontal uplift, which are discussed below, which may overshadow the mixing. Nevertheless, the latent heat released during the mixing is frequently an important source of energy for sustaining the motion.

The radiative cooling mechanism for condensation depends directly on radiation exchanges at the surface. When net radiation is negative, particularly on a calm, clear night, the air in contact with the ground is cooled. If there is sufficient moisture in the air, or the cooling is sufficient, the air will be cooled below its dew point and ground fog, usually called a *radiation fog*, will result. This will begin to form, very close to the ground, an hour or two after midnight and will gradually thicken and deepen as the night progresses. Soon after sunrise net radiation will become positive, heating of the air will commence and the liquid water droplets of the fog will evaporate back into the air.

Another cooling mechanism producing fog is associated with horizontal movement of air (advection). If a warm airstream starts to blow over a cooler surface the air itself rapidly adjusts to the temperature of the new surface. Again, given sufficient cooling, or sufficiently moist air, a fog will result. This *advective fog* will persist as long as the moist airstream blows over the cooler surface (see e.g. Fig. 5.5).

Although these cooling mechanisms may be locally important, by far the most common mechanisms for cloud formation occur as a result of vertical motions in the atmosphere. In order to appreciate these it is first necessary to examine the consequences of allowing air to rise vertically.

### Vertical motion in the atmosphere

To develop the concepts associated with vertical motions we consider a 'parcel' of air. Although such a parcel can be of almost any size, it is useful initially to visualise it as being about the same size as a small cumulus cloud. We make the assumption, which is very good in practice, that the parcel's vertical motion is sufficiently fast to prevent energy exchange with its surroundings. Such a process is called *adiabatic*. In this case, as the parcel rises through the atmosphere, its pressure decreases along with that of the surrounding environment; so the parcel expands and cools. The reverse also holds. In an adiabatic descent, a parcel of air would warm as it contracts. A familiar example of this descending condition would be the increase in temperature of the valve on a bicycle pump when used to inflate a tyre. The relationships can be quantified through the first law of thermodynamics which states that

$$\text{energy added} = \text{increase in internal energy} + \text{work done} \qquad [3.5]$$

In the adiabatic case the heat added is, by definition, zero. The increase in internal energy is proportional to the temperature change in the parcel, while the work done is represented by the effort needed for the parcel to expand against the outside pressure. Thus equation [3.5] can be restated for our parcel as

$$0 = c_p \, dT - \frac{1}{\rho} \, dp \qquad\qquad [3.6]$$

where $c_p$ is the specific heat at constant pressure. Using the hydrostatic equation which relates changes in pressure to changes in height in the atmosphere such that

$$dp/dz = -g\rho \qquad\qquad [3.7]$$

we have

$$dT/dz = -g/c_p \qquad\qquad [3.8]$$

i.e. the rate of change of temperature with height has a constant value. This rate of change of temperature with height is the *dry adiabatic lapse rate*, $\Gamma_d$. Numerically it is equal to 9.8 K km$^{-1}$, and is a constant. Note that this value is a *decrease* of temperature with height.

If the air is saturated when lifting occurs this rate must be modified to account for the latent heat released during condensation. In this situation the first law of thermodynamics leads to

$$-L dw_s = c_p \, dT - \frac{1}{\rho} \, dp \qquad\qquad [3.9]$$

where $w_s$ is the saturation mixing ratio. Inclusion of this term leads to

$$\frac{dT}{dz} = -\frac{g}{c_p} - \frac{L}{c_p} \frac{dw_s}{dz} = -\Gamma_d - \frac{L}{c_p} \frac{dw_s}{dz} = -\Gamma_s \qquad\qquad [3.10]$$

This *saturated adiabatic lapse rate*, $\Gamma_s$, is slightly variable with height because $dw_s/dz$ (which it should be noted has a negative value) varies with height. Nevertheless, when considering the lower and middle troposphere a value close to 5.0 K km$^{-1}$ for $\Gamma_s$ is sufficient for most purposes.

### Cooling rate changes during ascent

Usually when a parcel starts to rise in the atmosphere it contains some moisture but is not saturated. Hence it cools initially at $\Gamma_d$ (Fig. 3.10). Eventually the dew point temperature is reached and condensation

(a)

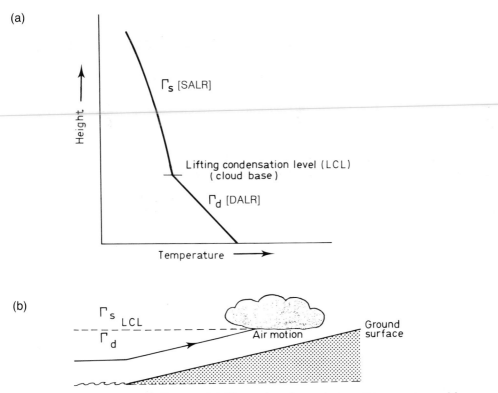

Fig. 3.10. Thermodynamic diagram (a) illustrating the variation of temperature with height as the air parcel is lifted over a raised ground surface as shown (b).

commences. This is the *lifting condensation level* and is the level of a cloud base. Continued uplift is accompanied by a temperature decrease at the saturated adiabatic lapse rate. Simple calculations of the lifting condensation level height are thus possible, provided it is borne in mind that there is a dew point lapse rate. This is approximately 1 K km$^{-1}$ and results from the influence of decreasing pressure on the condensation process. Complex and accurate calculations must be undertaken using one of the variety of *thermodynamic diagrams* (e.g. Fig. 3.15), which are not of immediate concern here.

There are a number of ways of creating the ascent of a parcel of air. Whenever the topography of the land surface dictates that air must rise, *orographic* uplift occurs. The angle of slope of the surface largely determines the size of the parcel that is uplifted and the type of cloud that results. Air flowing over a gently rising coastal plain, for example, will slowly be uplifted *en masse*, leading to stratus development. If the air reaches a larger obstacle, say a mountain, it may be forced to rise steeply and, if there is additional surface heating, cumulus clouds may result.

113

*Frontal uplift* occurs when two air masses of different temperatures come into contact. Although some mixing may occur, as described above, it is usual for the warmer, often moist, air to override the colder, being forced to ascend and thus to create clouds. The dynamics and characteristics of fronts, including the clouds associated with them, will be considered in detail in Chapter 5.

Uplift as the result of *confluence* between two airstreams generally results in widespread ascent. Such uplift is commonly associated with depressions in mid-latitudes and will be treated along with fronts in greater detail in Chapter 5.

So far we have considered parcels that are forced to rise. However, vertical motions leading to cooling can be created in a much more spontaneous manner through the process of *convection*. This process, which leads to cumulus clouds, is the result of *hydrostatic instability* of the atmosphere.

# 3.4 Hydrostatic stability

If the surrounding environment is cooler than the parcel, the parcel will be less dense than its surroundings, more buoyant, and thus will rise through the atmosphere, whilst continuing to cool at the appropriate adiabatic lapse rate. This simple statement summarises the concept of *hydrostatic stability*: the property of the atmosphere which controls the small-scale vertical motions within it.

Prior to considering hydrostatic stability in more detail we must introduce another lapse rate, the *environmental lapse rate*, $\gamma$. This is the observed temperature distribution with height at a given time and place. It is variable both in time and space. It must be clearly distinguished from the adiabatic rates since it bears no relation to parcels rising or falling. Nevertheless, the contrast between the environmental rate and the adiabatic rates controls the stability of the atmosphere.

### Stability conditions

Idealised stability conditions are illustrated in Fig. 3.11. In Fig. 3.11(a) it is assumed that a parcel, initially at the same temperature as the environment, rises at the saturated adiabatic lapse rate. Immediately upon uplift it becomes cooler than the environment and sinks back. Note that if it were a dry parcel, cooling at the DALR ($\Gamma_d$), the effect would be even more marked. These conditions in which the environmental lapse rate is such that it forces air to return to its original level are called *stable* conditions. Figure 3.11(b) indicates the opposite, *unstable* conditions. Here a parcel, whether saturated or dry, immediately becomes warmer than the environment and continues to rise. A third case, Fig. 3.11(c), illustrates *conditional instability*. Here the stability

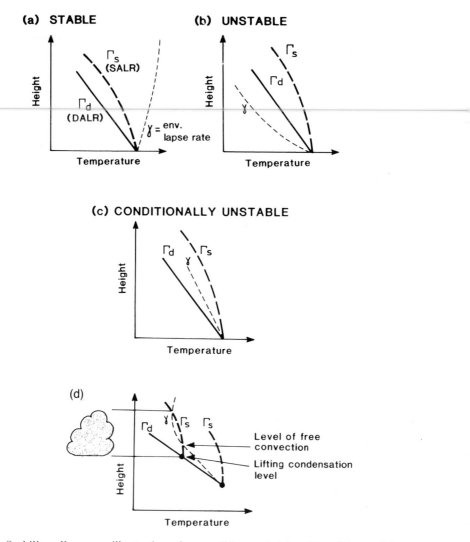

Fig.3.11. Stability diagrams illustrating the conditions of (a) stable, (b) unstable, (c) conditionally unstable atmospheres and (d) formation of cloud at the lifting condensation level (LCL).

condition depends on whether the parcel is dry or saturated.

Most parcels starting at or near the surface of the Earth contain some moisture but are not saturated. Frequently the atmosphere is conditionally unstable. These common conditions are illustrated in Fig. 3.11(d). As the parcel is forced upwards from its starting height it initially cools at the dry rate until saturation is reached at the lifting condensation level. Further upward movement is accompanied by cooling at the saturated rate. Usually this slower cooling rate in the conditionally unstable atmosphere eventually causes the parcel to change from being cooler to being warmer than the environment. At this point, the *level of free convection*, instability is established and vertical motions are continued without the necessity of the forced uplift needed to get the parcel to the level of free convection. Ascent will continue until the parcel once again becomes colder than the environment. At this point stability will be established and the cloud top reached.

When a deep layer of air is uplifted, the stability condition can change as it ascends. If the base of the layer is very moist while the top is comparatively dry (Fig. 3.12(a)), the bottom will reach saturation before the top. Thereafter it will cool more slowly than the top, steepening the lapse rate within the layer and eventually creating unstable conditions. This (Fig. 3.12(b)) is *convective instability*.

The environmental lapse rate is rarely constant with height. On the largest scale, as illustrated in Fig. 2.21, the various atmospheric layers have differing stabilities. In particular, the troposphere is conditionally unstable while the stratosphere is stable. Consequently few convective motions can penetrate into the stratosphere and the tropopause can be regarded as the practical upper limit of cloud formation. Not all clouds, of course, reach this height. However, a well developed cumulonimbus top is frequently very close to the tropopause.

### *Diurnal variations in the environmental lapse rate*

Within the troposphere the environmental lapse rate can change with location, time and height. At a given time and place several layers with differing stability may occur above each other (Fig. 3.13). As an example of temporal changes, we can consider what might occur near the surface during a cloudless day (Fig. 3.14). The results are an extension of the effects of the diurnal variation in the surface energy balance. Commencing soon after sunrise the heating of the ground causes a steep, unstable lapse rate in the lowest air layers. This persists and deepens throughout the morning and early afternoon. Once surface cooling is established after sunset, the lapse rate in the lowest layers is reversed. During the night this layer of reversed lapse rate deepens, only to be replaced by more normal conditions as the cycle starts again after dawn.

An environmental temperature profile where the temperature increases with height is an *inversion*. Such layers are very stable and allow little vertical mixing, a very important consideration in air pollution control (see sect. 6.7.1). Inversions can be formed in several ways in addition to the radiative cooling illustrated in Fig. 3.14. Many, such as that in Fig. 3.13, are the result of large-scale atmospheric motions which have advected warm, stable air over a cooler, less stable near-surface air mass.

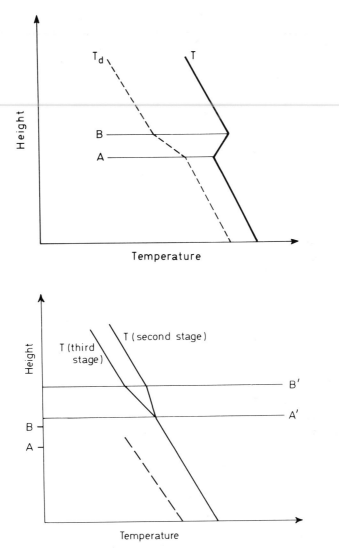

Fig. 3.12. Thermodynamic diagram illustrating convective instability in which the lifting of a large mass of moist air results in less rapid cooling of the base as it reaches saturation first, i.e. at height A before height B. This results in a slower cooling of the base (second stage) so that the lapse rate within the layer is steepened, finally creating the unstable conditions between heights A' and B'. ($T_d$ and $T$ are the dew point and air temperatures respectively.)

## *Triggers of convective processes*

In an unstable atmosphere convection will occur spontaneously, redistributing heat, and tending to create a *neutral* atmosphere, where $\gamma$ approaches $\Gamma_d$. If saturation occurs during convection cumulus clouds are formed whilst convective activity which does not lead to saturation

117

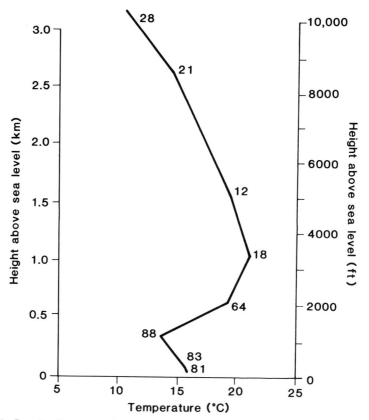

Fig. 3.13. Results from a typical radiosonde ascent taken on the Californian coast in summer (July 1957). The numbers beside the sounding curve give the relative humidity in percent. Note the change in sign in the lapse rate at about 0.35 km above sea level.

is known as *clear air turbulence* (CAT). In a stable atmosphere any tendency towards convection is rapidly damped and the stability retained.

Even with an unstable atmosphere there must be some 'trigger' mechanism to initiate the convection process. Rarely is there a shortage of such triggers. The turbulence inherent in any wind flow is almost always sufficient to move a parcel and thus create a temperature difference between it and the environment. Nevertheless, there are certain situations where there are well defined triggers which enhance the likelihood of convection at certain times or places. The orographic, frontal and confluence mechanisms treated above as forcing air to rise can all act to initiate convection. Similarly, differences in surface characteristics can start the process. A simple example would be the case of an island in a lake. During the day the island heats much more rapidly than the water, forming a heated air parcel over the island. Afternoon convective showers during the rainy season over the Lake

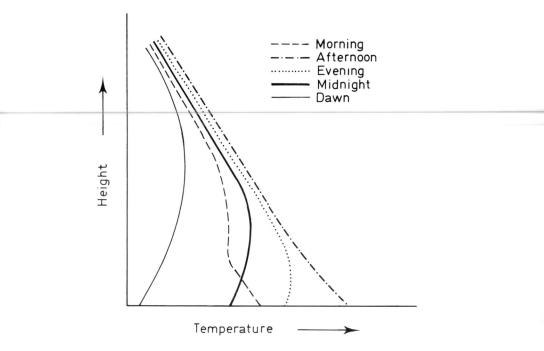

Fig. 3.14. Typical diurnal variation in atmospheric stability. Note the breakdown of inversion conditions established during the preceding cloudless night. The morning heating is seen to proceed more rapidly for the surface than for the atmosphere and the nighttime cooling is also more rapid for the surface than for the air.

Victoria coast of Uganda, for example, can frequently be traced to a source over the islands in the lake.

### Concept of potential temperature

The relatively simple diagrams of stability we have used so far provide only a qualitative picture of the atmospheric processes. Quantitative diagrams, called thermodynamic diagrams, however, can be developed. These depend for their basic construction on the concept of potential temperature. There are many atmospheric processes in which during ascent and descent the temperature changes at the dry adiabatic lapse rate. When this occurs a quantity called the *potential temperature* ($\phi$) is conserved. The potential temperature of a parcel of dry air is the temperature it would have if brought dry adiabatically to a surface pressure of $10^5$ pascals. $\phi$ is a function of $T$ and $p$ only. The second law of thermodynamics and the ideal gas law can be combined and rearranged to give

$$\frac{c_{\mathrm{p}}}{R}\ \frac{\mathrm{d}T}{T} = \frac{\mathrm{d}p}{p} \qquad\qquad [3.11]$$

119

Integrating upwards from surface conditions $p_o$ and $\phi$ to parcel position conditions $p$ and $T$ leads to *Poisson's equation*

$$(T/\phi)\ c_p/R = p/p_o \qquad\qquad [3.12]$$

This equation is usually solved graphically in meteorology using one of two types of thermodynamic diagram: either a *pseudo-adiabatic chart*, on which the axes are pressure and temperature, or a *temperature-entropy diagram*.

### Temperature-entropy diagrams

In temperature-entropy diagrams the coordinates are temperature and the logarithm of potential temperature (entropy). From Poisson's equation it can be shown that for a constant pressure process

$$\ln \phi = \ln T + \text{constant} \qquad\qquad [3.13]$$

An example of one of these diagrams is shown in Fig. 3.15. While considerably more complex than the simple diagrams we used above, this T$\phi$gram allows the determination of static stability in any and all conditions. In addition quantitative measures, such as the amount of vapour condensed during a particular uplift, or the temperature change resulting when a parcel is lifted over a mountain and descends the leeward side, can be calculated.

The effects of horizontal motions have virtually been ignored in this consideration of hydrostatic stability. A strong wind shear, i.e. a change of wind speed or direction with height, can also destabilise the atmosphere. This 'dynamic' instability can occur even in an atmosphere which is statically stable. This is the type of instability which is mainly responsible for clear air turbulence (CAT) and is also frequently associated with the jet stream which is described in the next chapter.

## 3.5 Clouds and climate

Clouds have two climatic roles to play in addition to being a source of precipitation: they are themselves a highly visible part of climate and they significantly modify the local radiation streams and thus influence, through the energy balance, the whole climate.

Despite the apparent ease of cloud observation, our knowledge of cloud amount and distribution around the Earth is by no means complete. Interpretation of both the traditional, surfaced-based observations and the more recent satellite measurements is difficult.

### Surface-based observations of clouds

The traditional method of observation, routinely used at a large

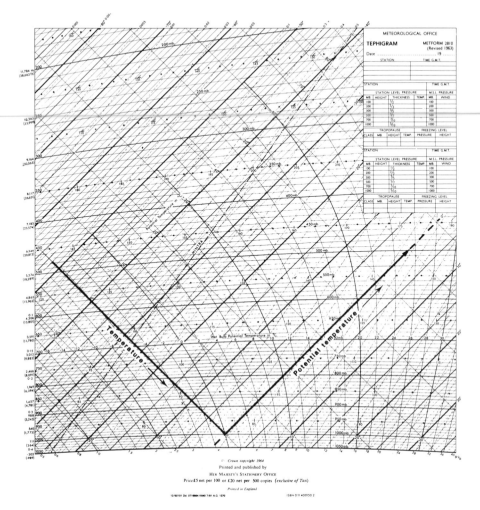

Fig. 3.15. An example of one type of thermodynamic diagram – a tephigram (Tφgram) which is used by the United Kingdom Meteorological Office. The direction of increasing temperature (*T*) and potential temperature (φ) which has a logarithmic scale are superimposed. The dry adiabats (lines of constant potential temperature) are straight. The chart has been rotated through 45° so that lines of constant pressure are approximately horizontal. (HMSO, 1964).

number of stations for many years, has been simply to divide the sky by eye into eight or ten parts and estimate the fraction covered (oktas or tenths) by clouds. In some cases the observation is refined to include cloud type and the amount at each of the three levels indicated for horizontal clouds in Fig. 3.8. This approach is rather subjective, and when estimates are made for several levels it is impossible to determine whether higher cloud exists but is hidden by lower layers. Nevertheless, some places now have detailed cloud climatologies. Although

121

these observations are mainly restricted to land areas the results give general indications of the average cloud cover of the Earth. These results can be related directly both to the formation processes discussed above and to the general weather and climate features of the major regions of the Earth to be considered in the next two chapters.

(a)

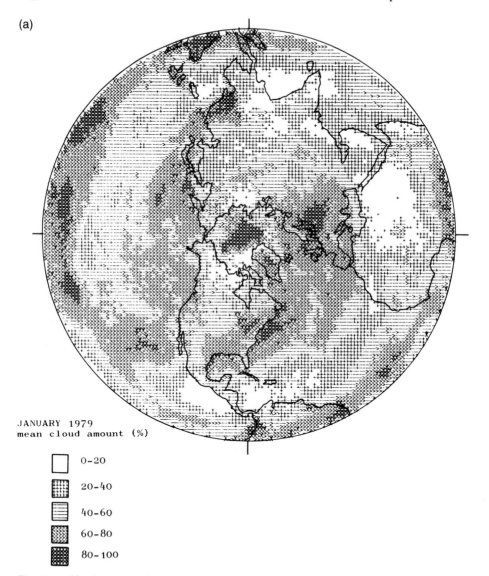

JANUARY 1979
mean cloud amount (%)

0–20

20–40

40–60

60–80

80–100

Fig. 3.16. Northern and Southern Hemisphere cloud amount: (a) and (b) for January 1979; (c) and (d) for July 1979; as compiled from a high resolution satellite-based global nephanalysis. Note the regions of extensive cloud cover, particularly associated with the mid-latitude depression belts and the intertropical convergence zone. (Courtesy: Dr. N. Hughes).

### Satellite-based observations of clouds

Satellite cloud observations have the global coverage that surface-based observations lack (Fig. 3.16). However, these observations, because of the sensor location, differ both in the type of cloud they see in any multilayer situation and in the total cloud amount observed (Fig. 3.17(a)). Hence it is very difficult to use the traditional and

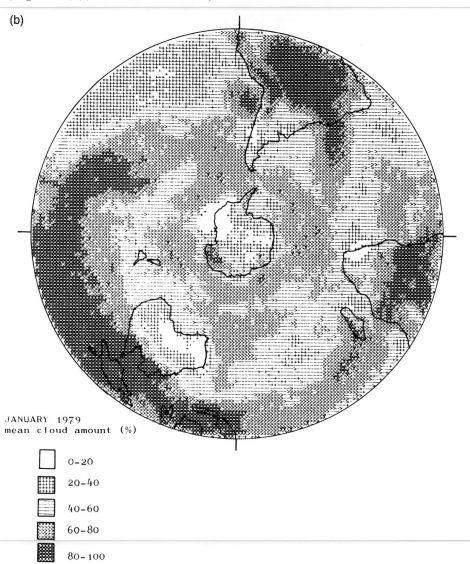

(b)

JANUARY 1979
mean cloud amount (%)

☐ 0–20

▦ 20–40

▤ 40–60

▨ 60–80

▩ 80–100

satellite data sets together to obtain a realistic cloud climatology for the Earth. Although such a single climatology might be intrinsically desirable, it is often preferable to treat surface- and satellite-based

(c)

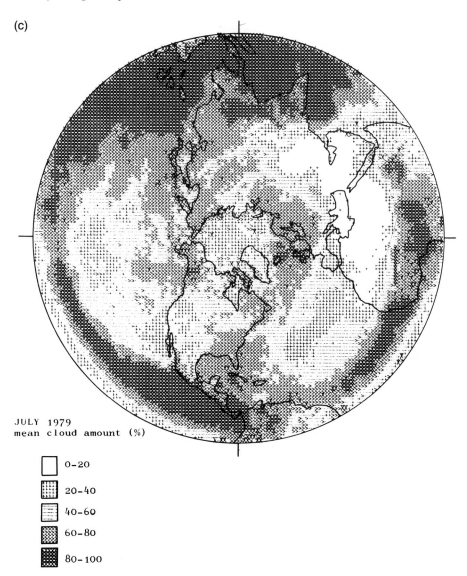

JULY 1979
mean cloud amount (%)

☐ 0–20

▦ 20–40

▤ 40–60

▨ 60–80

▦ 80–100

cloud observations as different facets of the same phenomenon. Surface values are useful in specifying the local climate and in considering what a person or plant 'sees', particularly in terms of the way in which clouds interact with solar radiation. Also surface observations can be used to estimate cloud base heights, a vital consideration for aviation (Fig. 3.17(b)). On the other hand, the satellite observations are much more useful when considering global energy flows, climate models, and the possible causes and effects of climatic change.

Although straightforward cloud images (e.g. Figs. 5.25 and 5.29) are

(d)

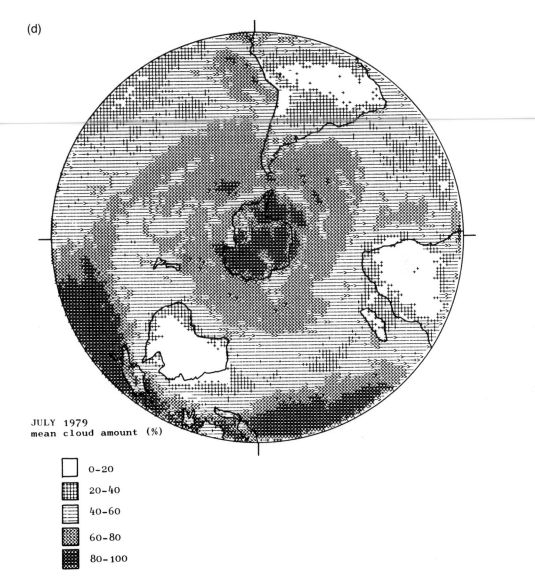

JULY 1979
mean cloud amount (%)

| | |
|---|---|
| ☐ | 0–20 |
| ▦ | 20–40 |
| ▤ | 40–60 |
| ▨ | 60–80 |
| ▩ | 80–100 |

an important tool in weather forecasting and are familiar to viewers of television weather programmes, extracting cloud information for climatic purposes from satellite measurements is not easy, since the information is based on measurements of albedo and outgoing long-wave radiation.

In the shortwave region of the spectrum the amount of reflection (the albedo) depends on the 'optical thickness' of the cloud. This is largely a function of the liquid water content which in turn depends on the thickness of the cloud. It also depends on the type of cloud, since

125

(a)

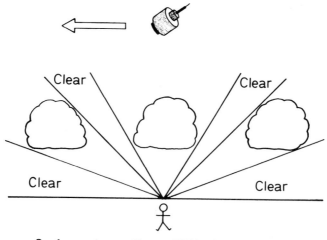

Surface observation = 75% ( 6 Oktas)
Satellite observation = 50% ( 4 Oktas )

Fig. 3.17(a). Schematic illustration of the way in which the same cloud configuration can result in different cloud amount observations when surface and satellite retrievals are compared. The satellite which passes rapidly over clouds obscuring about half the land surface reports 50% cloud cover whilst the surface observer, who sees the sides of the clouds and hence less than half the dome of the sky, reports 75% cloud cover.

(b)

Fig. 3.17(b). A cloud base height recorder is an automatic system consisting of a transmitter (shown in the figure), a receiver and a recording unit. The instrument shines a bright light upwards and then towards the receiver in a fixed time sequence. The time lag between emitting and receiving the light reflected from the cloud base permits calculation of the cloud base height.

**Table 3.2  Albedos and absorptances of four cloud types**

| Cloud type | Low | Middle | High | Cumuliform |
|---|---|---|---|---|
| Albedo | 0.69 | 0.48 | 0.21 | 0.70 |
| Absorptance | 0.06 | 0.04 | 0.01 | 0.10 |

cirrus for example is composed mainly of ice crystals. Values of albedo and absorptance for individual clouds vary tremendously, but generalised values are given in Table 3.2.

Longwave radiation emanating from a cloud top consists not only of the radiation emitted by the cloud itself but also radiation from lower levels transmitted by the cloud. The amount of absorption of radiation from lower levels again depends on the liquid water content of the cloud. Usually a cloud thicker than 1 km will absorb virtually all the radiation it receives. The transfer of radiation within the cloud is a function both of the liquid water content and the radiating temperature. Since there will be a decrease in temperature through the cloud, the amount of radiation finally emitted upwards is largely a function of the cloud top temperature.

The maps presented in Figs. 2.26 and 2.27 may be used to illustrate the method of interpreting radiation information. Generally areas with high albedo and small amounts of outgoing radiation are cloudy (see, for example, Fig. 5.29). Thus the towering cumulonimbus of the belt near the equator and the predominantly cloudy regions of the mid-latitude depression belts are readily discernible (see also Fig. 3.16). Other cloud areas are less easy to detect. For example, the low level stratus cloud that persists along the western coastal areas of North America, South America and Africa is not only a region of high albedo but also an area of relatively high outgoing radiation. This is because these clouds rarely have great vertical extent and are thus emitting at a temperature close to that of the surface. Usually it is possible to identify cloud areas at a particular time because of the distinctive pattern associated with many cloud systems. Consequently it has become possible to develop global cloud 'climatologies' for individual months (Fig. 3.16). Whilst it would clearly be unwise to claim that these cloudiness conditions are typical of other years, we are beginning to specify a realistic climatology for this vital element of the climate system.

## 3.6  Precipitation formation

The climatic role of clouds that we have yet to consider is that of precipitation production. When a cooling air parcel reaches saturation, condensation does not occur immediately. Indeed if the parcel were rising in pure, particle free air, relative humidities could reach several hundred percent before spontaneous condensation occurred. However, even clean air is rarely entirely free of particles, and certain particles act as *cloud condensation nuclei* (CCN), promoting condensation at relative humidities at or close to 100%.

Common CCN include dust, clay and organic particles derived from land surfaces, salt crystals derived from sea spray and particles created in the atmosphere by chemical actions usually called gas-to-particle conversions. An idea of the size and concentration of these particles can be gained from Fig. 3.18. Note that marine clouds tend to have a smaller concentration, but a larger range of droplet sizes, than continental clouds.

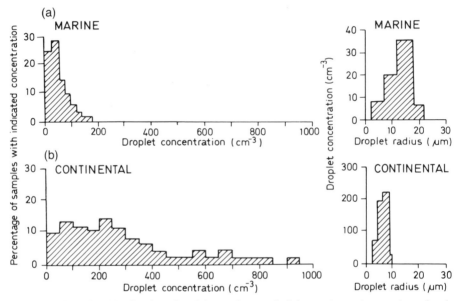

Fig. 3.18. Droplet distribution for (a) marine and (b) continental cumulus clouds showing (left-hand side) percentages of samples with given concentrations (droplets per cm$^3$) and (right-hand side) concentrations of different sized droplets. Note the scale change in droplet concentration between the two cloud types. The very much higher droplet concentrations in continental cumulus clouds suggests a much higher concentration of CCNs in the air mass from which they are formed.

### Growth of cloud droplets

The size of the particle has a great influence on its propensity for growth. Recalling that saturation is defined with respect to a plane pure water surface, water that condenses on a CCN is neither plane nor pure. Two effects come into play. The first is the *solute effect*. If a CCN is dissolved by the water, the air surrounding it may be saturated with respect to the resulting droplet at relative humidities <100% and further condensation is possible. Also, because we have a spherical droplet, the surface is not planar and surface tension creates the *curvature effect.* This second effect requires that the air be supersaturated before further growth can occur. In practice, these two effects occur simultaneously, so that droplet growth follows curves of the type shown by curves X and Y in Fig. 3.19. Curve Y starts with condensation on a CCN smaller than does X, and requires a higher supersaturation in order to continue growth. The value of supersaturation reached depends on the concentration and size of nuclei present and on the speed of the cooling process. Rapid ascent usually gives high values although rarely do they exceed 101%. If there are a large number of small CCN competing for water vapour, as with the continental cloud suggested by Fig. 3.18, the supersaturation may not achieve a high enough value to allow the droplets to pass over the maximum of the curve of Fig. 3.19. The droplets remain small and precipitation is

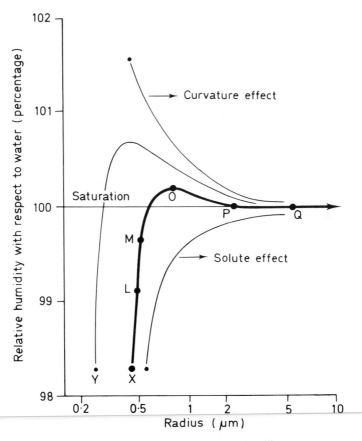

Fig. 3.19. An ordinary cloud condensation nucleus (X) will grow as a compromise between the solute and the curvature effect, the path of this process following the curve LMOPQ. The smaller condensation nucleus (Y) requires a large supersaturation in order to continue growth. (From Petterssen, 1969).

unlikely. If this happens at low levels, *haze* is formed. With a variety of CCN sizes, as found in marine clouds, some droplets will be able to pass over the 'hump' and continue to grow. They are then said to be *activated* and may continue to grow into raindrops.

## Growth of raindrops

Growth of activated droplets will continue by direct condensation. However, this process is very slow, much slower than the observed times needed for precipitation formation. Hence other mechanisms must be involved. To explore these mechanisms it is first necessary to consider the motion of drops within clouds.

Particles suspended in the atmosphere fall under their own weight according to *Stokes's Law*

$$V = 2 \ g(\rho_p - \rho_a) \ r^2 \ / \ 9 \ \eta \qquad [3.14]$$

where $V$ is the terminal velocity, $g$ is the acceleration due to gravity, $r$ the particle radius, $\rho_p$ and $\rho_a$ the densities of the particle and the atmosphere, and $\eta$ is the viscosity of air. The *terminal velocity* is the velocity a particle attains in free fall through still air, being the result of the balance between the gravitational attraction downwards and the frictional drag of the air itself. Stokes's law indicates that this velocity increases rapidly with particle size. For example, a $10 \ \mu m$ droplet will have a terminal velocity of about $10^{-3}$ m s$^{-1}$, while a 1 mm drop will have a terminal velocity of $\sim 10$ m s$^{-1}$. When clouds are formed, the air is rarely motionless, and in particular, there is likely to be an updraught in the cloud. This will mitigate against the downward movement so that the speed of the final movement towards the Earth's surface, the *fall velocity*, is given by

$$\text{Fall velocity} = \text{terminal velocity} - \text{updraught velocity} \qquad [3.15]$$

Only if the terminal velocity is greater than the updraught velocity will the drop fall. In practice the small cloud condensation droplets will have a terminal velocity that is smaller than the updraught velocity and will remain suspended in the cloud. Indeed, they may never grow to sufficient size to have a positive fall velocity and thus may never leave the cloud. In order for precipitation to occur, therefore, some mechanism is needed to increase the size of a cloud droplet to that of a raindrop. The mechanism which becomes active depends greatly on the temperature of the cloud.

## Collision and coalescence

In *warm clouds*, where the ambient cloud temperature is above 0 °C, all the condensation products are liquid water. The larger droplets will

130

have a higher terminal velocity than smaller ones and will fall through them, collecting them and increasing in size through the *collision and coalescence* processes. For collision to occur the smaller droplet must be close to the axis of fall of the larger drop, otherwise it will follow the air currents around the falling drop and there will be no impact. Generally the larger the falling drop, the more efficient is the collection, drops with a radius greater than 40 $\mu$m collecting most of the droplets they encounter. Even with a collision, growth will only occur if the two drops coalesce. This will occur most readily if the drops are of considerably different sizes. These factors thus re-emphasise the importance of a wide range of sizes for the original CCN for precipitation production.

### Bergeron–Findeisen mechanism

In *cold clouds*, where the ambient temperature is below freezing, the condensation products can be both liquid water and ice crystals. At temperatures below about $-40\,°C$ all the products are ice crystals and the cloud is said to be *glaciated*. Between $0\,°C$ and $-40\,°C$ ice and water coexist, giving a *mixed cloud*. In glaciated clouds there is some crystal growth by processes analogous to the direct condensation and the collision/coalescence processes of warm clouds, but rarely do these high, cirrus type clouds yield precipitation that reaches the surface. In mixed clouds the initial growth phase depends on the coexistence of ice and water. The process is known as the *Bergeron–Findeisen* process. At temperatures below $0\,°C$ the saturation vapour pressure with respect to water is greater than that with respect to ice (Fig. 3.5). Thus, in a mixed cloud, the air which is close to saturation with respect to water is supersaturated with respect to ice. Consequently ice crystals can grow much more rapidly than, and at the expense of, water droplets.

The individual ice crystals will grow and collect together to form snowflakes. The shape of snowflakes that are created depends on the temperature at which the condensation occurs. Since a crystal may move about in a cloud and experience a variety of temperatures, the shape of the snowflake which grows as a result of the accumulation of crystals can be very complex. In addition, the growing ice crystals may come into contact with supercooled liquid water droplets which will freeze onto or around the crystal immediately on contact, a growth process known as *riming*. Riming is a primary mechanism of hail formation. Crystals may also grow by aggregation, a process similar to collision/coalescence. Aggregation is most marked at ambient temperatures above $-5\,°C$, when crystal surfaces become moist and sticky.

The action of the Bergeron–Findeisen process is used in artificial *cloud seeding* (see section 6.8). Tiny particles of silver iodide or dry ice (solid carbon dioxide) are dropped into the tops of mixed clouds. Silver iodide particles serve as CCN with the correct crystal structure to act as a large freezing nucleus and initiate the growth of drops. Dry ice serves to cool the ambient air locally, enhancing the difference in saturation vapour pressure between ice and water and thus initiating the whole process.

An individual cloud need not be exclusively warm or cold. The lower portions of a cloud may be warm while higher parts may be mixed or even glaciated. Thus precipitation, created as snow, may melt before falling from the cloud. Furthermore, clouds can seed themselves naturally, when ice particles falling from a cirrus cloud, or from the glaciated portion of a towering cumulus, act as nuclei. In the same way ice particles or snowflakes from the mixed portion of a cloud can fall into the warm portion and provide the large nuclei needed for further growth.

# 3.7 Precipitation

The type and size of precipitation leaving a cloud base depends on the conditions within the cloud, but the precipitation that actually reaches the ground is modified by conditions in the air layer between the cloud and the ground. In general, the temperature structure determines whether the precipitation will arrive as frozen precipitation or as liquid water, while the humidity of the layer determines the amount of evaporation that will occur and hence the ultimate size of the precipitation particles (Fig. 3.20). In both cases the fall velocity will dictate the time over which the processes can act and hence how complete they will be. The major types of hydrometeors that can occur are summarised in Table 3.3.

### Precipitation type and duration

The intensity and duration of precipitation is determined largely by the type of cloud system involved. This in turn is intimately connected with the cloud formation processes considered above. In general, cumulus type clouds involve vigorous vertical motions, give large drops and intense precipitation for a short period. Usually their influence is restricted to a fairly small geographical area. Stratus and altostratus, in contrast, involve more persistent, less vigorous vertical motions over a much wider area. Hence prolonged, steadier, and usually less intense, precipitation results. This can be illustrated by reference to Fig. 3.21. Miami, Florida, USA is in an area dominated by cumulus type clouds. Short duration rainfalls are likely to be much more intense than in Seattle, Washington, USA, where precipitation from depressions is predominant. The difference in intensity decreases as the duration increases, but is still clearly discernible for 24-hour rainfall totals.

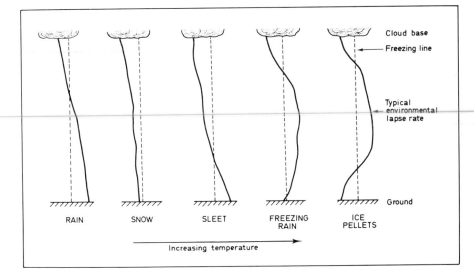

Fig. 3.20. The type of precipitation that reaches the ground depends on the temperature structure of the air layer between the cloud base and the ground. In the situations illustrated precipitation falls from a mixed cloud. The resultant precipitation type depends on the relation between the environmental lapse rate and the freezing temperature. In all cases typical profiles are shown and temperature increases from left to right.

## Table 3.3  Major types of hydrometeors (or precipitation)

| Hydrometeor | Description | Normal clouds from which precipitation can fall and reach the ground |
|---|---|---|
| Rain | Drops with diameter >0.5 mm (0.02 in) but smaller drops are still called rain if they are widely scattered | Ns, As, Sc, Ac castellanus, Cu congestus |
| Drizzle | Fine drops with diameter <0.5 mm (0.02 in) and very close to one another | St, Sc |
| Freezing rain (or drizzle) | Rain (or drizzle), the drops of which freeze on impact with the ground | The same clouds as for rain or drizzle |
| Snowflakes | Loose aggregates of ice crystals, most of which are branched | Ns, As, Sc, Cb |

| Hydrometeor | Description | Normal clouds from which precipitation can fall and reach the ground |
|---|---|---|
| Sleet | In Britain, partly melted snowflakes, or rain and snow falling together | The same clouds as snowflakes |
| Snow pellets (also known as soft hail and graupel) | White opaque grains of ice, spherical, or sometimes conical with diameter about 2–5 mm (0.1–0.2 in) | Cb in cold weather |
| Snow grains (also known as granular snow and graupel) | Very small, white, opaque grains of ice. Flat or elongated with diameter generally <1 mm (0.04 in) | Sc or St in cold weather |
| Ice pellets | Transparent or translucent pellets of ice, spherical or irregular, with diameter <5 mm (0.2 in). There are two types:<br>(a) frozen rain or drizzle drops, or largely melted and then refrozen snowflakes.<br>(b) snow pellets encased in a thin layer of ice (also known as small hail) | Ns, As, Cb<br><br>Cb |
| Hail | Small balls or pieces of ice with diameters 5–50 mm (0.2–2.0 in) or sometimes more | Cb |
| Ice prisms | Unbranched ice crystals in the form of needles, columns or plates. | St, Ns, Sc (sometimes falls from clean air, when it is just an advanced stage of ice fog) |

Figure 3.21 also indicates that intensity decreases as duration increases. World rainfall statistics suggest that intensity is approximately proportional to the inverse square root of the duration, but that there are many regional variations.

### Measurement of precipitation

Rainfall can be directly measured using a *rain gauge* (Fig. 3.22). This

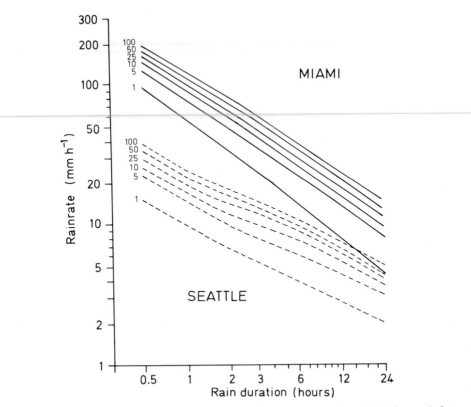

Fig. 3.21. Intensity duration rain curves for various return periods (1–100 years) for Miami, Florida (solid lines) and Seattle, Washington (dashed lines). The curves indicate that only once in 10 years is Seattle expected to have rainfall averaged over 12 h which will reach or exceed 5.5 mm h$^{-1}$.

is essentially a bucket with a horizontal orifice of known size exposed just above the surface of the Earth. The water is caught in the bucket and funnelled into a measuring cylinder and the depth of water collected at the end of a period of interest is then noted. In order to maintain uniformity, most nations have established standards for the size, exposure and measurement times. There are several problems inherent in this simple measurement technique. In windy conditions turbulence in the air flow is created by the gauge itself. In particular this increases the speed of the flow across the top of the orifice, decreasing the catch and leading to unrepresentative results. This underestimate in measured precipitation is much greater for snow than for rain because the fall velocities and momenta of snowflakes are much less than those of raindrops. The decrease varies with wind speed and, in sloping terrain, with wind direction. Some gauges are equipped with wind shields, but this only partially corrects the problem. Since wind

135

**(a)**

**(b)**

Fig. 3.22(a). Example of one type of rain gauge and (b) the chart recorder of a tipping bucket rain gauge. (c) shows a baffled surface erected to try to eliminate splashing into the gauge of rain which has fallen on to the ground. Raingauges are sited near the centre of the meteorological enclosure, not less than 3 m away from the Stevenson screen (Fig. 1.1). The gauge should be installed in horizontal ground in a vertical hole so that the rim is the specified height above ground (usually 30 cm or 12 inches).

Recording rain gauges are used to provide information above the time and duration (and hence intensity) of precipitation. The mechanism records each time the complete filling of a small water trough occurs. The trough tips when full, causing the water to be lost and permitting another cycle of filling to occur.

136

(c)

speed decreases rapidly as the ground is approached, a gauge close to the ground is desirable. However, the closer to the ground it is, the more likely it is to receive precipitation which is splashed into it in addition to the falling rain (cf. Fig. 3.22(c)). Hence it is difficult to measure with great accuracy the amount of rainfall that actually arrives at the Earth's surface. Nevertheless it remains the only instrument giving absolute measurements of precipitation, and there are numerous records of precipitation available from many locations around the world.

Rain gauges can be constructed so that a continuous record of receipt is made (Fig. 3.22(b)). A mechanism is inserted within the bucket which senses the change in water level as precipitation occurs and translates this into a trace on a chart located on a revolving drum. Such instruments allow not only the determination of total precipitation but also its intensity and duration.

Although these measurements give a good idea of the distribution of precipitation, especially when long-term averages are needed, they are spot measurements. Rainfall amounts, especially from a single storm, can have a very great spatial variation. Unless there is a very dense network of rain gauges these variations will not be identified. This can be a very serious consideration when intense storms that may cause flooding are common.

The lack of spatial coverage by rain gauges can be overcome to some extent by using information obtained by *radar* (Fig. 3.23). A radar emits a pulse of electromagnetic radiation which is reflected off any obstacle in the emission path and returned to the source. By measuring the time between emission and receipt, the distance of the obstacle from the source can be deduced. By choosing an appropriate wavelength for the emission, raindrops can become the obstacles. In sophis-

137

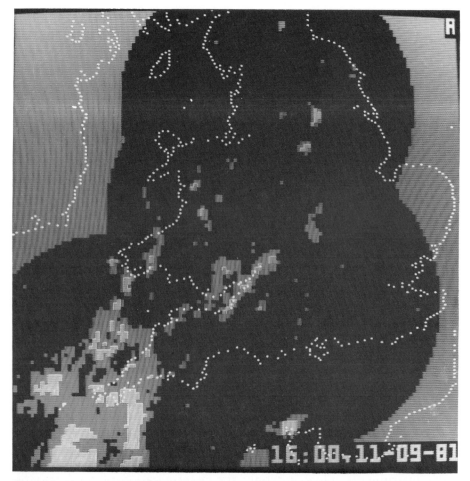

Fig. 3.23. This image, which is a composite formed from a network of four weather radars, shows where rain was falling in England and Wales at 1600 hours on 11 September 1981. An unstable southwesterly airstream covered the British Isles, producing heavy showers over Wales and the Midlands. An area of thundery rain can be seen approaching from the southwest, ahead of an upper trough. In this black and white photograph of a colour display, the dark grey represents light rain and pale grey heavy rain. Each picture element is a 5 km × 5 km square. (RSRE Malvern: British Crown copyright reserved).

ticated systems the direction and speed of motion of the raindrops can also be determined. The results can be translated into rainfall rates by calibration with conventional rain gauges. By using a full radar sweep, it is possible to determine rainfall over a wide area.

Another feature of precipitation that is routinely measured is snow depth. The procedure is again simple. A site where the depth of snow appears to be representative of the area, usually in the centre of a

138

region of open ground, is chosen and the depth measured with a normal measuring stick. For some applications this is the important information, but for others it is more useful to determine the water content of the snow which is related to snow density. This is particularly important when the snow pack is being used as a water reservoir, so that the spring meltwaters can be used for downstream water supply, a common practice in the western part of the United States. Water content can be determined by extracting a snow core of known diameter, melting it and weighing the meltwater. A more common, and much simpler, method is to assume that the snow is of average density and compaction and may be converted to a liquid water equivalent using the rule of thumb that 10 units of snow correspond to 1 unit of water. However, actual values range from about 6 : 1 to 30 : 1.

Two other sources of atmospheric water for the ground surface, in no way to be considered as precipitation, can be very important locally. Whenever cloud comes directly into contact with the ground, as in a fog, *direct water deposition* occurs. This can be an important water source; for example, in some highland regions of Central America where low level cloud is persistent. This cloud will not precipitate, but direct deposition on vegetation, especially trees, provides enough moisture to support a luxuriant vegetation.

The other non-precipitation water source is *dew*. This occurs, as noted earlier, whenever the flux of water vapour molecules towards the surface exceeds the flow leaving the surface. Normally this is a night time phenomenon, and the dew rapidly disappears soon after dawn. However, in regions with scarce precipitation it is possible to conserve and use the dew. For example, 'dew traps' are constructed in the vine growing regions on the island of Lanzarote in the Canaries. Volcanic ash, an abundant resource there, is used as the building material. This both insulates the soil below from the direct effect of solar radiation during the day and cools rapidly during the night, thus enhancing dew production. Even in the absence of rainfall, as happened in 1975 and 1976, the vines can live on entrapped dew alone.

## 3.8 Global precipitation distribution

We have seen that both the intensity and duration, and thus the amount of precipitation, in an individual event depend on the processes acting to create the precipitating clouds, and that the areal extent of the precipitation depends on the same factors. Since particular processes tend to dominate particular areas of the globe, we can make several pertinent generalisations about the global precipitation on an annual basis.

The area of maximum annual precipitation, over 2000 mm per year, extends in a band through the equatorial regions (Fig. 3.24). The

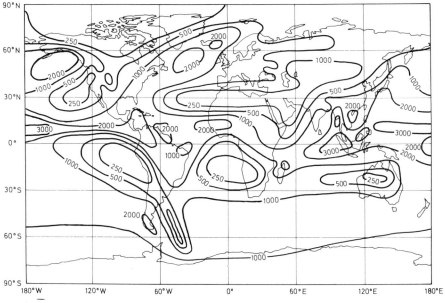

Fig. 3.24. Global mean annual precipitation (mm).

subtropical deserts and the polar regions have values below 250 mm. The mid-latitude regions have intermediate values, being in general about 1000 mm per year.

### Tropical precipitation

Precipitation in much of the tropics is associated with convective activity. Strong vertical motions occur in a fluctuating band near the equator. These release the abundant water vapour to create a regime of intense, short-lived storms from cumulus clouds. Rainfall rates in excess of 100 mm h$^{-1}$ are not uncommon. Although the location of the storms is partly controlled by local topographic features, storms tend to recur sporadically, so that precipitation does not occur at a particular place every day even though there may be a storm in the area each day.

More widespread uplift is associated with monsoonal circulations. Such circulations are particularly well developed over tropical Asia. Although this is a strongly seasonal precipitation regime, the effects of convective uplift, dynamical uplift and topographic forcing combine to produce high annual rainfall totals. Locally rainfall rates may be very high but generally the monsoonal condition is characterised by longer lasting, less intense precipitation.

### Mid-latitude precipitation

In mid-latitudes much of the precipitation production is associated

140

with depressions and fronts. The result is widespread uplift giving extended periods of gentle rain over a broad area. Rainfall rates can vary greatly, although 1–2 mm h$^{-1}$ can be regarded as a typical value. The intensity is partly controlled by the amount of water vapour available, which in turn depends on the source of the air which is being uplifted. Air derived directly from the subtropical oceans, where evaporation rates are high, is likely to lead to higher precipitation rates. If the source is the tropical deserts, the air is likely to be much drier and it is not uncommon in these conditions for dust and sand particles to form the condensation nuclei and hence be deposited in large quantities with the rain.

Convective activity in the mid-latitudes is primarily a summer phenomenon. It can be as intense, but is usually less regular, than in the tropics. A rainfall of 31 mm in 1 minute (1860 mm h$^{-1}$) was recorded in Maryland, USA, in 1956, and a fall of 126 mm in 8 minutes (945 mm h$^{-1}$) in Bavaria, Germany, in 1920. The strong upwelling of air which must be associated with such phenomena can lead to some surprising results. On 9 February 1859 an area of about 1000 m$^2$ centred on Aberdare, Wales, received precipitation filled with small fish (sticklebacks and minnows). Similarly, the Yachting Olympics in October 1968 at Acapulco, Mexico, experienced a heavy storm which deposited live maggots from 5 to 25 mm in length.

### Low precipitation regions

The regions of low precipitation in the subtropics result mainly from a lack of mechanisms for creating uplift and bringing the air to saturation. Certainly over the oceans, and to a large extent over the land deserts as well, there is no lack of moisture in the atmosphere. In contrast, over the polar regions the low precipitation totals are as much associated with lack of atmospheric moisture (Fig. 3.7) as with a lack of uplift mechanisms.

### Rain days

The spatial distribution of rainfall can also be viewed in terms of the number of *rain days* per year. A rain day is usually defined as a 24-hour period, usually starting at 0900 Z, during which 0.2 mm or more of precipitation falls. The climatic average of rain days varies from over 180 per year in humid coastal regions, such as Washington State in the USA, to less than one per annum in very arid regions. In general there is a close relationship between the number of rain days and the total precipitation. For example, the bases of the Hawaiian Islands are in a very different climatic regime from the mountain peaks. While the base has low precipitation and few rain days, the peak of Mount Wai-ale-ale on the island of Kauai could claim to be the wettest place in the world with an average annual total precipitation of 11,455 mm and an average of 335 rain days per year. However, seasonality can influence the relationship between rainfall totals and rain day numbers. Places influenced by the Asiatic monsoon may have annual totals approaching that of Wai-ale-ale, but, because of distinct wet and dry seasons, only half as many rain days.

The relationship between rainfall and rain days therefore depends strongly on the climatic regime and on the nature of the precipitation producing systems. For many purposes the total rainfall in a given period is the most useful measure of precipitation, but in some cases the number of rain days is more appropriate. It is perhaps unfortunate that most travel brochures and travel agents, when they consider climate, provide monthly average rainfall statistics; whereas the average number of rain days in a particular month would be of much more use to the prospective holidaymaker.

# 3.9  Local scale precipitation

Although the global scale considerations discussed in the previous section explain the major features of precipitation in particular regions, at any specific point there is liable to be a variety of types of hydrometeors that fall, as well as variations in the frequency, intensity and length of occurrence of precipitation events. While results for specific locations and conditions will be considered in later chapters, two general considerations pertinent to local precipitation remain to be considered. The first is discussion of the thunderstorm. The second is the use of precipitation information, since it is mainly at the local scale that precipitation information is most useful.

### 3.9.1  Thunderstorms

One of the most spectacular examples of processes associated with water in the atmosphere is the *thunderstorm*. This phenomenon is a local scale feature which is possible virtually anywhere in the world. Examination of its features can also be regarded as a summation of much of the material contained in this chapter. A thunderstorm consists of a series of cumulonimbus clouds or cells, each cell going through a sequential development, often called a life cycle, in about half an hour (Fig. 3.25(a)).

In order to initiate a thundercloud the basic prerequisite is an unstable, usually a highly unstable, humid atmosphere. In these conditions vertical motions are easily triggered. The initial cloud consists entirely of a rising cumulus 'bubble'. Once condensation begins, the release of latent heat provides additional energy and the updraught velocity increases with height, often reaching $10 \text{ m s}^{-1}$ near the cloud top. As a result of the rapid ascent, supercooled liquid water drops can be

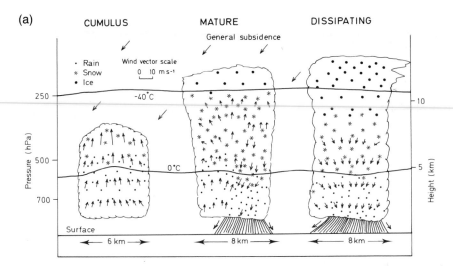

Fig. 3.25(a). Typical life cycle of a thunderstorm cell showing the cumulus stage, the mature stage and the dissipating stage.

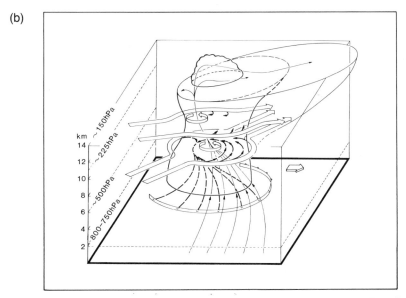

Fig. 3.25(b). Three-dimensional depiction of a cumulonimbus cloud. The flat arrow on the upper right of the surface shows the direction of travel. The thin, solid inflowing and ascending streamlines represent the trajectory of moist air originating in low levels (surface to ~750 hPa). The heavy dashed streamlines show the entry and descent of potentially cold and dry middle-level (700 to 400 hPa) air feeding the downrushing and diverging downdraught. The surface boundary between the inflow and downdraught is shown as a barbed band. The internal circular arrows show the net updraught rotation.

143

carried far above the freezing level and precipitation-sized particles are rapidly formed.

When precipitation starts to fall out of the cloud the cell has reached the mature stage. This precipitation creates a downdraught by entrainment of the adjacent air (Fig. 3.25(b)). The downdraught occurs initially close to the leading edge of a moving cell and brings snowflakes or hail pellets below the freezing level. This downdraught becomes increasing dominant and locally decreases the instability, so that eventually the updraught is suppressed completely. Meanwhile the cloud top is approaching the tropopause and beginning to spread laterally. Maximum upward velocities are now found in the middle of the cloud.

Once the updraught is removed the cell enters the dissipating stage. Without the updraught the cloud droplets cannot grow. Hence precipitation soon ceases and the remaining cloud droplets evaporate back into the air. However, another cell may have formed ahead of the dissipating cell. This is frequently the result of the cold downdraught air undercutting warm surface air ahead of the cell and releasing the latent instability there. Consequently the storm may continue as another cloud. Usually subsequent clouds have somewhat less vigour than the initial one, simply because each cloud serves to redistribute the vertical temperatures and thus decrease the instability.

In the type of thunderstorm just described it has been assumed that there is little vertical variation in wind speed. This is typical of *air mass thunderstorms*. These are the common type in the tropics and are the isolated summer thunderstorms typical of mid-latitude afternoons. However, most of the severe thunderstorms of mid-latitudes occur when there is significant vertical variation in wind speed. Although there are numerous possible types in this situation, we can consider the *squall line* as a representative example.

### *Squall line precipitation*

A squall line is a series of thunder cells aligned at right angles to the direction of motion. Although individual cells go through a life cycle akin to that for the air mass thunderstorm, the results are different because the wind speed of the environment increases with altitude, and the cells move at a speed approximately equal to the wind speed in the middle troposphere. Thus at low levels ambient air is being overtaken by the cell and incorporated into it (Fig. 3.26). Outside the storm, the ambient air is usually capped by a weak inversion, which prevents any spontaneous convection occurring outside the cell. The incorporated air is undercut by the cold downdraught air and lifted to the level of free convection. It then moves upwards towards the tropopause. At these higher levels the wind is moving faster than the cell and the cloud top is drawn out into an 'anvil'. The updraught

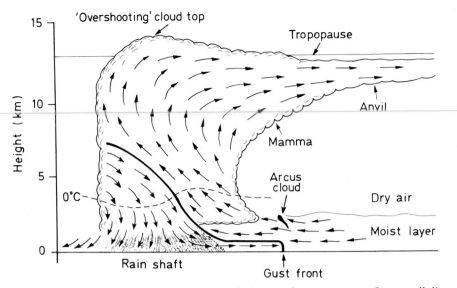

Fig. 3.26. Cloud outline and air motions relative to the movement of a squall line storm (from left to right).

creates the precipitation which now falls to the ground through the trailing edge of the moving storm, creating the downdraught area. Part of the downdraught travels along the ground under the updraught, creating a 'gust front'. This gust front undercuts the warm air, forcing it to rise and join the updraught. Thus in the squall line storm the updraughts and the downdraughts are complementary, not in opposition as in the air mass storm. This means that the storm can persist much longer and can reach much higher intensities. Indeed, many of the most intense short-term precipitation events recorded have been associated with this type of storm.

### Lightning

Lightning is the result of charge separation within a thundercloud. Although the mechanism of charge separation is not well understood, the effect is to produce a gradient of electrical potential within the cloud and between the cloud and the ground. Once this gradient exceeds a threshold value a lightning stroke results. A stroke is initiated by a 'stepped leader' moving in a jerky, but rapid, motion. This is usually directed from the cloud to the ground but may be in the opposite sense from an upstanding object such as a tall building. This stepped leader is met by a 'travelling spark', with the 'return stroke' flowing in the channel created by the two. This gives the bright flash that is visible as the lightning stroke which may be followed by one or two 'streamers' and by a second return stroke. In fact most

145

lightning flashes are made up of three or four strokes about 50 ms apart.

The return stroke raises the temperature of its channel by about 30,000 K. This occurs so fast that the air has no time to expand so that the pressure within the channel increases to 10, or sometimes 100, times its normal value. This sets up a shock wave emanating from the channel which is converted into the sound wave we hear as thunder. While lightning can be seen for long distances, thunder can usually only be heard for about 25 km from the lightning site, partly because of the damping effect of the air itself and partly because the temperature structure of the air usually ensures that the sound wave is refracted upwards away from the Earth's surface.

### 3.9.2 Applications of precipitation information

Three examples have been chosen to illustrate various aspects of the use of precipitation information. The solution of the types of problems posed would in actual practice follow the steps similar to those outlined for the solar energy study considered in Chapter 2. The examples illustrate a class of problem where the required information is the *probability of occurrence of an event*. These probabilities, determined from the climatic record, give no information about when the event will occur, only about its chance in a certain period. A short-term weather forecast would have to be used for guidance concerning the time of the actual event.

### *City snow clearance strategies*

The first example emphasises the distinction between using climate probability information to develop a long-term strategy and weather forecast information to put it into action. The 'problem' is snow removal from city streets. Long-term planning for snow removal is needed since equipment must be purchased and funds put aside to ensure adequate resources to operate the equipment and pay the crews. The city authorities must first determine how much snow they can tolerate lying on the streets before initiating cleaning operations. Thereafter the climatologist can use the historical record to determine how frequently snow above this threshold value occurs, in terms of the number of times per winter. Standard statistical methods can then be used to assess the probability of a certain number of snowfalls in the coming season. Armed with this climatological information the appropriate authorities can plan their strategies accordingly.

Frequently the determination of the number of snowfalls from the historical record is not easy. Long-term observations are likely to be available only for the local airport, usually some distance from the city centre. Estimates of how much snow fell in various parts of the city will therefore have to be made, requiring a detailed knowledge of the

146

local city climate. This may mean that any temporal trends found in the data for the airport will have to be converted to trends in the city, and they may not be the same. Further, the time of occurrence of a snowfall may be important, since an early morning rush hour fall may require a different response from one which occurs in the late evening. Again, the data may not provide an answer directly, and estimates based on knowledge of local climatic processes may be needed. Nevertheless, the climatologist should be able to provide to the city authorities the information they require. Thereafter it is the job of the local weather forecaster to provide the short-term warnings that are needed to put the strategy into operation.

### Power-line icing

The second example involves a situation where no appropriate measurements are available, and concerns the probability of ice formation on overhead power lines. Such icing can create loads of several tonnes per kilometre on the lines, leading to their subsequent collapse. This results not only in loss of service for the customers, but also repair costs and lost revenue for the utility. A light coating of ice can be overcome by boosting the power to create self heating within the lines, but regular, severe icing can only be overcome by laying the cable underground: a costly undertaking. Hence the climatological problem is similar to that for the snow removal: how frequently and over how wide an area will icing occur, and how severe will it be? The major difference is that here there are very few data upon which to base an assessment: no one routinely measures ice accumulation on a wire. Experiments and theory indicate that accumulation depends on precipitation type and rate, wire and air temperature, wind speed and direction, and humidity. With so many variables it has proved almost impossible to produce a 'predictive equation' which could use the more commonly measured climatic parameters to give the probability estimate needed. Instead the approach that has been adopted in some places has been to use the power company's own records of when icing problems occurred to identify the general meteorological situations responsible. Thereafter the pertinent parameters have been investigated to isolate the atmospheric causes further, eventually leading back to an assessment of the probability of occurrence of the particular meteorological situation. This provides some guidelines, but is obviously not as satisfactory as approaching the problem directly. Thus this example must be left as one where a rather unsatisfactory climatological answer has been given and where further investigation to gain insight into the problem is needed.

### Urban drain design

The final example of a climatic probability approach to problems

147

associated with water concerns urban drainage design. Storm drains are generally designed to be large enough to accommodate the largest storm runoff expected during the lifetime of a drain although some (unlikely) risk is usually acceptable. The obvious approach of searching records for the largest storm on record is not adequate, since the record is only a sample of all possible conditions and there is no guarantee that it will contain the largest possible event. Fortunately there are established statistical methods, called extreme value analyses, which allow one to estimate the 'largest possible' event from a set of records. Thereafter a set of intensity/duration curves can be developed (e.g. Fig. 3.21), for a variety of return periods. (An event with a return period of $n$ years has a probability of occurrence in any one year of $(100/n)\%$.) The drainage designer can then choose the appropriate return period, making a reasonable compromise between construction cost and size that ensures that the drain is adequate for all but the most unusual, extreme storm event. The designer will also be concerned with the areal variation of the intensity/duration curves. As with the snow removal example, it is unlikely that there will be records from a dense network of rain gauges available to provide the required information. Hence a knowledge of the precipitation producing mechanisms must be combined with the records that are available to produce the types of estimates that the designer needs.

## 3.10   The water balance

Now that we have returned the water to the surface of the Earth, it remains for us to complete the cycle that we started with evaporation from the surface. In the evaporation section we noted that rates of evaporation were partly controlled by water availability which, over most land surfaces at least, is largely controlled by precipitation. The difference between precipitation and evaporation in any area can loosely be called the *water balance* of the area. The concept of the water balance is of most immediate interest over land areas, since any excess of precipitation over evaporation represents water that runs off the surface or is stored in the soil or lakes. It therefore represents the water that can be used by mankind. For ocean areas the water balance is receiving increasing attention, after long neglect, since it is an important component in the air–sea interactions that have profound effect on climate.

The way in which we define the water balance, and the use to which the concept can be put, depends greatly on the space and time scales of interest. The most precise definition is needed for the smallest scales, where the local water balance of a land area has several important applications, notably in agriculture. On a regional scale the land surface water balance has a significant impact on the distribution of

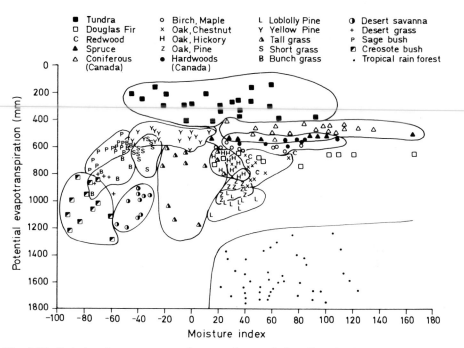

Fig. 3.27. Relation between natural vegetation and the climatic factors of potential evapotranspiration and a moisture index at selected stations in the United States, Canada and the tropics. The moisture index is $100[(P/PET) - 1]$ where P and PET are the annually averaged precipitation and potential evapotranspiration respectively.

natural vegetation. This relationship has been expressed in numerous ways. A typical example relates potential evapotranspiration and a moisture index to specific vegetation types (Fig. 3.27). The moisture index used here is equal to $100[P/PET) - 1]$, where P and PET are annual average values of precipitation and potential evapotranspiration respectively.

Ultimately, on a global scale there must be a balance between the water evaporated and that which is redeposited as precipitation. The total mass of water vapour in the atmosphere is approximately equal to one week's rainfall over the globe. Since this mass does not change appreciably with time it seems that the average time for a complete water cycle is about one week. Although we have emphasised the vertical component of the cycle, there are also, of course, significant horizontal motions for water vapour during the time that it resides in the atmosphere. In fact, there must be horizontal motions in order to maintain the global water balance (Fig. 3.28). There is a large movement of water vapour from the primary source area in the subtropics both into the equatorial regions and towards the middle and high latitudes.

149

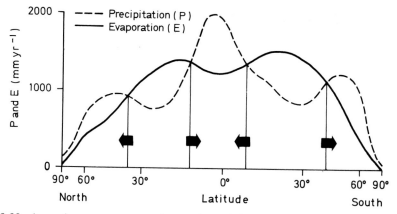

Fig. 3.28. Annual average evaporation and precipitation per unit area in mm yr$^{-1}$ as a function of latitude. The arrows show the sense of the required water vapour flux in the atmosphere.

## Latitudinal hydrological imbalance

These latitudinal motions involve not only the transport of water vapour, but also the transport of energy. Energy contained in the water vapour that is evaporated in the subtropics is released during cloud formation in the equatorial regions. Indeed, this energy transfer is vital for driving the circulation in the tropics. Similarly the poleward transfer from the subtropics plays a significant role in creating the circulation in mid-latitudes.

In order to understand and describe these circulations and, hence, the various climatic regimes of the globe, we have to combine the hydrological cycle we have discussed in this chapter with the energy exchanges considered in Chapter 2. This will be the focus of the next two chapters.

# Chapter 4
# The General Circulation and Global Climate

4.1   The function of the general circulation

4.2   Atmospheric pressure

4.3   Air movement around a rotating planet

4.4   Barotropic and baroclinic conditions

4.5   The general circulation of the atmosphere

4.6   The mid-latitude baroclinic zone

4.7   Large-scale effects of the surface boundary

   4.7.1   Oceans

   4.7.2   Cryosphere

   4.7.3   Continents

# Chapter 4
# The General Circulation and Global Climate

To describe, understand and explain the global climate fully we must add consideration of horizontal motions to the details of the energy and water systems we have discussed in the last two chapters. In the present chapter we are primarily concerned with the *general circulation of the atmosphere*: the large-scale air flow pattern of the whole globe. In addition to producing the wind and thus controlling horizontal motions such as cloud movement, the general circulation serves to redistribute energy and moisture, modifying their latitudinal imbalances and thus establishing the climate. The basic description of this circulation comes mainly from the numerous surface-based observations that have accumulated over the years. Our knowledge of the processes operating is rapidly expanding as the result of satellite observations which encompass the globe.

It is frequently very convenient to divide the general circulation into two components: the *primary circulation features*; the persistent large-scale features which cover large areas of the globe, which, whilst varying in detail, exist at all times; and the *secondary circulation features*; the short-lived, rapidly moving cyclones (depressions) and the much slower moving anticyclones which are superimposed on the former and are responsible for day to day weather changes over large portions of the Earth.

In the oceans there is a somewhat similar 'general circulation' which assists the atmosphere in the redistribution of energy and moisture. We shall consider this oceanic circulation when necessary for our understanding of the climate system. In many ways understanding of it is still at the descriptive stage, the characteristics being fairly well known, whilst the processes operating are less well known. Their investigation constitutes a rapidly expanding area of research.

We shall start our discussion with a brief overview of what the general circulation must do to maintain our present climate. This will

153

be followed by a section detailing the forces acting to produce horizontal motions in the atmosphere. The results of these two will be combined to approach the general circulation first in a simplified way in order to establish the basic features and then in the more complex way that more closely resembles reality. This will give us insight into the processes that control the climate of the globe as a whole as well as its major regions. Throughout this chapter we shall emphasise the processes acting to create the climate. A more descriptive discussion of the results of the processes in creating regional and local climates will be given in the succeeding two chapters, and finally, in Chapter 7, we will use the information to consider more fully the controls on the climate system and their implications for future climates.

# 4.1   The function of the general circulation

The latitudinal imbalance of absorbed and emitted radiation discussed in Chapter 2 and latitudinal variations in the components of the atmospheric water system detailed in Chapter 3 both indicate that horizontal motions are necessary to maintain the present climate. Thus a prime role of the general circulation, both atmospheric and oceanic, is to provide this redistribution in such a manner that the present climate is maintained. Indeed, almost all information that we have about past climates indicates that the general circulation has been operating in the same way for millennia, possibly aeons, climate changes being modifications of the basic pattern rather than radical departures from it. Hence we can take the present-day circulation as a model which, once understood, not only explains the present climate system but also provides insights into the past and possible future climates.

### Latitudinal energy imbalance

The discussion of Fig. 2.24 emphasised that the positive imbalance of energy in tropical and equatorial regions was compensated by a net negative energy budget at higher latitudes. Latitude zones poleward of about 30° emit more radiation than they absorb, while at lower latitudes the absorbed solar radiation exceeds the emitted infrared radiation. This compensation must be accomplished by the general circulation. However, the resulting circulation pattern is complicated by the seasonal variation in the energy budget components. This is shown in Fig. 4.1, using data for the period June 1974 to February 1978.

The outgoing longwave radiation (Fig. 4.1(b)) exhibits many of the same characteristics as the albedo (Fig. 4.1(a)) but in an inverse sense. This is because cloud cover generally extends into the middle and upper troposphere giving rise to the condition of high albedo and low outgoing longwave radiation as discussed in Chapter 3. Snow and ice give rise to the same relationship.

The variation of absorbed solar energy (Fig. 4.1(c)) shows a pronounced annual variation outside the zone 5–10 °N, the maximum of absorbed energy occurring during the summer months of each hemisphere, regardless of latitude. The phase of the variation clearly follows the course of the Sun during the year, as seen by the solar declination (the latitudinal position of the overhead, midday Sun) plotted in Fig. 4.1(c).

The net radiation (Fig. 4.1(d)) also exhibits a pronounced annual variation with a phase relationship similar to the absorbed solar energy, i.e. maximum of net radiation during the summer months of both hemispheres. The maximum and surplus of energy also follows the solar declination. It is likely that the annual variation of net radiation is dominated by the variation in absorbed solar energy, which has a much greater amplitude than the variation of outgoing longwave radiation. The net radiation achieves its greatest surplus, in excess of 100 W m$^{-2}$, over the subtropical high pressure zones of the Southern Hemisphere.

### Role of the general circulation

The profound role of the general circulation in our climate can be illustrated by reference to Fig. 4.2. Without the horizontal motions the temperatures at each latitude would be dictated by radiation alone. Thus summers would be rather warmer than observed for much of the globe. Winter temperatures in the tropics would also be higher, but would drop lower than the actual ones very rapidly as the poles are approached.

The fluxes that modify these radiative temperatures are shown in Fig. 4.3. These fluxes have already been discussed separately. Here we have combined the effects of the latitudinal imbalance of radiation (Fig. 2.24) with the imbalances of the hydrological cycle (Fig. 3.28), added the effects of oceanic transport and expressed them all in energy units. Although the individual fluxes have seasonal variations, as is also implied by Fig. 4.2, we shall temporarily concentrate on the annual conditions.

The major role of the oceanic flux is movement of some of the sensible heat away from the equator through the action of the ocean

Fig. 4.1. (Overleaf) Top of the atmosphere radiation characteristics displayed as a function of month and latitude as observed by the scanning radiometers on board the NOAA polar orbiting satellites for the period June 1974 to February 1978. (a) albedo; (b) outgoing longwave radiation (W m$^{-2}$); (c) absorbed solar radiation (W m$^{-2}$); (d) net radiation (W m$^{-2}$). The seasonal variation in these characteristics can be compared with the annual average global distributions illustrated in Figs 2.26–2.28. (Ohring, G. and Gruber, A. 1983, *Adv. Geophy.*, **25**).

**(a)**

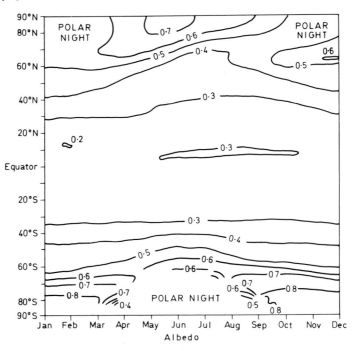

Albedo

**(b)**

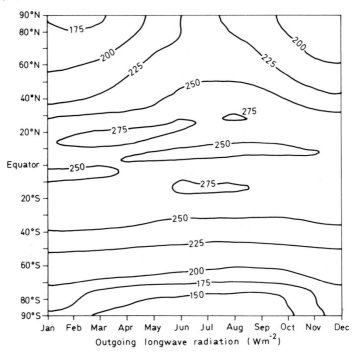

Outgoing longwave radiation ( Wm$^{-2}$ )

156

**(c)**

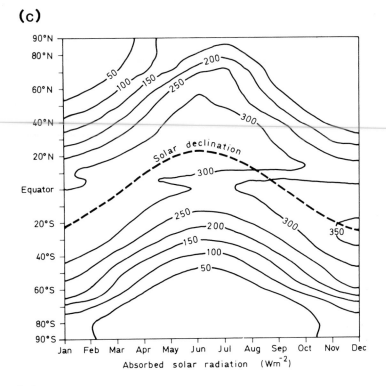

Absorbed solar radiation $(Wm^{-2})$

**(d)**

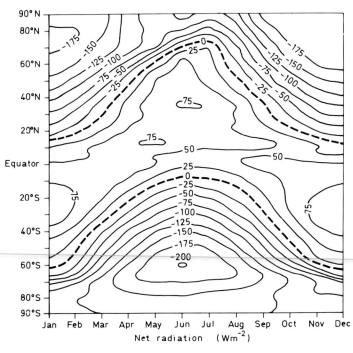

Net radiation $(Wm^{-2})$

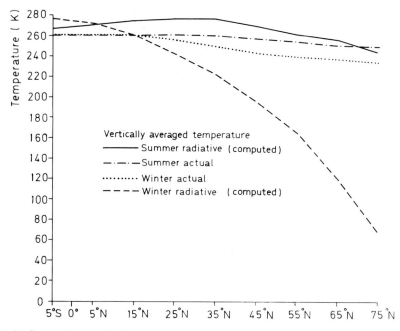

Fig. 4.2. Comparison of theoretically derived radiative equilibrium and observed vertically averaged temperature profiles for the summer and winter. Without energy transfer from low to high latitudes the equilibrium temperatures in mid and high latitudes are extremely low.

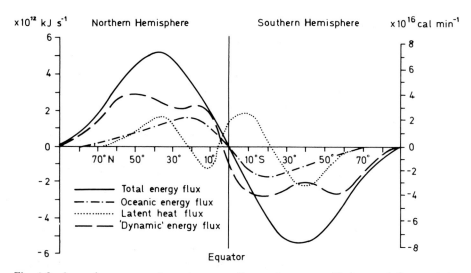

Fig. 4.3. Annual mean northward energy fluxes shown as: (i) the total flux and its three components; (ii) oceanic; (iii) latent; and (iv) dynamic (sensible).

currents. The remaining sensible heat is transferred by the atmospheric circulation. Again this is mainly a simple equator to pole transfer, although reaching a maximum rate farther poleward than the oceanic flux. The latent heat flux is rather more complex. The major source regions for the water vapour are the subtropical oceans, where net radiation is high (Fig. 4.1(d)), so that the flow is both equatorward and poleward from source regions around 10°. The relative magnitude of the three fluxes is very roughly:

oceanic flux – 25%

atmospheric dynamic (sensible heat) flux – 60%

atmospheric latent heat flux – 15%

Hence the oceanic and latent heat fluxes have about the same magnitude whilst the dynamic flux transfers about twice their combined amount. All three combine to produce a flux close to zero at the equator and a maximum near $5 \times 10^{12}$ kJ s$^{-1}$ at about 40 °N and 35 °S.

### Controls and constraints on the general circulation

Although it has been emphasised above that the general circulation is primarily a response to the energy imbalances, introduction of consideration of water vapour movement indicates that there are 'constraints' under which the general circulation must act if the present climate is to be maintained. The atmosphere must act to maintain the global water balance, preserving in an approximate way the present distribution and amounts of precipitation and evaporation. Similarly, it must maintain a balance of atmospheric mass. Finally, it must also maintain the planet's angular momentum balance. Since there is frictional coupling between the atmosphere and the rotating Earth there is a possibility that the rotation rate can be altered by the general circulation. In essence, a constant angular momentum requires an approximate equality between eastward and westward components of the wind. These simple constraints must be borne in mind as we develop our description and understanding of the general circulation.

## 4.2 Atmospheric pressure

Horizontal air motion is a response to horizontal variations in atmospheric pressure. Pressure is created because the atmospheric gas molecules are in constant motion. They therefore exert a force whenever they impact upon a surface. The total force produced per unit area is the pressure.

159

## Sea-level pressure

Variations in pressure, both in time and space, result from changes in the energy and number of molecular impacts, which are primarily determined by the density and temperature of the gas. The most fundamental density variation is its decrease with height in the atmosphere. Thus at sea level the pressure is commonly around 1013.2 hPa, which is taken as 'standard' sea-level pressure. At 3000 m it is about 70% of this value and at 10,000 m the pressure is around 300 hPa. Near the surface a change in height of about 100 m leads to a decrease in pressure of about 10 hPa. Such rapid vertical changes can mask horizontal changes unless all pressure observations are taken at a uniform height. Since over land this is a practical impossibility, observations are modified using known relationships between height and pressure, so that all refer to sea level. This modification process is known as *reduction to sea level*.

Once reduced to sea level, observations from a network of observing stations reveal distinct horizontal *pressure patterns*: discrete regions or bands of high and low pressure. These are readily seen whenever a pressure, or *isobaric* map is inspected (e.g. Fig. 4.4). An isobar is simply a line joining places with equal atmospheric pressure. In the atmosphere the surface pressure is likely to be low in regions of high temperature, in accordance with the universal gas laws. It is also likely to be low in regions of ascending air where molecules are being removed from the surface. Although the geographical distribution of pressure need not concern us until later in the chapter, Fig. 4.4 clearly indicates certain regions of persistent high and low pressure. It is these features which are responsible for the primary circulation features of the general circulation. Certain zones have frequent frontal activity and it is in these regions that the secondary circulation features are particularly important.

## Pressure variations higher in the atmosphere

Horizontal pressure variations also occur in the free atmosphere away from the surface. It is possible to represent these on a map in two ways. The first is directly analogous to the isobaric chart already introduced, but with the surface replaced by a specific level in the atmosphere. Pressure changes at a constant altitude are thus displayed. However, it is usually much more convenient to use an alternative approach and display changes of height over a constant pressure surface. The resultant constant pressure surface map looks rather like an isobaric chart, but the lines are height contours rather than isobars (Fig. 4.5). When we discuss winds associated with pressure variations, it will be seen that in fact the two types of maps can be interpreted in a similar way.

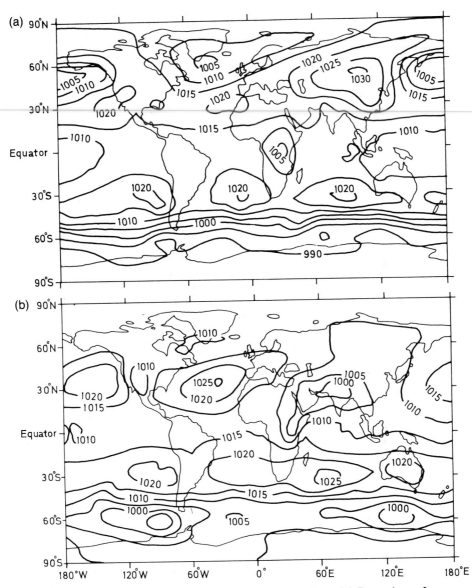

Fig. 4.4. Mean sea-level pressure (hPa) averaged for: (a) December, January, February (1963–73); and (b) June, July, August (1963–73). Note the movement of the position of the Intertropical Convergence Zone (the belt of lower pressure near the equatorial belt) and the subtropical high pressure areas.

## Measurement of atmospheric pressure

Surface pressure is measured by a *barometer*: an evacuated tube sealed at one end which is inverted into an open dish of mercury (Fig. 4.6). The level of the mercury in the tube adjusts itself until the pressure

161

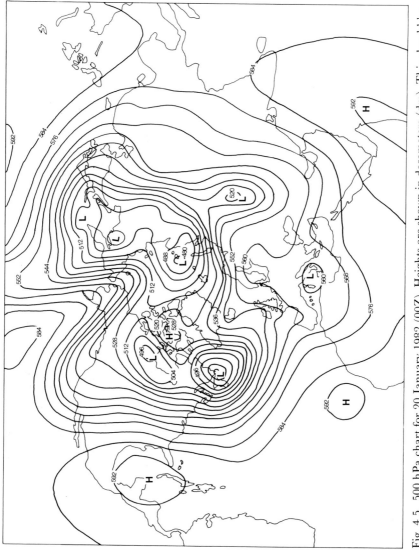

Fig. 4.5. 500 hPa chart for 20 January 1982 (00Z). Heights are shown in decameters (dm). This should be compared with the associated surface chart shown in Fig. 5.26. In particular note the close association between the position of Rossby waves over the UK and the North Atlantic and the family of depression systems in Fig. 5.26. The lows over the Mediterranean and south of Greenland are seen on both the surface and upper air charts.

it exerts on the mercury in the dish exactly balances that of the atmosphere. The height of the mercury column, after adjustment for expansion and contraction caused by temperature variations, is thus a measure of atmospheric pressure. This, of course, has led to pressure being expressed in terms of 'mm of mercury', or even just in term of height, rather than the fundamental pressure unit (Pa) used here.

In the free atmosphere pressure measurements are made during radiosonde ascents. Although the method adopted depends on the particular type of radiosonde used, it is common to employ an *aneroid barometer*. This is a partially evacuated semi-rigid metal bellows which flexes as pressure changes. The change in box configuration is transmitted by a mechanical linkage to a pointer or recorder which has been calibrated to relate the mechanical changes to pressure changes. Similar instruments, calibrated somewhat differently, are the basis of many aircraft altimeters. The instruments are also familiar to most people, since they are the type of barometer displayed in many homes. Such home observations of pressure changes, being for a single point, will not reveal all the complexities of the atmospheric motions.

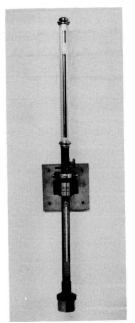

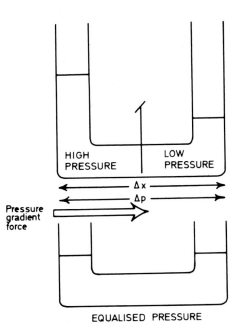

Fig. 4.6. Photograph of a barometer showing the mercury column and the vernier scale used to read the height of the column. The instrument reading is usually scaled to sea-level pressure.

Fig. 4.7. Schematic diagram illustrating the pressure gradient force. As the valve between the two sides of the container is opened the pressure gradient force operates to move the liquid until the pressures are equalised.

However, because pressure is so intimately linked to winds and weather, such observations remain the single most useful measurement which, if judiciously used, can reveal a great deal about the current weather and its likely future course.

# 4.3   Air movement around a rotating planet

### *Pressure gradient force*

The driving force for all air motions is variations in atmospheric pressure. For horizontal variations in pressure, a force is created acting from high to low pressure (Fig. 4.7). This *pressure gradient force*, $P_f$, is given, per unit mass, by

$$P_f = - \frac{1}{\rho} \frac{\Delta p}{\Delta x} \qquad\qquad [4.1]$$

where $\Delta p$ is the pressure change over distance $\Delta x$. Thus $\Delta p / \Delta x$ is the pressure gradient. The negative sign indicates that the force operates from high to low pressure, i.e. that $p$ decreases as $x$ increases. The atmospheric situation is illustrated by analogy in Fig. 4.7. If the valve separating the two liquid columns is opened, the fluid attempts to readjust to achieve equalised pressure.

Although there is a similar pressure gradient force in the vertical, acting upwards, this is almost exactly balanced by the downward acting force of gravity. Only in certain circumstances, such as during cloud formation, is vertical motion important. Generally the horizontal force exceeds the vertical one by about three orders of magnitude for most scales of atmospheric motion and the horizontal wind speed is very much greater than the vertical speed. For our present analysis it is a very convenient simplifying assumption to regard wind as having only a horizontal component.

On a non-rotating planet, air would flow under the influence of the pressure gradient force towards low pressure, a simple result of Newton's second law of motion. Its speed would depend on the magnitude of the force, which in turn is proportional to the spacing of the isobars or height contours on a chart such as Fig. 4.5.

### *Coriolis force*

On a rotating planet the speed of the wind is still governed by this pressure gradient force, but the rotation causes a change in the direction of the flow. In terms of Newton's second law, another force, created by the rotation, acts on the moving air parcel. This is the

*Coriolis force.* Any projectile given a velocity across the surface of the Earth will experience an apparent force which tends to turn it to the right in an anticlockwise rotating frame (e.g. the Northern Hemisphere as viewed from space) (see Fig. 4.8) and to the left in a frame that is rotating in a clockwise direction (the Southern Hemisphere as viewed from space). This additional force acts upon all projectiles moving within the rotating frame of the Earth. The deflection can be most easily understood by thinking of the surface of the Earth as being approximated by a disc at the location of interest. This disc has a spin about the centre (the local vertical) of $\Omega \sin \theta$ where $\Omega$ is the rotation rate of the Earth and $\theta$ the latitude of the location. As seen in Fig. 4.9, any projectile appears to move to the right because of the rotation of the disc. Thus to a stationary observer standing at the centre of the disc the projectile coming from the edge, which is moving more rapidly around the centre than the inner disc, appears to be deflected to the right. The extra apparent acceleration given to all air motions around the Earth is given by

$$\text{Coriolis acceleration} = (2\Omega \sin \theta)\, u \qquad\qquad [4.2]$$

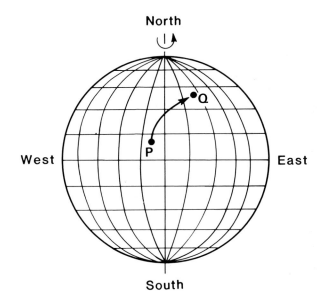

Fig. 4.8. Schematic diagram illustrating the direction (P to Q) which a projectile will take due to the rotation of the Earth, given an initial velocity towards the North Pole. When apparently stationary at point P the projectile already has a large angular velocity about the axis of rotation of the Earth as does the piece of land upon which it lies. Points nearer the pole have a smaller angular velocity and so, once moving, the projectile appears to turn towards the right as it travels poleward.

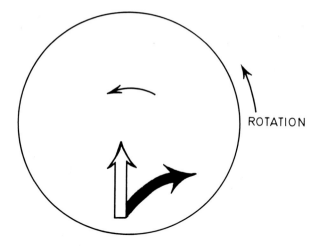

Fig. 4.9. Deflection (black arrow) to the right of a projectile (whose initial direction is shown by the open arrow) moving across the face of a disc which is spinning anti-clockwise. The projectile has a larger tangential speed than the inner part of the disc over which it is moving by virtue of its initial position at the edge. Hence as it moves towards the centre there is a turning to the right. Similarly a projectile whose initial position was at, or close to, the disc centre has a lower tangential speed than the outer part of the disc over which it is passing and hence it, too, is deflected to its right.

where $u$ is the speed of the projectile. The quantity known as the Coriolis parameter, $f$, where

$$f = 2\Omega \sin \theta \qquad\qquad [4.3]$$

is constant for a given latitude. Since the angular speed of the Earth is $2\pi$ radians, in 24 hours we have

$$\Omega = 2\pi / (24 \times 60 \times 60) = 7.27 \times 10^{-5} \text{ s}^{-1} \qquad [4.4]$$

Hence $f$ is always small, varying from $1.5 \times 10^{-4}$ s$^{-1}$ at the poles to zero at the equator. However, it is this force which imparts an east/west component to meridional atmospheric motions.

### Geostrophic flow

When isobars are straight and parallel and we are considering motion in the free atmosphere away from the effects of surface friction, only the pressure gradient and the Coriolis forces act on a parcel of air. The pressure gradient force initiates the motion and immediately the Coriolis force commences its deflecting action. The two forces rapidly come into equilibrium and there is a balanced flow with two equal forces. Thus, using equations [4.1] and [4.4] and rearranging

$$V_g = -\frac{1}{\rho f}\frac{\Delta p}{\Delta x} \qquad [4.5]$$

where $V_g$ is the *geostrophic wind*. This wind therefore blows parallel to the isobars with low pressure on the left-(right) hand side as you stand with your back to the wind in the Northern (Southern) Hemisphere (Fig. 4.10). Its speed is proportional to the isobaric spacing. The geostrophic wind, determined entirely from observations of the pressure distribution, is identical to the actual wind when isobars are straight and parallel and when the motion is far enough removed from the surface. In reality, of course, the actual situation rarely completely fulfils these constraints. For much of the time, however, the isobars are not highly curved and the geostrophic wind provides a very good approximation to the real wind. Certainly in these conditions it enables us to get a much better idea of the wind speed than can be obtained from the rather sparse and expensive direct observations. The geostrophic approximation can only be used with confidence poleward of about 30°, since in equatorial regions the Coriolis force tends towards zero and there is no strong deflection of the winds.

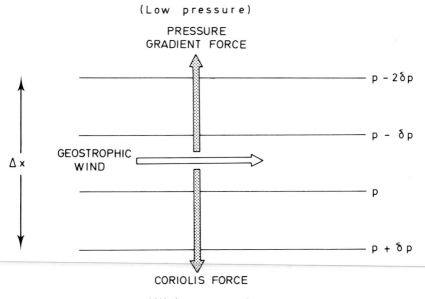

Fig. 4.10. Diagram showing the geostrophic wind, which is the balanced result of the two opposing forces (pressure gradient and Coriolis)

167

The major drawback to the use of equation [4.5] is that it includes the air density, which varies with temperature and height. This difficulty can be overcome by introducing the hydrostatic equation (equation [3.7]). Substituting for the density, the geostrophic wind equation becomes

$$V_g = (g/f)\ \Delta h/\Delta x \qquad\qquad [4.6]$$

where $\Delta h/\Delta x$ is the rate of change of height of a constant pressure surface. This can be obtained directly from a constant pressure chart such as Fig. 4.5. This equation is much simpler to use than equation [4.5], because only the Coriolis parameter, the acceleration of gravity and the slope $\Delta h/\Delta x$ are needed. The equation, however, can be interpreted in a similar way to equation [4.5], with the geostrophic wind blowing parallel to the height contours with low height on the left hand side (for the Northern Hemisphere) and at a speed proportional to the contour spacing. This is just one example of the simplification possible when constant pressure, rather than constant height, charts are used to depict conditions in the upper air.

Isobars are rarely straight. In most cases there is some degree of curvature, either *cyclonic* with air having a counterclockwise motion around a low pressure area, or *anticyclonic*, with clockwise motion around a high pressure system. [N.B. These directions are given for the Northern Hemisphere. They must be reversed when the Southern Hemisphere is being considered.] However, except for small-scale motions or motions associated with severe storms, the geostrophic approximation holds for such curved motions.

## Gradient wind

When there is marked curvature in the isobars a third force, the *centrifugal force*, must be introduced. This acts outwards from the centre in any curved motion. It can readily be demonstrated by whirling a stone on a string. The tension felt in the arm during this activity is an expression of this force. In the case of rotation around a high pressure area this force is in the same direction as the pressure gradient force and hence leads to an increase in wind speed over that calculated for the geostrophic wind. Around a low pressure centre the centrifugal force opposes the pressure gradient force and decreases the wind speed (Fig. 4.11). The wind that results from a balance of the three forces is known as the *gradient wind*. The direction of this wind, like the geostrophic wind, is parallel to the isobars. The speed of the gradient wind differs significantly from that of the geostrophic wind only when there is severe curvature at very large pressure gradients. For instance, in the case of a tropical hurricane the calculated geostrophic flow might be 500 m s$^{-1}$, but the gradient wind speed is only 75 m s$^{-1}$.

168

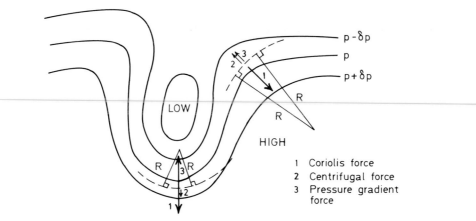

Fig. 4.11. The three-way balance between the horizontal pressure gradient force (3), the Coriolis force (1) and the centrifugal force (2) in atmospheric flow along curved trajectories (dashed) with radius of curvature R. The resulting winds are gradient winds. Only three isobars are shown for simplicity (p−δp, p, p+δp) although the concept of the gradient wind only becomes important in situations of large pressure gradients.

## *Near-surface wind*

As we approach the surface of the Earth the influence of surface friction is increasingly felt. This *frictional force* acts directly against the airflow, leading to a reduction in wind speed. Since the Coriolis force is a function of wind speed it is also reduced and the flow, even with straight parallel isobars, is no longer balanced (Fig. 4.12). A cross-isobaric flow, directed towards low pressure, is induced. The angle at which the air crosses the isobars depends on the magnitude of the frictional force. A fairly smooth water surface rarely produces inflow to low pressure at an angle greater than about 8° to the isobars, while a land surface of rolling terrain may lead to an angle somewhat in excess of 25°. Very rough terrain is much more likely to create its own circulation patterns in the lowest air layers than simply to modify the geostrophic wind. The frictional force is at its maximum right at the surface and gradually decreases with height until it becomes insignificant and the geostrophic wind approximation holds. This decrease with height also leads to a clockwise change in wind direction with height, which is sometimes called the *Ekman spiral*. The layer where friction is effective is usually known as the *friction layer*, with the free atmosphere above it.

So far when considering the free atmosphere we have assumed that the motions have been horizontal, or at least quasi-horizontal. Certainly for any given level the geostrophic approximation gives a good indication of windflow patterns. However, interactions between levels, particularly when spatial temperature variations are present,

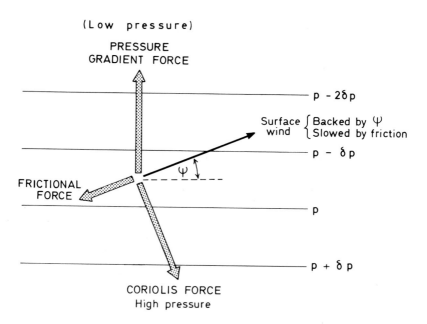

Fig. 4.12. The effect of surface friction is to 'back' (move in an anticlockwise direction) the surface wind compared with the direction of geostrophic flow and to reduce its speed.

can lead to vertical motions that modify the simple analysis presented here.

# 4.4  Barotropic and baroclinic conditions

We can commence our discussion of factors which influence the vertical variations of horizontal wind, and sometimes themselves lead to vertical motions, by considering the relationship between temperature and pressure in the vertical. The height difference between any two given pressure levels is known as the *thickness*. Since density decreases as temperature increases, it follows that a warmer layer must cover a greater geometrical height to embrace the same mass of gas, so that thickness varies directly with temperature.

### Barotropic atmosphere

Using this relationship, the simplest situation we can envisage is one where there is horizontally uniform temperatures at all levels. Hence there is no spatial change of thickness. We can introduce a pressure gradient and therefore allow horizontal motion. This situation, where there is a pressure gradient but no temperature gradient, is a *barotropic* situation. There is no change in wind speed or direction with height and there is no chance for disturbances to grow in the airflow.

170

### Equivalent barotropic state and the thermal wind

If we introduce a temperature gradient such that the isotherms are parallel to the isobars, we generate an *equivalent barotropic* atmosphere (Fig. 4.13). If we assume, as in Fig. 4.13(a), that the low pressure area is cold and the high pressure area is warm, the increasing thickness as we move into the warmer air will lead to a steepening of the pressure gradient with height. This will cause an increase in wind speed with height, but the wind direction will not change. Figure 4.13(b) indicates the opposite conditions, with the warm air over the near-surface low pressure region. Now the wind speed decreases with height, eventually becomes calm, and then increases with height, but moving in the completely opposite direction. Since we can treat this change as simply the wind blowing in the same direction, but with a negative velocity, we can state that in an equivalent barotropic atmosphere there is no change in wind direction. As with the true barotropic situation, there is no chance for disturbances to grow. The difference in wind speed between the top and bottom, the vertical shear in the calculated geostrophic winds, is proportional to the horizontal gradient of the mean temperature of the intervening layer. Such a thermally induced gradient in wind speed is termed a *thermal wind*.

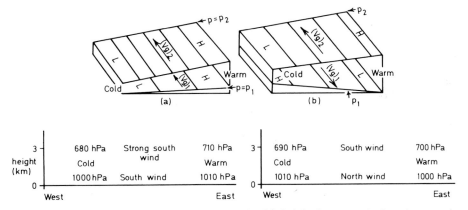

Fig. 4.13. The change of the geostrophic wind with height in an equivalent barotropic flow in the Northern Hemisphere. (a) $V_g$ increasing with height and (b) $V_g$ reversing in direction with height.

### Baroclinic atmosphere

When the isotherms are not parallel to the isobars a *baroclinic* situation occurs (Fig. 4.14). In this case the temperature, and thus the thickness, varies along an isobar. Consequently the pressure pattern changes with height, as does the wind speed and direction. The vector difference between the speed and direction at the lower and upper levels is also a thermal wind. It has a direction parallel to the isotherms and a speed

171

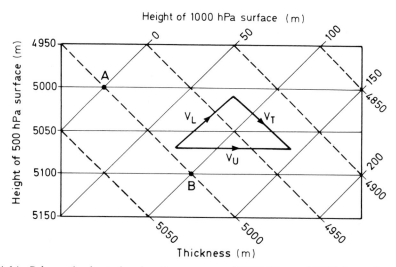

Fig. 4.14. Schematic chart showing the contours of 1000 hPa and 500 hPa surfaces and the thickness of the layer. Lines of constant values for the height of the 500 hPa surface are horizontal and those of the height of the 1000 hPa surface are skewed (both unbroken in diagram). Thus, for example, at point A, the height of the 500 hPa surface is 5000 m and the height of the 1000 hPa surface is 0 m, the difference between these two surfaces (of 5000 m) being known as the *thickness*. Similarly, at point B the thickness equals 5100 − 100 = 5000 m. The locus of such points is thus a line of constant thickness and a set of such lines is shown dotted in the diagram.

The lower level wind, $V_L$, blows parallel to the lines of constant values of the height of the 1000 hPa surface; and similarly the upper level wind, $V_U$, is parallel to the lines of constant values of the height of the 500 hPa surface. It can be seen from the diagram that the vector difference between upper and lower level winds, $V_T$ blows parallel to the lines of constant thickness and is known as the thermal wind. (From Petterssen, 1969).

proportional to the isotherm spacing, again blowing with low temperature to its left (in the Northern Hemisphere). The governing equation is similar to that of the geostrophic wind, so that its value can easily be calculated. The thermal wind is a very useful concept in meteorology since it allows the calculation of the actual wind at any level once the pressure distribution at one level and the horizontal and vertical temperature distributions are known. In practice the sea-level pressure distribution is obtained from the numerous surface-based observations, while the temperature distributions can be obtained from upper air soundings (radiosonde ascents) or from satellite measurements.

In baroclinic conditions the wind is blowing across the isotherms, which leads to *advection* of energy into or out of an area. Thus a flow from a colder to a warmer region leads to cold advection. When the wind is *backing* (changing in a counterclockwise direction) with height, cold advection is occurring, while a *veering* wind indicates warm advection. Figure 4.14 indicates a veering wind and warm advection. Backing or veering can often be observed when several different cloud layers occur simultaneously, although the phenomenon can easily occur without the presence of clouds.

This energy advection associated with baroclinicity plays a vital role in the creation of disturbances in the atmospheric flow patterns. We can start with an equivalent barotropic situation which has west to east, or zonal, airflow and isotherms parallel to the isobars. If for some reason, such as the presence of a topographic barrier, the zonal flow is disturbed, a baroclinic situation will be generated (Fig. 4.15). The flow at point A is carrying cold air southward while at point B warm air is being moved northward. The latitudinal temperature contrasts will continue to increase as the advection continues. Eventually the contrasts will be so large that spontaneous vertical motions will be generated. In energy terms, this implies that the increase in baroclinicity increases the potential energy in the system, which is then released as the kinetic energy of motion. Southward moving cold air is associated with sinking motions, northward moving warm air with rising ones. These motions are induced by the dynamics of the flow and will occur whether or not the atmosphere is hydrostatically stable in the sense introduced in Chapter 3.

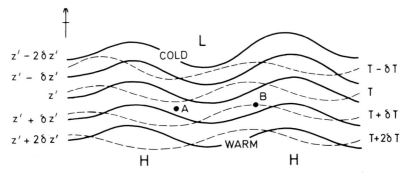

Fig. 4.15. Plan view (on a constant pressure surface) of the variation of temperature (dashed) and geopotential height (solid) in a developing baroclinic wave in the Northern Hemisphere.

The vertical motions associated with waves in the horizontal flow are enhanced by the curvature of the waves themselves. Recalling from the discussion of the geostrophic and gradient winds that for a given pressure gradient the wind speed will increase as the anticyclonic curvature increases, there will be accelerations in the airflow as it moves through a wave. In particular, as the air passes through point A (Fig. 4.15) its curvature is becoming more cyclonic and the air is decelerating, while at point B the opposite is occurring. Assuming that there is no significant lateral motion, this leads to further vertical motions.

173

### Divergence, convergence and vorticity

It is a simple extension of the necessity to conserve mass that if a constant volume of fluid extends its horizontal dimensions, or *diverges*, its vertical extent must decrease. Similarly a vertical column experiencing *convergence*, for example through a decrease in downstream velocity, is constricted in the horizontal and must therefore be extended vertically. Figure 4.16 is a schematic representation of the vertical velocities associated with the situations of convergence and divergence at the surface and aloft.

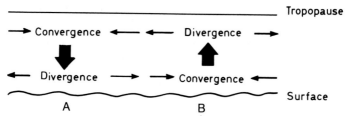

Fig. 4.16. Schematic illustration of the balance of surface convergence/divergence by divergence/convergence aloft.

In the case of our baroclinic wave occurring in the free atmosphere, convergence aloft at point A (Fig. 4.15 and 4.16) reinforces the sinking motion while high level divergence near point B enhances the upward motions. Thus the near-surface conditions are of sinking and diverging air at A and converging and rising air at point B.

The rising air near the surface at point B tends to be associated with a low pressure region at the surface. Air converges across the isobars

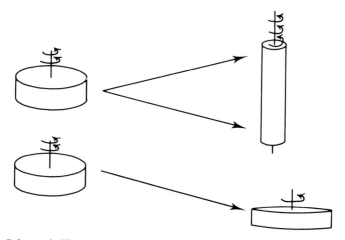

Fig. 4.17. Schematic illustration of the variation in vorticity caused by the conservation of angular momentum. As the upper cylinder decreases its radius, vorticity increases, whilst for the lower cylinder the increased radius leads to a vorticity decrease.

174

within the friction layer. This increases the *vorticity*, defined as twice the angular velocity of the rotation, of the air in this region. Figure 4.17 illustrates the principle. It is also clearly displayed by a skater who draws in arms and legs closer to the body, causing personal 'convergence' and an increase in vorticity as expressed by the increase in spin rate. Hence at point B in Fig. 4.16 a cyclonically curved spiral is generated as the air moves in towards low pressure.

The effects of the baroclinic instability, divergence and vorticity, all combine to create disturbances within the baroclinic zone. Detailed consideration of the processes involved is vital for day to day weather forecasting, and is important in computer modelling of the general circulation, but is not warranted in the context of climatology. It must be emphasised, however, that baroclinic conditions leading to disturbances occur only in restricted regions of the globe and even in these regions only for part of the time. Equivalent barotropic conditions are by far the dominant conditions over the globe.

## 4.5 The general circulation of the atmosphere

Characterisation of the general circulation of the atmosphere has been a central problem in both meteorology and climatology since the subjects emerged as sciences. Observation and theory have here gone hand in hand. Once surface-based observations became sufficiently widespread to give approximately global coverage, the *three-cell model* of the general circulation was developed as a theory fitting the known facts (Fig. 4.18). Although this model is now known to be a vast oversimplification, it still provides a useful conceptual tool.

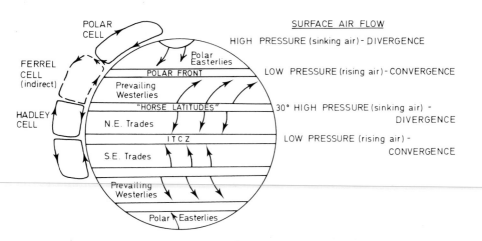

Fig. 4.18. The three-cell circulation of the atmosphere showing the resultant meridional pattern – high and low pressure areas and the direction of surface wind flow.

The model was developed from the observation that there were zonal belts of low pressure around the equator and, in a more diffuse form, around 60° latitude. High pressure dominated around 30° latitude and at the poles. Since low pressure is associated with convergence and ascending air and high pressure with descent and surface divergence, it was a relatively simple matter to create the three cells. The tropical and polar cells were supposedly being driven by the effects of surface heating and were called 'thermally direct' (i.e. ascent over a warm area and descent over a cold area), while the mid-latitude one was a response to them and was 'thermally indirect'. Ascent creates clouds and precipitation, while descent gives dry, cloudless conditions, in accordance with low-latitude observations. The air motions at the surface, after incorporation of the Coriolis effect, also correspond reasonably well with observations. It can also be seen that the model fulfils the basic functions of the general circulation to redistribute energy and moisture without changing the angular momentum or mass balance of the planet.

### Empirical data contradict the three-cell circulation model

As further observations were made, particularly in the upper atmosphere, it became obvious that the three-cell model was incomplete in all areas. It is also incorrect in at least one area. In mid-latitudes the upper airflow is westerly (Fig. 4.19), not easterly as the model predicts. Indeed, it can be seen clearly that mid-latitude upper level winds in both hemispheres and both seasons are characterised by extremely high velocity *westerlies*, centred on a core, or *jet*, which changes position during the year. These westerlies take on a wavelike motion, the waves being known as the *Rossby* or *planetary scale* waves. These are a function of the zonal temperature gradient and the angular velocity of the motion and are influenced by the characteristics of the underlying surface.

Thus the thermally indirect mid-latitude cell is much more complex than was supposed. Indeed it is no longer appropriate to think of it as a cell. Instead the necessary poleward energy transports are performed by horizontal wavelike motions and their embedded disturbances. The two thermally direct cells remain, although notions of their driving mechanisms have changed.

The most significant features of the general circulation are illustrated clearly in Fig. 4.20. They are the upper level, high velocity westerly flow in both hemispheres (the Rossby waves) which dominate the zonal wind pattern (Fig. 4.20(a)) and the strong cellular circulation near the equator clearly visible in the vertical wind field shown in Fig. 4.20(c). However, the circulation must be viewed as a single system, with all parts connected, a view strongly emphasised when satellite-derived information is incorporated into our picture. Despite

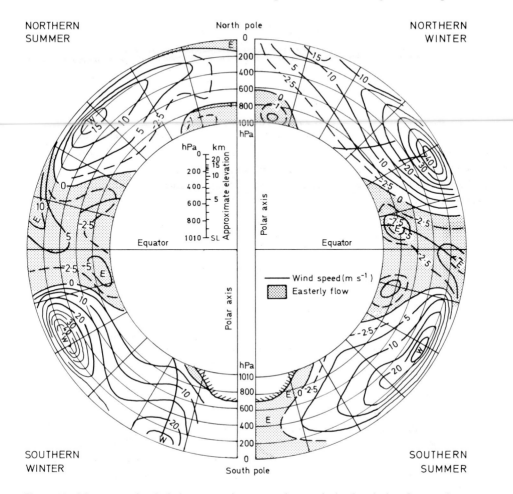

Fig. 4.19. Mean zonal wind (east–west) averaged over latitude circles for northern summer (left) and northern winter (right). Winds are in m s$^{-1}$ and easterly winds are shaded. Note the very strong upper westerly flow which contradicts the three-cell model shown in Fig. 4.18. A conversion scale between pressure (hPa) and height (km) coordinates is inset.

the necessary modifications, however, the three-cell framework, or perhaps better, the three-division framework (Hadley cell, Rossby waves, polar cell), is still evident in our diagram of the general circulation and is clearly reflected in the various regional climates.

The circulation in mid-latitudes seems to be more complex than in the tropical or polar regions. This is almost certainly the case, although it should be borne in mind that we know more about the mid-latitudes than the other regions and so are more likely to recognise complexity. The differences arise largely because of the different mechanisms responsible for energy and water redistribution. Meridional transport,

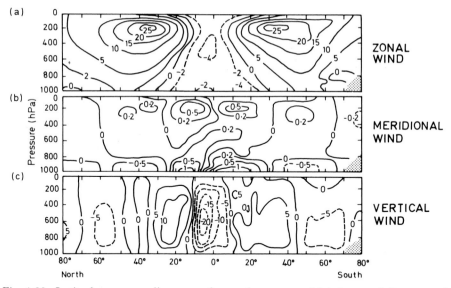

Fig. 4.20. Latitude pressure diagrams of annual averages of (a) the zonal (i.e. around a latitude zone) wind (m s$^{-1}$); (b) the meridional (i.e. along a longitude zone) wind (m s$^{-1}$); (c) the vertical wind ($10^{-5}$ hPa s$^{-1}$). Diagram (a) clearly shows the position of the two jet streams at about 40° latitude. Evidence for the Hadley circulation cells in the lower latitudes can be seen in (b) and still more clearly in (c). The units in which the vertical wind is given are the result of computing the rate at which isobaric (constant pressure) surfaces are crossed in unit time.

the cellular like motion, dominates in the tropical regions, while eddy transport, the horizontal wavelike motions, dominates in mid-latitudes (Fig. 4.21). Thus the relatively simple cellular description is adequate for general consideration of tropical and possibly polar processes (though the latter are at present not well understood) but is completely inadequate for the mid-latitudes.

### Hadley cell regime

The thermally direct tropical, or *Hadley*, cell generally exists within an equivalent barotropic atmosphere. At these low latitudes, without a distinct thermal winter and summer, continentality effects are minimal and so the near-surface temperature distribution is roughly uniform zonally, with a temperature decrease poleward. The basic pressure distribution is also zonal. The surface, heated by solar radiation absorption, provides a heat source for the lower atmosphere. Latent heat released by convective motions within the ascending air of the *intertropical convergence zone* (ITCZ), where the cells in each hemisphere meet, provides a second heat source. The heat sinks occur at the top of the atmosphere and at the poleward limb of the cell.

The Hadley circulation is driven by the continuous supply of heat

178

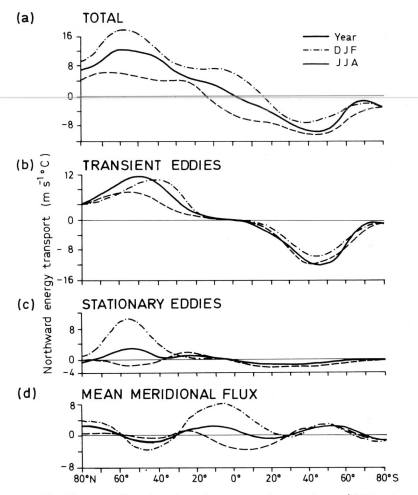

Fig. 4.21. Meridional profiles of northward transport of energy in m s⁻¹ °C for the cases of (a) total energy transfer, (b) transfer by transient eddies, (c) transfer by stationary eddies, in which the longitudinal component of the motion is responsible for the northward flux, and (d) the mean meridional flux. Note that the meridional flux is largest near the equator while transient eddies perform most of the energy transport in the mid-latitudes.

to the equivalent barotropic atmosphere. The horizontal components of the flow are only slightly affected by the Coriolis deflection at these low latitudes, so that the resulting winds, the *trade winds*, do not blow parallel to the isobars, but have a marked cross-isobaric component, blowing equatorward at low levels and poleward in the upper troposphere. Thus the wind patterns allow a direct transfer of energy by meridional flow away from the equator. In some areas variations in surface characteristics may lead to the superimposition of other circulation patterns, particularly at low levels. For example, surface heating and topographic effects lead to distinct circulations known as monsoons, while surface temperatures over the desert regions commonly occurring around 30° are frequently higher than temperatures around the equator. These differences, together with small-scale atmospheric disturbances, lead to local and regional variations in weather and climate. Nevertheless, from the global perspective the tropical conditions are adequately described and explained by the Hadley circulation.

### Polar regions

In theory the polar cell is similar to the Hadley one. A surface covered with snow and ice produces a uniform temperature distribution which leads to equivalent barotropic conditions and the cellular motions. In this case, of course, they are completely deflected to zonal flow by the Coriolis force. The resultant surface easterlies are relatively weak throughout the year in the high latitude Arctic but are a fairly strong feature over Antarctica in the southern summer (Fig. 4.19). There is general atmospheric subsidence. During winter in both hemispheres there is evidence for a local velocity maximum at 75–80° which has a jet-like structure that is particularly well developed in the Southern Hemisphere. This feature, which can be regarded as an extension of the mid-latitude westerlies, dominates the polar circulation. Hence the 'pure' polar cell is rather weakly developed and there is relatively little meridional transport (Fig. 4.21). In any event its geographical extent is very small and in both hemispheres it can be regarded as the area to which energy must be transported and hence is not itself required to play a great role in energy transport.

### Baroclinicity in the mid-latitudes

The mid-latitude region is frequently a region with an equivalent barotropic atmosphere and thus there is a tendency for cell-like meridional overturnings and meridional energy transport. However, the unique characteristic of mid-latitudes, creating complexity in both the energy transfer mechanisms and the weather and climate, is the frequent establishment of baroclinic conditions. The causes of this complexity will be explored in the next section.

The present configuration of the general circulation of the atmosphere into three major regions appears to be a relatively stable one. This has been established both by numerical results and through laboratory simulations of atmospheric motions using *dishpan* (or *annulus*) experiments. The dishpan is a fluid-filled annulus which can be rotated at various rates and which can be heated at the bottom and outer edges (Fig. 4.22). Thus it is a type of scale model of the Earth's atmosphere. With a small amount of heating at the base and outer edge

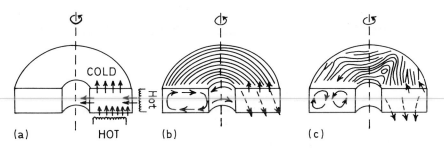

Fig. 4.22. A differentially heated rotating annulus experiment showing (a) the heating regime; (b) the symmetric flow, i.e. a Hadley circulation, in the case of a low rotation velocity; and (c) the Rossby wave regime set up in the case of a higher circulation velocity.

and relatively low rotation rate a single cellular circulation is established, very similar in nature to the Hadley circulation. As the rotation rate and heating gradient are increased to simulate more closely the existing atmospheric conditions, this single system breaks up. The Hadley circulation becomes confined to the 'tropics' near to the outer edge and a Rossby-like circulation develops in mid and high latitudes. In some cases a very weak polar circulation is also established. This configuration is maintained through a relatively wide range of rotation rates and heating gradients, suggesting that the present atmospheric configuration is 'stable', at least for the rotation rates and energy gradients likely to be encountered in the foreseeable future. The results, of course, do not indicate an unchanging climate, but rather one that can change within the limits imposed by the configuration of the general circulation of the atmosphere.

## 4.6  The mid-latitude baroclinic zone

### Rossby waves

The Rossby waves are probably the most important feature of the mid-latitude circulation. This flow pattern (Fig. 4.23) influences the whole depth of the troposphere and covers most of the mid-latitude region. However, the waves are usually best developed, and well defined, in a relatively narrow band at restricted latitudes. It is in this band that the major poleward temperature gradient is established and thus that baroclinic conditions are likely to occur. This has several consequences. A marked horizontal temperature gradient is set up at the surface, leading to the development of the *polar front* (Fig. 4.24). This major temperature discontinuity, although not continuous around the hemisphere, usually covers a significant portion of it. It extends upwards, as a *frontal surface*, through much of the depth of the troposphere. In this region the atmosphere is at least equivalent barotropic and may

181

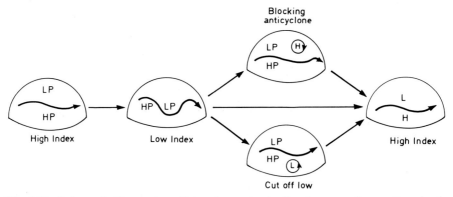

Fig. 4.23. Schematic illustration of the change in the Rossby wave flow pattern in the transition from high to low index and then from low to high index regimes.

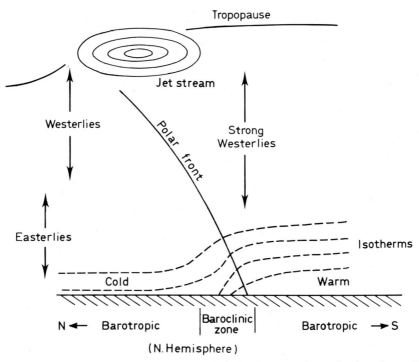

Fig. 4.24. Schematic cross-section of the mid-latitude atmosphere showing the position of the polar front, the jet stream and the baroclinic and barotropic zones.

be baroclinic. In any event, there is an increase in wind speed with height which leads to the development of a *jet stream* at or just below the tropopause. This is a ribbon-like belt of rapidly moving air, a few hundred metres in depth, a few kilometres wide and hundreds, or even thousands, of kilometres long.

The location of these three associated features, the Rossby waves, the polar front and the jet stream, varies with time. In general the Rossby waves tend to move eastward about 15° of latitude per day, although they may be retarded, or even 'anchored', by a large topographic barrier such as the Rocky Mountains. In addition the fluid mechanics of the flow will dictate that they change in amplitude and configuration with time. Further, the jet stream, being dynamically unstable, may also create meanders of its own, which may also lead to a change in the Rossby wave pattern. Similarly, the polar front may change position and affect the other two features.

### *Index cycle*
In general there is a tendency for a pseudo-cyclic change in the Rossby wave amplitude (Fig. 4.23). This can be conveniently characterised by the zonally averaged pressure difference between two latitude circles, a difference known simply as the *index*. The latitude circles usually chosen for the Northern Hemisphere are 35° and 55°, latitudes which approximately correspond to the southern and northern limits of the main Rossby flow and which have an adequate spacing of observing stations. As the wave amplitude changes there is an *index cycle*. Starting with a strong zonal flow, there is a high index and a small number of relatively smooth Rossby waves. As the amplitude of the waves increases and the index decreases the main axis of the strong westerlies is displaced southward and may become discontinuous. Distinct high and low pressure centres are formed (Fig. 4.25). Thereafter the amplitude may slowly decrease again, or a 'cut-off' may occur, leaving either a *blocking anticyclone* (see sect. 5.5) as a stationary atmospheric feature, or a *non-frontal depression* as a feature that usually moves eastwards slowly. The whole cycle may take 20 to 60 days to complete, but it is very irregular in length and in the speed with which each stage progresses.

Changes in the polar front position will arise because of the baroclinic nature of this feature. In such baroclinic conditions vertical motions associated with convergence and vorticity will be developed. In particular, these will lead to the development of *frontal depressions*, features with cyclonic motion around a low pressure centre. These probably originate near the 500 hPa flow. As they develop and extend down through the atmosphere to become 'surface features', they will not only move in response to the Rossby wave motion, but will also become sufficiently large to affect this flow themselves. They are not only responsible for rapid weather changes in the areas over which they pass, but they also play a significant role in the poleward energy transport. They are responsible for energy transport by transient eddies, supplementing the energy transport provided by the Rossby waves (Fig. 4.22). In general terms, a high index circulation with strong zonal

183

(a)

(b)

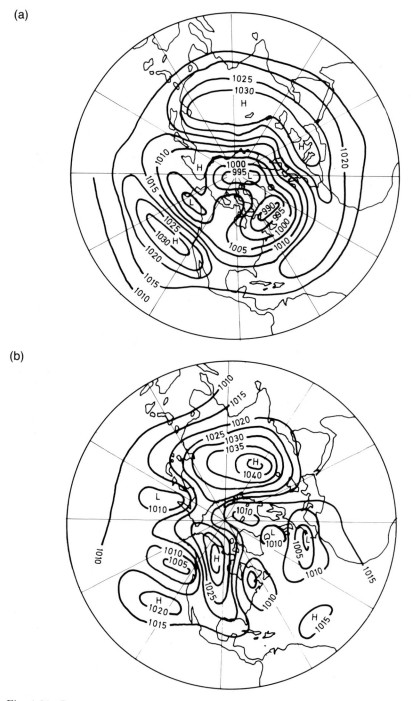

Fig. 4.25. Pressure distributions (hPa) at sea level during two 7 day periods showing (a) a high index and (b) a low index.

flows is likely to be a period with deep, vigorous depressions which are the main transporters of energy northward. With a low index the depression circulation is much weaker and the Rossby waves transport the energy directly.

Away from the main line of the Rossby waves are areas where conditions are barotropic. Such regions serve as the sources for *air masses*, regions of horizontally uniform and relatively calm conditions. The air in these regions remains over the same surface for several days and can take on the temperature and moisture characteristics of that surface. Eventually these air masses will move, either because they have become sufficiently strong to influence the general circulation in mid-latitudes, or because changes in the wave circulation allow them to move into a new area. In either case they take their inherent characteristics with them into the new area and thus affect the weather there.

# 4.7 Large-scale effects of the surface boundary

The Earth's surface forms a highly discontinuous boundary to all atmospheric processes and therefore its effect on the general circulation must be taken into account. We have already shown that practically all of the energy input to the atmosphere comes from the surface, either as the result of atmospheric absorption of longwave radiation emitted by the surface or as the result of sensible and latent heat transfer. Surface friction changes the near-surface wind direction and also serves as a sink for kinetic energy, since the energy extracted from the wind as its speed decreases is transferred to the surface. This exchange of kinetic energy, while much smaller than the sensible and latent heat exchanges between the surface and the atmosphere, is significant in maintaining the global angular momentum balance and in producing motion in the oceans. Hence both the thermal and the frictional character of the surface have profound influence on the atmosphere. In this context we can divide the Earth into three basic surface types: water, snow and ice, and land; and look at each in turn.

## 4.7.1 Oceans

The single most important characteristic of the ocean for atmospheric circulation is the large heat storage capacity it represents. By comparison, atmospheric storage of heat is an order of magnitude smaller (Fig. 4.26). In terms of mass, the atmosphere is equivalent to a layer of water approximately 10 m deep. If we consider only the *mixed layer* of the ocean, that top layer where vertical stirring is possible, which is about 70 m deep, then, since the specific heat capacity of water is 4.2 times greater than air, this upper layer of the water can store approximately 30 times more heat than can the atmosphere. Thus for a given change in heat content, the temperature change in the

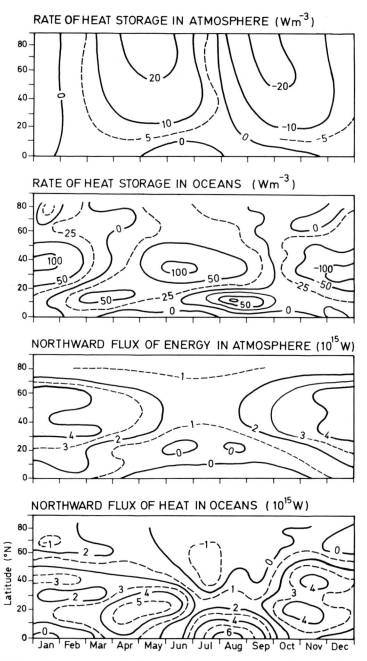

Fig. 4.26. Rate of heat storage in the atmosphere and oceans and the northward transport of energy in the atmosphere and oceans as a function of latitude and time of year. The atmospheric heat storage peaks in high latitudes in early summer and is matched by a heat loss in autumn. The oceanic heat storage, by contrast, is greatest at ~30° in midsummer. The fluxes of energy are similarly mismatched with the largest atmospheric flux being in the mid-latitudes in winter while the maximum oceanic flux is in lower latitudes in the summer.

atmosphere will be around 30 times greater than that in the ocean. Since the emission of thermal energy is a function of temperature, the ocean will lose its heat by radiation much more slowly than will the air. The overall result is that temperature changes are much less, and much less rapid, in the ocean than in the air. Heat storage by land surfaces is intermediate between air and water, so, as noted previously in our discussion of continentality, temperature changes for land surfaces are larger than those for water, but smaller than changes for air.

The second major characteristic of the oceans is their ability to transport sensible heat. The magnitude of the flux on an annual basis was indicated in Fig. 4.3. A monthly and latitudinal comparison between atmospheric and oceanic transport (Fig. 4.26) indicates that the oceanic transports are comparable to those of the atmosphere in many places at many times.

### Atmosphere–ocean interactions

Vertical interactions between the ocean and the atmosphere involve fluxes of moisture, heat, momentum and gases. Of major concern here is the momentum flux. Momentum is transferred to the ocean whenever the wind blows over it. The kinetic energy in the wind is effectively transformed into the kinetic energy associated with ocean currents. On a small scale, waves are produced. On a larger scale, when strong winds blow, the drag on the ocean surface is sufficient to permit the movement of the warm surface waters, which are then replaced by upwelling colder water from greater depth. Thus a specific sequence of events can be initiated by a warm water surface. The warm water tends to create unstable atmospheric conditions which are favourable to the development of storms and strong winds. These in turn lead to upwelling of cold water, which serves to stabilise the atmosphere and dampen the winds: an example of a 'negative feedback' within the climate system.

### Walker circulation and El Niño

A large-scale example of this type of action is provided by the *Walker circulation* (Fig. 4.27). The normal atmospheric circulation mode is a longitudinal cell readily observable across the Pacific Ocean. Near the coast of South America the winds blow offshore and result in an upwelling of water which tends to be some 5 K colder than the waters in the western Pacific. The air is stabilised by the cold water and cannot rise and join the normal Hadley circulation. Instead it flows westward, forming the southeast trade winds across the South Pacific, to the warm western Pacific, where it gains moisture and heat, and rises. Some of the rising air then flows eastward to complete the cell.

A disturbance occurs in this circulation pattern every 2–7 years. The

187

NORMAL

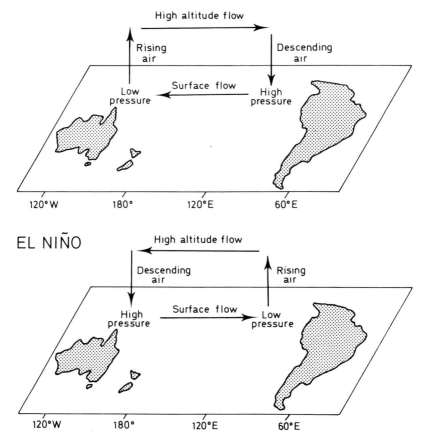

Fig. 4.27. The two modes of the Walker circulation which form part of the Southern Oscillation. In the period of El Niño the circulation is opposite to the normal flow.

ocean surface temperature off the Peruvian coast suddenly rises by about 4 K, killing fish, especially anchovies, and thus disrupting the local economy. This effect is called El Niño (The Christchild) because of its near coincidence with the Christmas period. It is suspected that this unusual outbreak of warm water comes from the western Pacific in the form of equatorial currents. In any event the result is a reversed but weaker Walker circulation (Fig. 4.27), but an enhanced Hadley circulation. This increased Hadley circulation will increase the surface trade winds, which will keep the warm water from flowing across the Pacific, thus ending El Niño. When the colder water is re-established in the eastern Pacific, the Hadley cell weakens and conditions are set up for the return of the warm water currents.

There seem to be linkages between the oscillation in direction of the

Walker circulation and other anomalies in the climate of the Southern Hemisphere, but they are not fully understood. For example, drought in Australia seems to be associated with negative anomalies in sea-surface temperatures in the west Pacific Ocean which occur at the times of El Niño circulation pattern. Certainly the strength of the Walker circulation, which varies in response to changes in ocean temperatures, forms an important component of the *Southern Oscillation.* This phenomenon is characterised by an exchange of air between the Indonesian low and the southeast Pacific area of high pressure and occurs at intervals of between 1 and 5 years with an average period of about 2.5 years. The Southern Oscillation provides one of the few well documented associations between atmospheric circulation patterns and surface and deep ocean conditions. At present we can do little more than suggest that the occurrence of sea-surface temperature anomalies seems to coincide with disrupted atmospheric patterns around the globe and that Rossby wave positions appear to be very sensitive to such anomalies.

### Modelling oceanic circulation

An understanding of the processes relating atmospheric and oceanic circulations offers hope for the development both of seasonal weather forecasts and of predictions of the timing and location of oceanic upwelling, which constitute the major regions of oceanic biological productivity. However, links between the atmospheric state and circulation and the ocean state and circulation are very difficult to observe and still more difficult to model. These difficulties arise because the oceanic response to any changes tends to be slower than that of the atmosphere. Thus an oceanic change may have been triggered by an anomaly, say in cloud cover or prevailing surface wind, which was persistent but very slight. Two modelling approaches are being pursued to try to simulate and understand such atmosphere–ocean linkages. The first specifies various sea-surface temperature distributions so that their effects on the atmospheric circulation can be investigated. The other approach simulates the development of ocean currents by using an atmospheric model to drive an ocean model. Figure 4.28 illustrates the result of one such simulation of this type. This figure indicates that oceanic flow can be simulated with reasonable accuracy when the observed wind systems are used as input, but that we still have a long way to go when we try to link modelled atmospheric flow to ocean circulations.

### Satellite observations of the ocean surface

Any understanding of the links between the ocean and the atmosphere depend, of course, on adequate observations of both. Of particular importance is the specification of sea-surface temperature. Until very

189

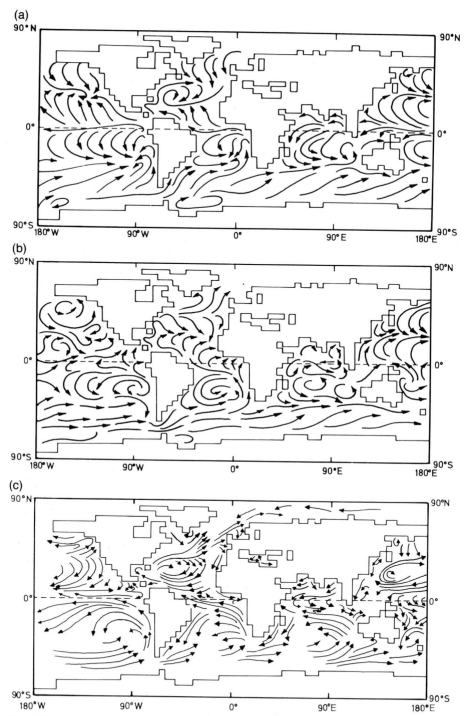

Fig. 4.28. Comparison of the oceanic circulation resulting from a coupled atmosphere-ocean model driven by: (a) observed atmospheric flow; and (b) the computed atmospheric flow from a general circulation climate model; (c) illustrates the observed oceanic circulation in January. (Perry, A.H. and Walker, J.M., 1977, *The ocean–atmosphere system*, Longman).

recently the only methods of measuring such temperatures were the use of fixed and drifting buoys or direct measurement by ships. Buoys, whilst likely to produce accurate results, are difficult and expensive to deploy and monitor. Measurements made from ships are, naturally, only available at the locations of the ships. Thus they only include the major shipping routes and even then regions and times of bad weather are excluded. Furthermore, ship observations are usually made in the intake area of the engine and are therefore temperatures of the ocean at a depth of approximately 5 m, rather than true water surface temperatures.

Satellite observations of the world's oceans have no such restrictions. The ocean is probably the simplest surface to which remote sensing techniques are applied. This is because the albedo, temperature and emissivity vary little over large areas. Unlike buoy and ship data, the satellite-derived measurements of ocean temperatures are homogeneous in space and regular in time. Figure 4.29 shows the distribution of sea-surface temperature for November 1983 derived by the multi-channel sea-surface temperature method. In this method, three channels, with wavelength ranges 3.55–3.93 $\mu$m, 10.3–11.3 $\mu$m and 11.5–12.5 $\mu$m, are

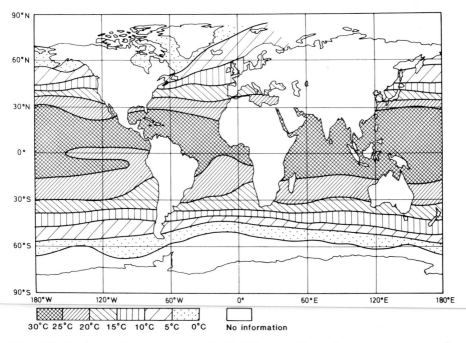

Fig. 4.29. Ocean surface temperatures derived from satellite radiance measurements for November 1983. These satellite retrievals have been shown to be in excellent agreement with more traditionally observed temperatures and, in addition, provide an homogeneous global data source.

used. The latter two channels sense only thermal infrared radiation whilst the 3.7 μm channel senses a mixture of reflected solar radiation and emitted thermal infrared radiation during the day (e.g. Figs 2.4 and 2.5 in Ch. 2), but only emitted radiation at night. Empirical equations have been established from which sea-surface temperatures can be derived accurately for both the daytime and nighttime overpasses of polar orbiting satellites. These equations are

$$\text{ocean temp. (day)} = 1.0346\ T_{11} + 2.5\ (T_{11}-T_{12}) - 283.21 \qquad [4.7]$$

$$\text{ocean temp. (night)} = 1.0170\ T_{11} + 0.97\ (T_{3.7}-T_{12}) - 276.58 \quad [4.8]$$

where $T_{3.7}$, $T_{11}$, $T_{12}$ are the temperatures (K) calculated for the three wavelength channels listed above. Results from equations [4.7] and [4.8] have been used in the production of Fig. 4.29. Such satellite-derived ocean surface temperatures can be compared with buoy data. In general the distribution of temperatures is very similar and the average differences are ~0.2 K between satellite retrievals and drifting buoys and ~0.5 K for moored buoys. It is clear that data as accurate as these which are available continuously for all the world's oceans will permit a much better assessment of oceanic climatology in the future.

### 4.7.2 Cryosphere
The Earth's cryosphere consists of snow on continents and ice over both land and sea. The existence and persistence of the cryosphere depends upon subfreezing temperatures and thus it occurs mainly at high latitudes and at high altitudes. In addition there must be adequate precipitation to maintain the snow and ice supply (Fig. 4.30). The observed snow line in equatorial regions is lower than in the subtropics probably because of the extra precipitation in the former.

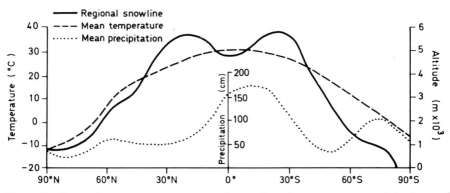

Fig. 4.30. Generalised curves showing the fluctuation of temperature, precipitation and regional snow line with latitude. Note that in the equatorial zone the snow line falls, despite increased temperature, apparently because of increased precipitation.

The perennial cryosphere covers 8% of the Earth's surface (Fig. 4.31). The very large seasonal fluctuations in the climate of the high latitudes are underlined by the fact that the seasonal cryosphere covers an additional 15% of the surface in January and an additional 9% in July. Antarctica has a complete cover but the Arctic Ocean, although frozen all year, has an ice mass which is not complete. It consists rather of numerous large ice floes, with an average thickness around 4 m, which are in continuous motion within the Arctic Basin. The year to year fluctuations in seasonal sea-ice extent are considerable in both hemispheres. For example, an anomalous southward extension of the Arctic ice between Iceland and Greenland in 1968 permitted polar bears access to Iceland for the first time in over 50 years.

The cryosphere affects the general circulation because its cold surface has a stabilising effect on the atmosphere. This is really significant only in polar regions, where the cryosphere has its major extent. Thus the poles tend to be high pressure regions with low level inversions and relatively calm conditions.

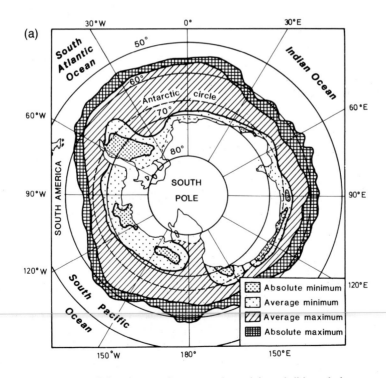

Fig. 4.31. The extent of the present-day cryosphere (a) and (b) and the extent of the Northern Hemisphere cryosphere at the height of the Pleistocene (c).

193

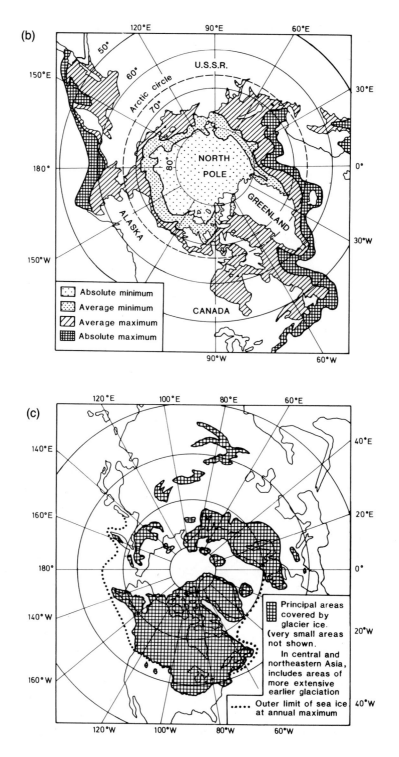

## Cryospheric changes as a trigger for climatic changes

The cryosphere has a high surface albedo and thus makes an important contribution to the planetary albedo. It is possible that the sudden onset of glacial periods may be due to what is termed 'the ice–albedo feedback mechanism'. The hypothesis is that a trigger, such as a fluctuation in solar luminosity, could cause an increase in overall glacial extent. If it is assumed that the areas which remain unglaciated do not suffer a significant change in cloud amount, the resultant increase in mean global albedo may lead to a new, colder climate which is sufficiently stable to persist for a long period of time.

## Present-day cryospheric variations

Most current concern, however, is with the opposite effect, resulting from a possible global warming. The West Antarctic ice sheet is a massive ice shelf grounded below sea level. At present the midsummer temperatures at the outer limits of the ice sheet are only a few degrees below freezing point. Thus there is the possibility that the effects of increasing $CO_2$ in the atmosphere could lead to an ocean temperature increase great enough to begin to melt this ice sheet. The immediate consequence would be to raise world sea level by some 5 m.

### 4.7.3  Continents

The continental surface is characterised by a vast array of topographic features and surface types, all of which, to varying degrees, influence the general circulation of the atmosphere. The most obvious and pervasive influence arises because of the contrast in thermal properties between land and sea leading to the continentality effect. This contrast can readily be illustrated by displaying a global map of temperature but omitting the coastlines (Fig. 4.32). The thermal contrasts created give rise to pressure contrasts, which influence the secondary circulation features of the general circulation and eventually create distinct regional variations in continental climates.

A generalised scheme of continental climate regions can be sketched (Fig. 4.33) using the information we have already established about latitudinal energy budgets (and hence temperatures); evaporation and precipitation and geostrophic wind direction. The major features of each zone are listed in Table 4.1. The isopleths of temperature are displaced from zonality by the effects of continentality. The dry regions reflect continentality effects in the continental interiors, being far from moisture sources. The west coast deserts are areas under the descending limb of the Hadley cell.

These regional climates are further differentiated as a result of the differences in the vertical fluxes associated with land and water. The Indian Ocean is typical of the tropical oceans in that about 90% of the available energy is used for evaporation. In arid land regions, such

195

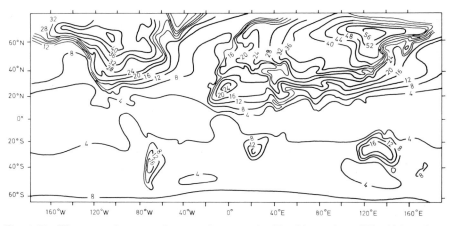

Fig. 4.32. The annual range of temperature at the Earth's surface (K). Although no land–sea boundaries are depicted on this map, the positions of the continents are clearly apparent.

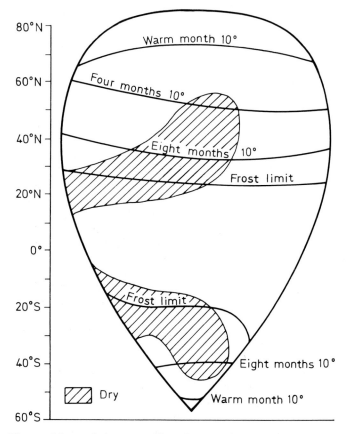

Fig. 4.33. The positions of the controlling features (temperature and aridity) of the regional climate on a stylised continent stretching from the North Pole to 60°S.

**Table 4.1  The major climatic zones**

| Climatic zones | Latitude | Description |
|---|---|---|
| High polar zone | 80–90°N, 70–90°S | Little precipitation all year round, variable winds, predominantly east. |
| Subpolar zone | 60–80°N, 55–70°S | Thick cloud and heavy precipitation all year round, strong, variable winds. |
| Temperate zone | 40–60°N, 35–55°S | Varying clouds and moderate precipitation all year round, prevailing west winds. |
| Subtropical winter rainy zone | 30–40°N, 30–35°S | Predominantly in western region of continents – fair weather in summer with weak winds – varying clouds and precipitation in winter, prevailing west winds. |
| Subtropical dry zone | 20–30°N, 20–30°S | Absent in eastern portion of continents; fine all year round with weak dry winds. |
| Tropical summer rainy zone | 10–20°N, 5–20°S | In summer cloudy with showers, variable to west winds. Fair and dry winter trade winds from the east (monsoon region). |
| Equatorial rainy zone | 5°S–10°N | Very cloudy all year round, thunderstorms, weak west winds. |
| Monsoon | – | During Northern Hemisphere summers, maximum heating in Northern Hemisphere drives ITCZ northward, brings low pressure northward; during Northern Hemisphere winters, maximum heating in Southern Hemisphere drives ITCZ southward and thus descending motion takes over with fair weather. Additionally, colder continental surface stabilises air in winter, high pressure builds, drives low pressure zone southward. In summer the reverse occurs. |

as part of Asia and Australia, most of the energy goes directly into warming the air. In the moist mid-latitudes and the tropical jungles both sensible and latent heat are removed from the surface, but most of the available radiative energy is used for evapotranspiration. Finally in the polar regions the average energy flux is from the air to the surface in the form of sensible heat flow. In each case the fluxes influence the amount of clouds and, to some extent, the amount of precipitation in the area. They are all, of course, influenced by the horizontal air motions mentioned earlier. These motions are themselves influenced by topography, which varies from continent to continent, leading to different distributions of climate regions over each continent.

Thus the effects of the various types of surface boundary for atmospheric motions are combined with the wind flow patterns and pressure distributions dictated by the general circulation of the atmosphere to produce distinctive patterns of climate over the surface of the Earth. These patterns are discussed in the following chapter.

## *Summary*

In this chapter we have considered the movement of two quantities fundamental to the climate system: energy and water. By combining the requirement to conserve angular momentum in the rotating frame provided by the Earth, the poleward and equatorial flow of water from tropical regions and the overall poleward flow of energy have been shown to give rise to mass fluxes which are strongly meridional in low latitudes and highly zonal in mid-latitudes. These four fluxes are summarised in Fig. 4.34. Thus from a knowledge of the zonal net imbalance of energy and water vapour, and incorporating the effect of the rotation of the Earth, we have been able to construct a theoretical picture of the general circulation of the atmosphere in accordance with the considerable array of surface and satellite observations and with the constraints placed on the system by the necessity to retain balances of the fluxes. Longer term climatic variations have been seen to be linked to ocean temperature and circulation patterns, and the possibility of climatic change, perhaps in response to cryospheric changes, has been noted. The interlinked role of the general circulation and the continental surface in producing regional climates has also been introduced. This will be amplified in the next chapter.

---

Fig. 4.34. Summary of the fluxes associated with the general circulation of the Earth's atmosphere. Diagrams show streamlines of zonal mean flow of (a) energy, (b) relative angular momentum, (c) mass and (d) water substance averaged for the two seasons December/January/February and June/July/August. There is seen to be a net energy deficit in the December–February period which is balanced by a positive balance during the June–August period. The seasonal movement of the ITCZ and the Hadley cells can be particularly well seen in the diagram showing the mass flux.

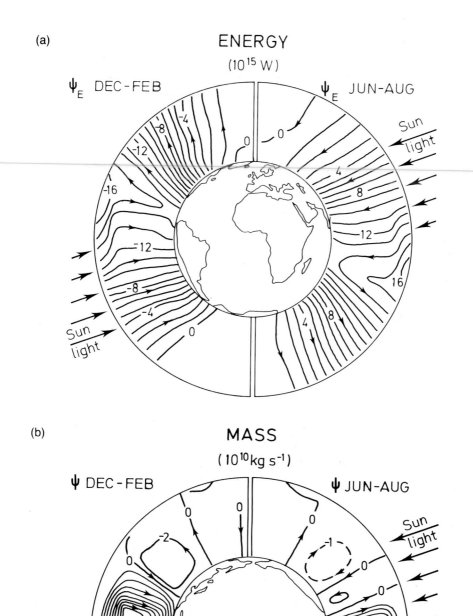

(a)

ENERGY

($10^{15}$ W)

$\Psi_E$ DEC-FEB     $\Psi_E$ JUN-AUG

Sun light

Sun light

(b)

MASS

($10^{10}$ kg s$^{-1}$)

$\Psi$ DEC-FEB     $\Psi$ JUN-AUG

Sun light

Sun light

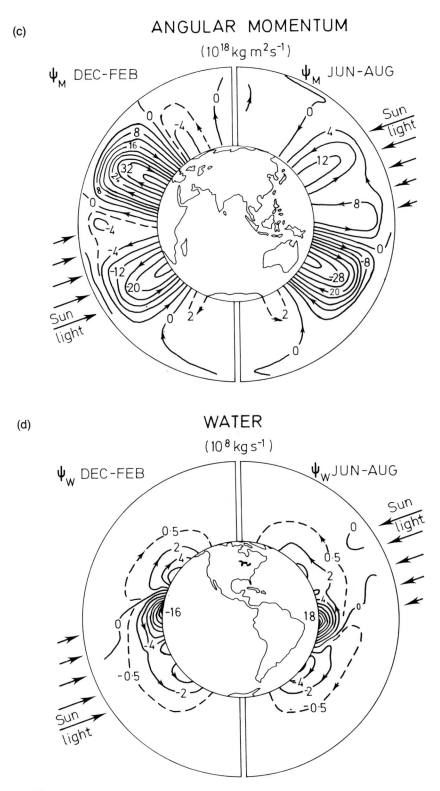

(c)

# ANGULAR MOMENTUM

$$(10^{18} \text{kg m}^2 \text{s}^{-1})$$

$\psi_M$ DEC–FEB        $\psi_M$ JUN–AUG

Sun light

Sun light

(d)

# WATER

$$(10^8 \text{kg s}^{-1})$$

$\psi_W$ DEC–FEB        $\psi_W$ JUN–AUG

Sun light

Sun light

# Chapter 5
## Regional Climates

5.1   Climate classification

   5.1.1   Development of thunderstorm regions for the United States: an example of climatic classification

   5.1.2   Methods of classification of climate

5.2   The empiric approach

   5.2.1   The Köppen classification

5.3   Tropical climates – the Hadley cell

   5.3.1   Hurricanes

5.4   Tropical climates – the monsoon regime

5.5   Mid-latitude weather

5.6   Mid-latitude climate regions

5.7   Polar climates

5.8   Changes in regional climates

# Chapter 5
# Regional Climates

The surface of the Earth can be divided into a series of regions which have similar climates. In this chapter we present a quantitative description of these regional climates, and combine the information with that obtained from the process orientated approach of the previous chapter, to develop a *regional climatology* of the Earth. The objective of such a climatology is to develop succinct descriptions of climatic conditions likely to be encountered at any point on Earth, together with an explanation of their causes and indications of their stability and variability with time. This serves two major purposes. Firstly, the information is useful for anyone with more than a passing interest in a particular place. It can provide information, for example, pertinent to the types of crops which could possibly be grown, or the heating and cooling requirements of housing in the area. Thus it provides an estimate of the climatic 'resources' of an area. Secondly, the identification of regions is also important for climatology itself. If patterns in the spatial distribution of regions are found they may provide insight into the processes that are acting to create those regions. Furthermore, examination of these processes can lead to estimates of whether a particular region is likely to be modified over time as the result of natural or anthropogenic changes to the atmosphere or surface.

It has long been recognised that climatic conditions have an impact on human activity. Climate regionalisation schemes have frequently served to provide the information needed to use climate as a resource. Many schemes have been developed over the years, either emphasising a particular set of climatic parameters perceived as important to a particular activity, or specifying in more general terms the overall world regional climate. The development of regionalisation schemes of both types has historically gone hand in hand with the understanding of climatic processes. The early recognition that several widely divergent areas on the Earth's surface have similar climates stimulated a search to identify the processes acting to create them. This search played a vital role in advancing our understanding of the nature of the general circulation of the atmosphere and thus stimulated study of most of the processes we have so far discussed in this book. Conversely, as understanding of the basic processes increased, we were able to refine our understanding and definition of the regions. This interaction is continuing. It is becoming increasingly possible to use satellite observations to specify processes, especially the energy budget, on a regional basis. As a consequence, we can start to investigate not only the controlling mechanisms for a particular region, but also the possibilities for change, whether natural or Man-made, in regional climates.

## 5.1   Climate classification

There are an infinite variety of climates over the Earth, every place being slightly different in some aspect from all others. Consequently the first step in developing a regional climatology must be to develop a classification scheme which allows us to identify major, or in some aspect 'significant', differences in climate. As with classification in any branch of science, such a scheme must aim to simplify and clarify the variations in order to enhance comprehension and understanding. The classification scheme produced automatically leads to the creation of a series of 'climate types'. Provided the scheme leads to a manageable number of these, they can be mapped to produce the climate regions. The major problem in developing a climate classification scheme is in defining climate. Many elements are involved. If only one is used, it hardly qualifies as a 'climate classification', although the areal distribution of a single element can provide much useful information. On the other hand, if we try to use all elements, the resulting complexity defeats the purpose of classifying for simplicity and clarity. Hence usually two or three elements are used.

The elements are usually chosen because they are perceived to be important in the context of the use to which the classification scheme is to be put. It is also within this context that the method of expression of the elements must be chosen. It is possible to express precipitation, for example, in terms of the number of rain days or the total rainfall, while various averaging periods, such as monthly or annual, could be used.

Once a decision about the elements and their method of expression is made, the next stage in developing a classification scheme is to identify threshold values which specify an important change in the impact of the parameter. It might be decided, for example, that a temperature of 18 °C was an important threshold, since above this temperature no residential heating is required. However, for a scheme emphasising agricultural applications, this value may have little meaning. Instead, a threshold of 5 °C may be more useful, since many plants commence growth once this temperature is reached.

Once the classification scheme, with appropriate threshold values, has been developed, the climate types are automatically established. Thereafter it is conceptually simple to develop a map of the resulting climate regions. Data for the whole area of interest are examined and standard cartographic techniques used to produce the regions. There are, however, several problems inherent in producing such a map. Climate is a spatially continuous variable but observations are available only for discrete points. Hence regional boundaries must be established by interpolation. The accuracy of such interpolation depends greatly on the density of the network of observation stations. Some land areas have a sparse network and any boundaries developed can only be approximate. Over the oceans the problem may be even more severe since most observations are at island stations which probably do not represent true ocean conditions. For some parameters, notably temperature, satellite observations are enabling us to develop oceanic climate regions with more confidence. The problem of drawing accurate boundaries is also acute in mountainous terrain. Here climatic conditions can vary dramatically within a short distance. Not only is there usually a paucity of observational information about these variations, but the rapid changes present challenges to cartographic representation. In many classification schemes this problem is avoided by specifying a category 'mountain climates', with a definition which emphasises the extreme rapidity of spatial variations in these areas. Finally, there is a problem associated with the boundaries themselves. In most cases there is not an abrupt boundary between climate types, but rather a transitional region, so that almost all boundaries that appear on maps of climatic regions should be interpreted as transitional zones.

### 5.1.1 Development of thunderstorm regions for the United States: an example of climatic classification

The development of a thunderstorm climatology is here used as an example of a method of developing a classification and regionalisation of a particular climatic element which could have specific applications. In addition, it indicates the climatological insight needed to develop a realistic regionalisation and also suggests some insights that can be gained as a result of the study.

The timing and frequency of thunderstorm occurrence is of significance in several segments of society. As already noted, for example, hail and its associated property and crop losses, is commonly associated with such storms. In addition, aviation operations are affected by them. In this case it is particularly important to identify when, as well as how often, they occur.

### Observation of thunderstorms

The time of beginning and ending of thunderstorms is recorded by approximately 450 observing stations of the US National Weather Service, and data are available for about 30 years. Thus it is possible to develop a climatology based on both time and frequency of occurrence. A simple distribution of frequency is easy to construct. Figure 5.1 shows such a distribution for total annual number of thunderstorms. Similar figures could be produced for the number of thunderstorm days on an annual, seasonal, or monthly basis, the choice essentially depending on the purpose of a particular study. A regionalisation of Fig. 5.1(a) is also relatively easy. In broad qualitative terms there is an absolute maximum around the Gulf Coast and relative maxima in the mountainous areas of the southern Appalachians and the Rockies. The central mid-west has high values, while the west coast and all northern areas have relatively low ones.

Thunderstorms can be generated by various mechanisms and so theoretically can occur anywhere at any time of the day. Nevertheless, consideration of the climatic characteristics of the various areas of the United States suggests that different mechanisms are likely to be dominant in different areas, leading to a spatial variation in the time of occurrence of the maximum number of storms. In the particular analysis illustrated here giving rise to Fig. 5.1(b) two types of threshold were set. The first was a value of total number of storms below which they are sufficiently infrequent to be ignored from the standpoint of,

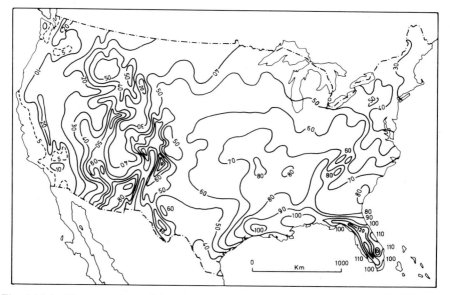

Fig. 5.1(a). Mean number of thunderstorms in a year for the USA.

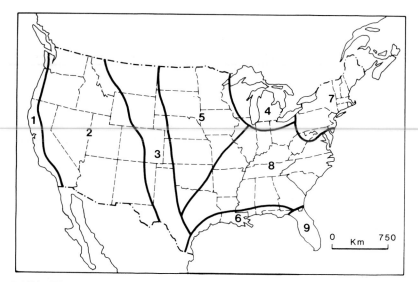

Fig. 5.1(b). The diurnal variability of thunderstorm activity in the conterminous United States (Easterling, D. R., 1984. M.A. Thesis, University of North Carolina, Fig. 16. p. 40).

*Key*

|  | 1 West coast | |  | 2 Montane and inter-montane west | |  | 3 Western plains | |
|---|---|---|---|---|---|---|---|---|
|  | Timing | Conc. |  | Timing | Conc. |  | Timing | Conc. |
| Spring | A | H | Spring | A | H | Spring | A/N | H/M |
| Summer | – | – | Summer | A/N | H | Summer | A/N | H/M |
| Autumn | A/N | L/M | Autumn | A | H | Autumn | A/N | H/M |
| Winter | A/N | M | Winter | – | – | Winter | – | – |

|  | 4 Great Lakes | |  | 5 Central US | |  | 6 Gulf coast | |
|---|---|---|---|---|---|---|---|---|
|  | Timing | Conc. |  | Timing | Conc. |  | Timing | Conc. |
| Spring | N | L | Spring | N | L | Spring | M/A/N | L |
| Summer | N | L | Summer | N | L | Summer | M/A | M/H |
| Autumn | N | L | Autumn | N | L | Autumn | M/A | M |
| Winter | N | L | Winter | – | – | Winter | N/M | L |

|  | 7 Northeast | |  | 8 Southeast | |  | 9 Florida peninsula | |
|---|---|---|---|---|---|---|---|---|
|  | Timing | Conc. |  | Timing | Conc. |  | Timing | Conc. |
| Spring | A | M | Spring | A | L/M | Spring | A | M |
| Summer | A | M | Summer | A | H | Summer | A | H |
| Autumn | A | M | Autumn | A | M | Autumn | A | H |
| Winter | – | M | Winter | N | L | Winter | A | L |

| | Timing | | Conc. |
|---|---|---|---|
| M | = Late morning | L | = Storms at any hour of day |
| A | = Afternoon | M | = Storms mainly at time of maximum |
| N | = Nocturnal | H | = Storms exclusively at time of maximum |
| – | = none | – | = few or no storms |

for example, aircraft operations. The second threshold separated those areas where storms were highly concentrated at a particular time of day from areas with a broad range of storm times. The resulting information could then be mapped and a small number of regions produced (Fig. 5.1(b)). It was found that the regions identified were stable throughout the year, but that the frequency and timing varied seasonally.

## 5.1.2   Methods of classification of climate

Using the approach to climate classification outlined above, it is obvious that classifications are possible on all time and space scales. However, the formal development of classifications and regions has traditionally been on the global or continental scale, producing regions approximately the size of a European nation or a US state. Similarly, there has usually been emphasis on the climatic 'normal' conditions, although year to year variations are frequently incorporated. It is this scale with which we are concerned in this chapter. Given the number of choices that have to be made in developing a classification, it should not be surprising that numerous classification schemes, even on this restricted scale, have been proposed. For convenience the approaches can be divided into two types:

genetic – relating to the origin of the features (emphasising atmospheric dynamics);

empiric – relating to the observation of features (emphasising use of observed climatic normals).

Each has particular strengths and weaknesses.

*Genetic* classifications are based on the activity and effects of the general circulation of the atmosphere. As such they emphasise the role of the climate controls in creating climate and its various regional expressions. By emphasising the factors causing climate they can be exceedingly useful in understanding climate and instrumental in investigating the nature and impacts of climatic change. Indeed, in a somewhat informal way, Chapter 4 provides the basis for this type of classification.

Emphasis on the dynamic nature of climate, however, makes it exceedingly difficult to produce a succinct summary of the climate of a region whilst retaining the dynamic flavour. It can succeed provided the user is already familiar with the way in which synoptic climatology translates into the more familiar climatic elements such as cloud amount, temperature, or precipitation. Even then, it provides no direct quantitative information about precipitation amounts or actual temperatures. Furthermore, regionalisation, in the sense of creating distinct boundaries between regions, is not appropriate for this type of classification.

The objective of an *empiric* classification scheme is to produce a quantitatively defined series of regions which specify the overall climate without emphasis on any specific application and without regard to the causes of the climate. Our discussion of classification above implicitly concerned empiric schemes. The parameters most frequently used are temperature and precipitation, or variations on these, such as evapotranspiration or soil moisture. The result is a strictly defined scheme based on two or three variables. This type of classification has a long tradition and the names given to some regional types, such as 'Mediterranean climate' or 'humid continental climate', have become widely known and evocative of particular conditions. Indeed, the major strength of this approach is the succinct way in which a great amount of information can be conveyed. In addition, reasonably clear regional boundaries can be established.

Empiric classifications use summarised data, such as monthly average temperature or monthly total precipitation, as the basic input. Hence these schemes do not deal with climate in its role as a collection of individual weather events. Furthermore, climate is restricted to two or three parameters, so that a whole range of climatic phenomena are not included.

The empiric and genetic approaches should be viewed as complementary. Although the regions defined by each may not coincide, when used together they give a quantitative indication of the average conditions for specific elements, the variations about the average likely to be encountered and the sequence of weather events that are likely to occur. Indeed, the approach we adopt in this chapter is first to introduce empiric schemes and then use one, together with the material from Chapter 4, as the basis for investigating the regional climates thus identified.

## 5.2   The empiric approach

Numerous empiric schemes for climatic regionalisation have been proposed. This is not surprising given the complexity of climate and the wide choice of parameters and threshold values available. Attempts have been made, for example, to develop classifications based on the surface energy and moisture fluxes considered in earlier chapters. The resulting regions emphasise the climatic controls operating in a particular area. However, because climate is a continuously varying phenomenon in time and space, it is extremely difficult to identify a set of threshold values which are appropriate for climate *per se*. Hence most schemes have been designed with a particular application in mind, so that the application largely dictates the threshold choices to be made. Furthermore, the application frequently dictates the complexity of the scheme. At one end of the scale are simple regionalisations based on a single parameter. These are rarely thought of as true classifications, but rather as regional maps for a specific purpose. Figure 2.34, showing the spatial variability of the length of the frost-free season, is an example with applications in agriculture. At the other end of the scale are schemes involving numerous parameters. One of the best known of these is the 'rational' classification created by Thornthwaite. He postulated that the surface water balance was the single most important characteristic of climate in any area, a contention which would certainly find support when agricultural productivity is the main concern. This water balance depends not only on precipitation and evaporation at a particular time, but also on their seasonal variability. This led to the creation of a 'moisture index' as one of the pertinent variables in the scheme. Significant threshold values were then derived using information similar to that contained in Fig. 3.27. The resulting classification is somewhat complex and is not very appropriate for arid areas. Hence it has not been extensively used on a global scale. Nevertheless maps of climate types for the mid-latitude continents convey in symbolic form a tremendous amount of information pertinent to agriculture.

A group of schemes intermediate in complexity between these two extremes has been developed following the work of Köppen, who first proposed a classification scheme in 1918. We shall explore this scheme in a little more detail, since it forms a very convenient framework for our division of the globe into climatic regions. In the detailed discussions of the various regions, however, we shall not be overly concerned with rigorous definitions of either the scheme or the boundaries of the resultant regions.

### 5.2.1  The Köppen classification
The Köppen classification is almost certainly the most widely known climate classification scheme. Originally devised as an aid to understanding worldwide vegetation distributions, it has been modified and generalised by several workers. Now, therefore, it is widely accepted as the apotheosis of empiric classifications, although it still betrays its application origins in the choice of parameters used.

Monthly and annual normals of mean temperature and total precipitation are the input variables. The scheme divides these into a series of categories, the boundaries of which represent some vegetation based threshold value (Table 5.1). Each region is categorised, in symbolic form, by a series of two or three letters. The first letter initially separates dry from moist climates and then, for the latter, divides them on the basis of temperature. A second letter then refines this to define the degree of aridity for the dry climates and the temporal distribution of

## Table 5.1

**Criteria for classification of major climatic types in modified Köppen system (based on annual and monthly means of precipitation in mm and temperature in °C)**

| Letter symbol | | | Explanation |
|---|---|---|---|
| 1st | 2nd | 3rd | |
| A | | | Average temperature of coolest month 18 °C or higher |
| | f | | Precipitation in driest month at least 60 mm |
| | m | | Precipitation in driest month less than 60 mm but equal to or greater than $(100 - r)/25$[a] |
| | w | | Precipitation in driest month less than $(100 - r)/25$ |
| B | | | 70% or more of annual precipitation falls in warmer six months (April through September in the Northern Hemisphere) and $r/10$ less than $2t + 28$[a] |
| | | | 70% or more of annual precipitation falls in cooler six months (October through March in the Northern Hemisphere) and $r/10$ less than $2t$ |
| | | | Neither half of year with more than 70% of annual precipitation and $r/10$ less than $2t + 14$ |
| | W | | $r$ less than one half of the upper limit of applicable requirement for B |
| | S | | $r$ less than upper limit for B but more than one half of that amount |
| | | h | $t$ greater than 18 °C |
| | | k | $t$ less than 18 °C |
| C | | | Average temperature of warmest month greater than 10 °C and of coldest month between 18 and 0 °C |
| | s | | Precipitation in driest month of summer half of year less than 40 mm and less than one third of the amount in wettest winter month |
| | w | | Precipitation in driest month of winter half of year less than one tenth of the amount in wettest summer month |
| | f | | Precipitation not meeting conditions of either s or w |
| | | a | Average temperature of warmest month 22 °C or above |
| | | b | Average temperature of each of four warmest months 10 °C or above; temperature of warmest month below 22 °C |

| Letter symbol | | | Explanation |
|---|---|---|---|
| **1st** | **2nd** | **3rd** | |
| | | c | Average temperature of from one to three months 10 °C or above; temperature of warmest month below 22 °C |
| D | | | Average temperature of warmest month greater than 10 °C and of coldest month 0 °C or below |
| | s | | Same as under C |
| | w | | Same as under C |
| | f | | Same as under C |
| | | a | Same as under C |
| | | b | Same as under C |
| | | c | Same as under C |
| | | d | Average temperature of coldest month below −38 °C (d is then used instead of a, b or c) |
| E | | | Average temperature of warmest month below 10 °C |
| | T | | Average temperature of warmest month between 10 and 0 °C |
| | F | | Average temperature of warmest month 0 °C or below |
| H | | | Temperature requirements same as E, but due to altitude (generally above 1500 m) |

[a] In formulae $t$ is the average annual temperature in °C, $r$ is average annual precipitation in millimetres

precipitation for the moist ones. A final letter is used to characterise the seasonal variations for mid- and high-latitude climates.

Simply as an example of this scheme, we can choose the 'Cfa' climate. In brief this is a mild humid climate with a hot summer but no dry season. More specifically, we can say that precipitation exceeds evaporation on an annual basis. From a botanical standpoint, this implies, to a first approximation, that there is sufficient moisture for tree growth as the natural vegetation. The 'C' climate is defined as one with the average temperature in the coldest month between −30 °C and 18 °C and at least one month with average temperatures above 10 °C. Thus there is a distinct summer and winter. The final 'a' indicates that this warmest month is, in fact, above 22 °C. The middle 'f' represents a climate where precipitation in the driest month exceeds 30 mm, so that there is no dry season.

Using this scheme it is possible to produce a map showing the major climate regions of the Earth (Fig. 5.2). Even a cursory examination of the map indicates that patterns of climate are repeated from continent to continent, reflecting the overall control that the general circulation places on the climate and following the basic pattern shown in

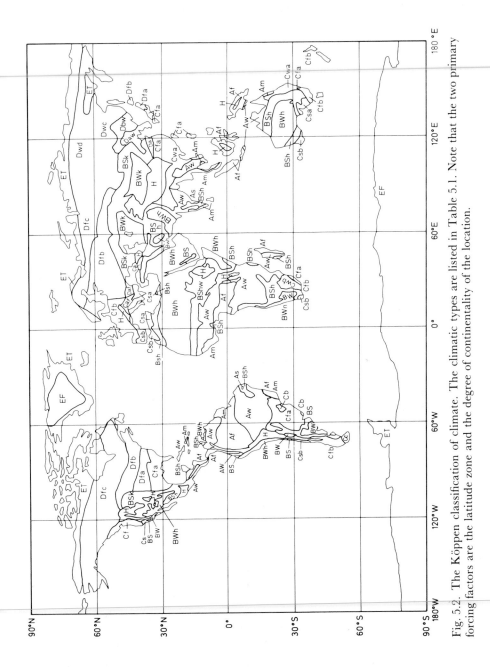

Fig. 5.2. The Köppen classification of climate. The climatic types are listed in Table 5.1. Note that the two primary forcing factors are the latitude zone and the degree of continentality of the location.

213

Fig. 4.33. The details differ between continents, of course, depending on the particular configuration and topography of each.

Using this type of empiric system thus gives the first feel for the type of climate to be experienced in any part of the world. Much information is conveyed, as indicated in Table 5.1 and as shown when the Cfa climate was considered above. This type of information is often just what is needed. We have all seen tourist brochures which quote monthly temperatures and precipitation. However, only part of the climate is considered. Questions such as 'What is the average number of rain days?', 'Does it rain as a continuous drizzle or as a short downpour?', 'Is the average temperature misleading because half of the days are hot, half are cold and none are near the average?' are not answered. They might be important when we are using tourist brochures to plan a vacation. Obviously these questions, and many others, could be answered from the surface-based observational record. To do so, however, increases the complexity of the classification, probably unnecessarily. Hence it must be remembered that an empiric scheme such as the Köppen system can convey a great amount of information, but it must also be recognised that there are limitations to the amount that can be given without confusion arising.

In the sections that follow we shall use the framework of the Köppen system to explore regional climates in more detail. Climatological data from the stations shown in Fig. 5.3 will be used to illustrate the climate types. However, we shall combine this empiric with a more genetic approach both to describe and explain those regional climates.

# 5.3 Tropical climates – the Hadley cell

Tropical climates are usually characterised as climates where there is no true temperature distinction between summer and winter. They normally occur between 30°N and 30°S. It is very convenient to divide this tropical area into two distinct, but interlinked, climatic types. Firstly, there are those regions where the Hadley cell circulation dominates for much of the year. These will be discussed in this section. Secondly there are the regions where monsoonal circulations are of overriding importance in some seasons. These will be the subject of the next section.

The basic Hadley cell circulation (Fig. 4.18) occurs in each hemisphere. Air flows equatorward at low levels in each cell. These airflows converge at the Intertropical Convergence Zone (ITCZ) and rise to create cloud and precipitation. The latent heat released during this ascent provides much of the energy needed to continue the whole cellular circulation. The air ascends towards the tropopause where it diverges and begins to flow poleward. As it flows, it cools by longwave radiation loss, increases in density and descends. Although the descent

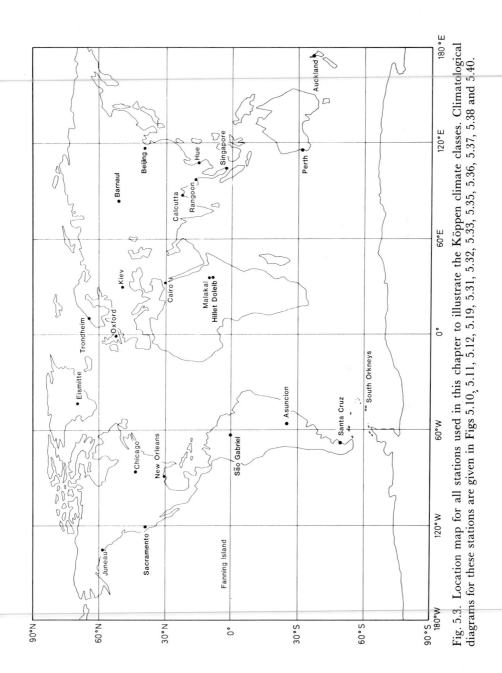

Fig. 5.3. Location map for all stations used in this chapter to illustrate the Köppen climate classes. Climatological diagrams for these stations are given in Figs 5.10, 5.11, 5.12, 5.19, 5.31, 5.32, 5.33, 5.35, 5.36, 5.37, 5.38 and 5.40.

covers a wide area it is particularly concentrated around 30° latitude. The descending air is adiabatically warmed, so that it arrives back at the surface as a dry, cloudless airstream. Once at the surface the air diverges, some flowing as the 'trade winds' towards the ITCZ to complete the cell. These surface winds, initially dry, become increasingly moist as the equator is approached. The trade winds are of course influenced by the Earth's rotation, so that in the Northern Hemisphere the near surface flow is the northeast trades. The airflow aloft, known as the counter trades, is from the southwest. In the Southern Hemisphere the corresponding winds are the southeast trades and the southwest counter trades.

In the above brief review it is clear that the major area of cloud formation in the tropics is around the ITCZ, which is a region of instability. Virtually all of the rest of the tropics are subject to more or less persistent subsidence inversions in which warm descending air 'traps' somewhat cooler air right at the surface (Fig. 5.4). This feature is usually best developed, and closest to the surface, at the outer, poleward limits of the Hadley cells. However, it persists throughout much of the trade wind area, although rising to a higher level as the equator is approached. This feature is known as the *trade wind inversion*. Any tendency towards instability caused by surface heating can rarely overcome this dynamically created feature. Consequently the surface trade

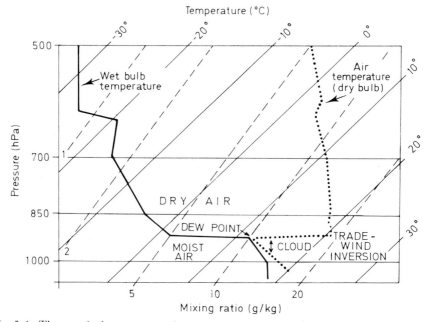

Fig. 5.4. The vertical structure of trade wind air at 30°N, 140°W at 0300 Z in July. This diagram is drawn on the thermodynamic diagram like that illustrated in Fig. 3.15. Note the very strong trade wind inversion.

winds are able to pick up a great deal of moisture as they flow over the ocean, but they cannot release it, or the associated latent energy, until they reach the ITCZ.

A small amount of cooling, however, can easily bring these moisture-laden winds to saturation. Fogs can occur when the air flows over a cool surface. Such fogs are common, for example, when the air passes over the cool ocean currents off the coasts of Peru and northern Chile and off the Namibian coast (Fig. 5.5), giving a frustrating sight for dwellers in these desert areas. Topographic barriers can also lead to orographic cloud and fog in the trade wind zone. However, significant vertical development will occur only when the barrier is sufficiently large to break through the inversion.

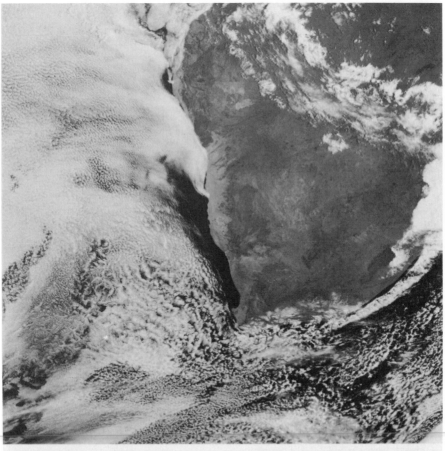

METEOSAT    1978 MONTH 10 DAY 11 TIME 0755 GMT (NORTH) CH. VIS 1/2
            NOMINAL SCAN/PREPROCESSED AREA B65    COPYRIGHT- ESA -

Fig. 5.5. Off-shore fog and low level stratus and stratocumulus adjacent to the Namibian coast (visible METEOSAT image for 11 October 1978). (METEOSAT image supplied by the European Space Agency).

Thus the only major precipitation forming mechanism of wide geographical extent in the Hadley cell is the ITCZ itself. The ITCZ occurs in the equatorial low pressure region, commonly called the *equatorial trough*. This has traditionally been characterised as a rather wide zone of cloudy, near calm conditions, the *Doldrums*, extending completely around the Earth. Satellite observations, however, have revealed that this region is one of gentle easterly winds, rather than calms. These observations have also shown that there is not a complete cloud band, but a series of fairly well developed cloud clusters separated by larger clear sky areas (Fig. 5.6) and that the cloud characteristics show great spatial and temporal variability.

METEOSAT   1979 MONTH  7 DAY  7 TIME 1225 GMT (NORTH) CH. VIS 1/2
NOMINAL SCAN/PREPROCESSED SLOT 25 CATALOGUE 1025010221

Fig. 5.6. METEOSAT image (visible channel) for 7 July 1979 showing clearly the position of the ITCZ identified by the band of cloud close to the equator over Africa and the adjacent Atlantic Ocean. (METEOSAT image supplied by the European Space Agency).

### Precipitation in the region of the ITCZ

Precipitation from the ITCZ clouds is almost entirely convective. It appears to be organised on two scales: small groups of convective cells which remain as small-scale features; and larger and better organised features which have the potential to become hurricanes. These latter tend to form in certain well defined areas whereas the former seem to be more random in their distribution within the ITCZ. The processes going on in the clouds of either type are very poorly understood. For example, it is often observed that heavy rain, or even hail at higher latitudes, is produced by some cells, but it is unclear how such large droplets can be produced in the environment of the cloud. Recent theoretical studies of vigorous cumulus growth suggest that an important mechanism is the initial growth of snowflake particles in an adjacent 'feeder cloud' which are then transported into the main cloud (Fig. 5.7). The veracity of this model is currently being tested by observation.

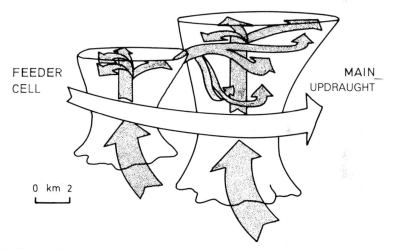

Fig. 5.7. Schematic representation of the hail formation process. Hail 'embryos' are generated in a feeder cell and transferred to the storm's main updraught core where rapid growth occurs.

### Easterly waves

Within the area of the ITCZ there are no major synoptic scale features comparable to the depressions of the mid-latitudes, largely because the Coriolis force is weak and cannot generate circular motions. Nevertheless, in certain areas organised flow patterns, known as *easterly waves*, have been identified (Fig. 5.8). (Such a feature can also be seen in Fig. 5.14.) These are troughs of low pressure, weakly defined at the surface but extending some distance into the atmosphere sloping away eastward with height, which slowly move westward. Behind the trough

219

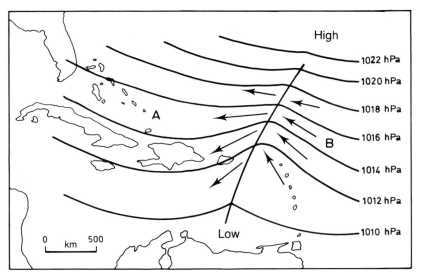

Fig. 5.8. Pressure distribution in an easterly wave in the Caribbean. Winds blowing from the direction of B converge towards the wave; they diverge in the region of A. (This feature is illustrated in the geostationary satellite image in Fig. 5.14.)

cumulonimbus clouds with intense rain and thundery showers occur. The origin of these waves is difficult to trace, but they seem to originate over the Caribbean when the trade wind inversion is weak. This weakness occurs during the period of maximum surface heating in summer and autumn, and the inversion can then be relatively easily broken. Further, it has also been suggested that they occur as the result of the penetration into the area of mid-latitude cold fronts, which also tend to be farther south in these seasons. There also appears to be a link between these waves, organised cloud clusters and hurricane formation. Again the links are unclear and further investigation is required.

### Seasonal movement of the ITCZ

The position of the ITCZ varies seasonally in rough accordance with the location of maximum solar heating. Thus, in July the ITCZ is likely to be around 25°N over the Asian continental interior and 5–10°N over the oceans, reflecting the continentality effect in the Northern Hemisphere. As a result of the relatively small amount of land in the Southern Hemisphere the average January position is at about 15°S over land and close to the equator over water (Fig. 5.9). Since this feature is effectively the only rain-producing system for most of the tropics, this seasonal variation is highly significant. Near the equator the ITCZ is always fairly close so its influence is felt in most months. Thus there are no dry periods. The detailed seasonal distribution may, of course, be influenced by the exact position of the ITCZ as well as by local topography. For example, the distribution at

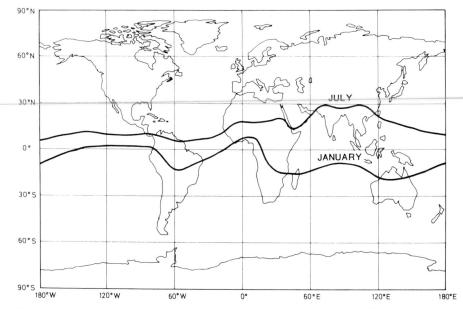

Fig. 5.9. Mean positions of the Intertropical Convergence Zone (ITCZ) in January and July

Fanning Island, a low lying island in the Pacific Ocean, reflects the dominance of the nearly overhead ITCZ in the Northern Hemisphere winter (Fig. 5.10(a)). In contrast, Singapore (Fig. 5.10(b)) has a more even precipitation distribution, partly as a result of orographic effects. At both stations temperatures remain high throughout the year, and the resulting climate type is Af.

Away from the equator, rainfall is concentrated in the summer, when the influence of the ITCZ can be felt. The further one progresses polewards the smaller the total precipitation and the shorter the length of the rainy season. A coastal station such as Calcutta (Fig. 5.11(a)) has a relatively long season and high totals partly because of the proximity to the ocean and partly because of its position relative to the ITCZ. Temperatures remain high throughout the year, although its location at 23°N leads to at least an embryonic summer and winter. Thus this is an A climate, with the distinct wet and dry seasons creating an Aw climate. Malakal (Fig. 5.11(b)) falls within the same group. Being closer to the equator, the seasonal temperature changes are not as marked, minimum temperatures occurring during the cloudy conditions when the ITCZ is most nearly overhead. It has a very well developed wet/dry regime. Only in the summer with the ITCZ nearly overhead does significant precipitation occur. During the other seasons there are no mechanisms for releasing any moisture within the trade winds, although in this particular situation the trades,

221

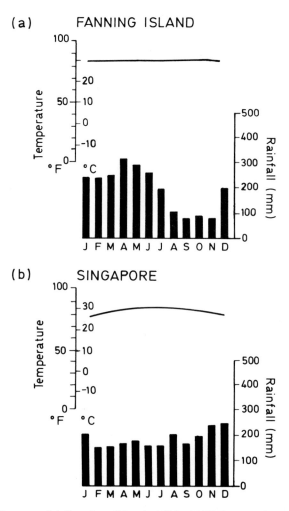

Fig. 5.10. Af climates. (a) Fanning Island (4°N, 160°W) and (b) Singapore (1°N, 104°E) (for location see Fig. 5.3).

having originated over the deserts to the north, are themselves rather dry. Indeed, Malakal is a station that is close to the desert margin, but it can still be regarded as an A climate since the summer precipitation is sufficient to maintain a positive annual water balance.

### Desert climates
Only a few kilometres away from Malakal, the station at Hillet Doleib (Fig. 5.12(a)) is classified as a desert climate, BSh. The atmospheric conditions of the two places are essentially the same; the only difference is a slight decrease in annual total precipitation at Hillet Doleib. This

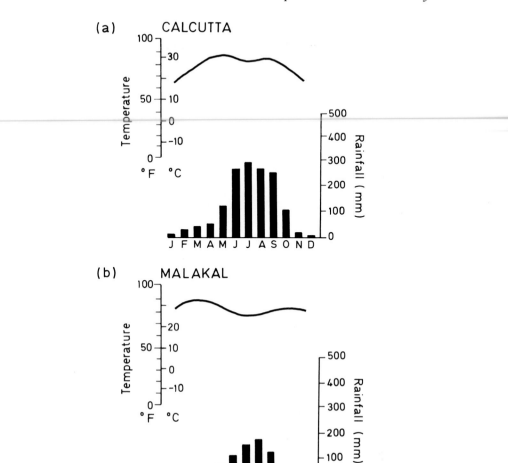

Fig. 5.11. Aw climates. (a) Calcutta (23°N, 88°E) and (b) Malakal (9°N, 31°E) (for location see Fig. 5.3).

is sufficient to create a net annual water balance which is negative, leading to the change in classification. Obviously the distinction here is more a function of the classification scheme and the rigid boundaries than a real change in climate between them. This underlines our earlier caution regarding the placement of faith in any one classification scheme. Certainly the outer limbs of the Hadley circulation, areas which are rarely influenced by the ITCZ, are desert areas. In the case of Hillet Doleib, there is some ITCZ associated rainfall and a BS climate results. As one moves poleward this influence continues to decrease and the true desert, BW, is encountered (Fig. 5.12(b)). The deserts created by the descending limb of the Hadley cell are generally

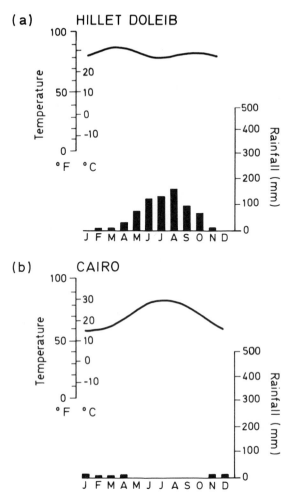

Fig. 5.12. Hot desert (BS and BW) climates. (a) Hillet Doleib (9°N, 32°E) and (b) Cairo (31°N, 31°E) (for location see Fig. 5.3).

hot deserts, BWh or BSh, and are concentrated around 30° latitude. Cold deserts, occurring farther poleward, may be spatially connected to the hot ones, but usually have a mechanism of formation which is not connected with the Hadley cell circulation.

As the precipitation amounts decrease as one moves away from the equator, so does the reliability of the precipitation. In the extreme case, a station such as Cairo (Fig. 5.12(b)) may record an average annual precipitation around 25 mm, but this is very misleading because there may be several years without rain followed by one which has a few intense storms and a rainfall of 100 mm or so. Within the wet/dry tropics receipt of precipitation depends on the movement of the ITCZ.

The amount of movement poleward, and the intensity of the zone as it moves, all vary from year to year. Thus there tends to be both a decrease in amount and an increase in year-to-year fluctuations as one moves away from the equator.

As yet the causes of these annual variations are by no means clear. Some insights are being gained as we increase our ability to model the climate system and its variations, and as satellite observations provide more information about the energy exchanges affecting individual regions. Meanwhile problems associated with the ITCZ movement continue. Recently they have been particularly acute in the southern margins of the Sahara Desert in Africa. Since the early 1970s there has been a series of years when the ITCZ has not moved as far north as usual and rainfall has been substantially below normal in the Sahel region (Fig. 5.13). In particular those areas close to the desert margin, with climates similar to Malakal, have only received about 50% of normal rainfall. Thus the already marginal climate for agriculture, in this case mainly cattle grazing, became much more unfavourable, a drought was created and with it much human suffering. Recently a similar problem has arisen, for similar reasons, in the eastern part of Africa.

Although our analysis of the Hadley cell circulations implied an almost closed circulation, connections with mid-latitude airstreams must occur in both the area of convergence at the top of the outer, descending limb and in the divergence region at the surface. Thus the

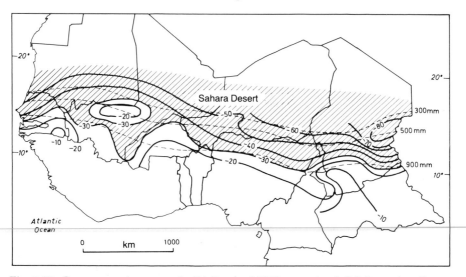

Fig. 5.13. Percentage departure (solid lines) of 1973 annual rainfall from the climatological normal over the African Sahel. Shaded region received less than 70% of the normal. Dotted lines are the mean annual isohyets.

225

climates resulting from the Hadley circulation do not occur in isolation from the rest of the globe. Interchange of air is a continuous process and specific connections between tropical and mid-latitude systems were cited in association with the Southern Oscillation/El Niño phenomenon (sect. 4.7.1). Two further connections require detailed treatment. The first, the monsoon circulations, cover a sufficient area to be treated in a separate section as distinct regional climates. The other is the hurricane, which originates in the tropics but frequently has a major impact on conditions in mid-latitudes.

### 5.3.1  Hurricanes

Hurricanes are intense circular vortices with winds over 34 m s$^{-1}$ (75 mph) spiralling around a low pressure centre (Fig. 5.14). Such systems are known as 'hurricanes' in the Caribbean and the Gulf of Mexico regions, 'typhoons' in the western north Pacific and 'cyclones' in the Bay of Bengal. 'Hurricane' has however become the accepted general name in meteorology. Similar storms having speeds below 34 m s$^{-1}$ are known as 'tropical storms'. The major characteristic of hurricanes is the strong circular flow around a low pressure centre (Fig. 5.15). Right in the centre is a calm region known as the 'eye'. Surrounding this is a bank of convective cloud, usually twisted by the strong winds into a piled mass of cumulonimbus cloud through which air ascends in a spiral motion. Adjacent to the eye this cloud bank is a continuous mass called the 'eye wall'. Additional convective bands occur further from the eye. These are also twisted into a spiral configuration by the winds. When the air from each of these ascending regions reaches the tropopause, it moves outwards, creating a veil of cirrus which may extend several hundred kilometres radially out from the hurricane centre. The whole feature thus has a very characteristic shape (Fig. 5.14).

### *Formation of hurricanes*

Hurricanes have their origin over the tropical oceans (Fig. 5.16) at latitudes between about 7°–15°. Equatorward of 7° the Coriolis force is too weak to initiate a circular motion, whilst poleward of 15° sea-surface temperatures are too low. Sea-surface temperatures greater than 27 °C (300 K) are required to ensure sufficient evaporation to provide the latent heat release within the storm needed to maintain its energy. It is usually in the late summer and early autumn that sufficiently high temperatures are achieved, creating a distinct 'hurricane season'.

### *Development of hurricanes*

The development of an organised cluster of tropical convective cells (Fig. 5.6) appears necessary for hurricane formation. These must be

Fig. 5.14. GOES East satellite visible channel image for 8 August 1980 at 1700 Z showing hurricanes Allen in the Gulf of Mexico and Isis to the west of Mexico in the eastern north Pacific. The concentric cloud pattern spiralling about the central eye is clearly discernible in both hurricanes. Allen was a particularly intense hurricane with aircraft measurements of wind speeds higher than 80 m s$^{-1}$. Note also the area of activity in the eastern Caribbean which is associated with an easterly wave (see Fig. 5.8). (*Weather Satellite Picture Interpretation*, Ministry of Defence).

organised in such a way that a distinct low pressure centre can be established over a sea surface. The low surface friction of the ocean prevents significant cross-isobaric flows, which would tend to fill in the central low pressure, but at the same time allows sufficient such flow to maintain the vertical motions within the storm. This central low frequently falls as low as 880 hPa. In fact, the lowest sea-level pressure ever observed was in the centre of typhoon 'Tip' on 12 October 1979, when a central pressure of 870 hPa was recorded.

227

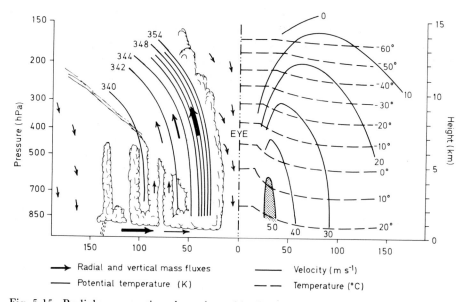

Fig. 5.15. Radial cross-section through an idealised axially symmetric hurricane. On the left the radial and vertical mass fluxes are indicated by arrows and equivalent potential temperature (K) by solid lines. On the right, isopleths of tangential velocity (m s$^{-1}$) are shown as solid lines (i.e. the speed of rotational flow around the 'eye') and temperature (°C) by dashed lines.

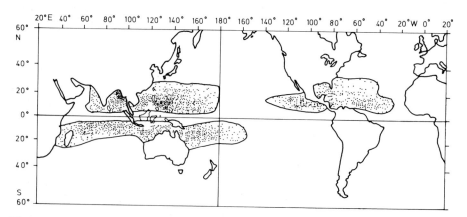

Fig. 5.16. Geographical distribution of the points of origin of hurricanes in the 20-year period, 1952–71. Each dot represents the first reported location of a storm which subsequently developed sustained winds in excess of 20 m s$^{-1}$. About two thirds of these storms eventually developed winds of hurricane force (>34 m s$^{-1}$).

From their source regions the storms usually start to move in a westward or northwestward direction (southwestward in the Southern Hemisphere). They contain sufficient energy to cut across the poleward limb of the Hadley cell and move into middle latitudes. Once there they tend to recurve under the influence of the prevailing westerlies and move towards the northeast (or southeast in the Southern Hemisphere). In theory, as they move away from their warm source waters they rapidly lose the energy source needed to sustain them. This should become particularly acute when they move over land, with its lower moisture supply and increased surface friction. In practice, although this is basically the case, they may travel a great distance, in a highly unpredictable direction, over mid-latitude areas (Fig. 5.17).

The number of hurricanes that actually form in any source region and move away is only a small fraction of the potential 'hurricane forming' situations there. Indeed, the identification of the conditions favourable for the formation of a hurricane which will move from its source region remains as a major meterorological problem. Hurricanes that do leave are usually named. The idea of naming them sequentially on an alphabetical basis originated in the Second World War, but it was not until 1978 that there was international agreement to use male as well as female names for these intense storms. The ocean source areas clearly differ since Atlantic hurricanes rarely get to the letter M (13 a year), whereas in the eastern Pacific the letter P (16 storms) is frequently used and in the west Pacific the end of the alphabet is often reached.

### *Effect of hurricanes*

Hurricanes are feared because of their destructive capabilities. The pressure gradients around the eye of the storm frequently produce sustained winds in excess of 50 m s$^{-1}$ (100 mph). This can have a direct destructive effect (Fig. 5.18). Equally, or more, destructive in coastal areas are storm surges resulting from local uplift of the sea surface under the central low pressure. Finally, the rainfall rate associated with the rapid vertical air motions often exceeds 0.03 mm s$^{-1}$, and the total rainfall as a storm passes overhead may be in excess of 1000 mm, with the associated possibility of flash flooding in many mountainous areas.

Although we usually think of hurricanes as agents of destruction, it must be remembered that they play a vital role in maintaining the general circulation and hence the present climate system. They are extremely efficient in moving excess energy away from the tropical regions towards the poles. Without this, there would either be a marked change in climate worldwide, or some other feature would be developed by the atmosphere to perform this transfer function.

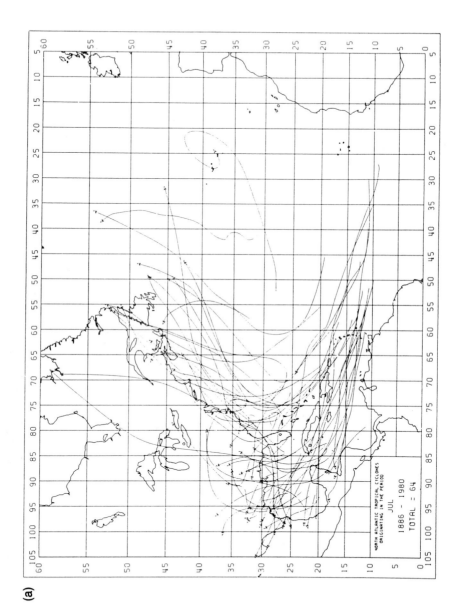

NORTH ATLANTIC TROPICAL CYCLONES
ORIGINATING IN THE PERIOD

JUL
1886 - 1980
TOTAL = 64

(a)

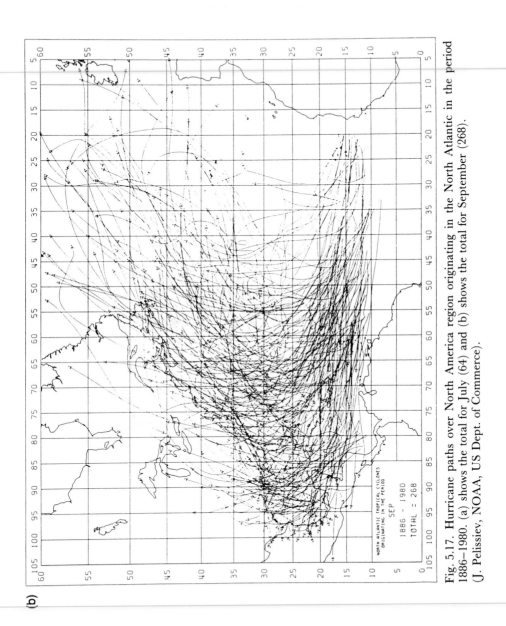

Fig. 5.17. Hurricane paths over North America region originating in the North Atlantic in the period 1886–1980. (a) shows the total for July (64) and (b) shows the total for September (268). (J. Pelissiev, NOAA, US Dept. of Commerce).

NORTH ATLANTIC TROPICAL CYCLONES
ORIGINATING IN THE PERIOD

SEP

1886 – 1980
TOTAL = 268

(b)

(a)

(b)

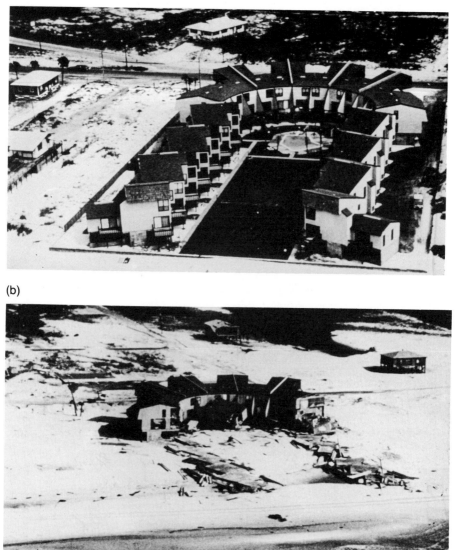

Fig. 5.18. Hurricane destruction: the Roundtowner Motel in Panama City Beach, Florida (a) before and (b) after it was hit by hurricane Eloise in 1975. (Courtesy: National Weather Service).

## 5.4 Tropical climates – the monsoon regime

Monsoon climates can be characterised as those where seasonal climate changes result from seasonal changes in the wind regime. In particular, the major monsoon areas exhibit twice yearly reversals in the prevailing wind direction. These lead to distinct wet and dry˙ seasons (Fig. 5.19). Although these graphs look similar to those of Fig. 5.11, the important difference is the amount of precipitation in the wet season. The very large amounts prevent any suggestion that these are dry climates, although there may be very definite dry periods. The similarity of the diagrams masks a further difference. The wet season for the Aw climates is one of the convective precipitation associated with the nearby ITCZ within the general Hadley circulation. For the monsoon (Am) climates the differing wind regimes lead to changes in all elements of the climate. The wet season is one of hot, on-shore humid winds and extensive cloud decks, while the dry season tends to be somewhat cooler with low humidity and off-shore winds. For Rangoon (Fig. 5.19(a)), temperatures are relatively low during the monsoon rains, largely because of the presence of clouds, and during the dry season, mainly because of the cooler winds. Temperatures are at a maximum at the intermediate seasons, when neither of these regimes is fully established. At Hue (Fig. 5.19(b)) there is no clear dry season, orographic effects providing some rainfall throughout the year. Overall temperatures are lower than at Rangoon, at the same latitude, and there is a distinct seasonal temperature trend created not so much by the wind and cloud conditions as by the solar altitude variations.

### Monsoon circulation

The prime cause of monsoon circulations is the thermally direct cell resulting from surface temperature differences (Fig. 5.20). This type of circulation can only become well established in approximately barotropic conditions and where there is a favourable distribution of land and sea. Thus, although a summer monsoon type of circulation will be developed over all tropical regions, it is in most areas overshadowed by other features and may not be accompanied by the necessary wintertime reversal of wind direction required for a true monsoon regime. In general it is only the continent of Asia that displays large areas of monsoon climate. Most of the western hemisphere is unaffected, although there are some indications of seasonal wind reversals near the mouth of the Amazon in South America. Similarly, only small portions of the west coast of Africa display a monsoon climate, although the whole tropical portion of the continent is potentially part of the regime (Fig. 5.21). Here, however, the Hadley cell dominates the circulation patterns.

### The Asiatic monsoon

The Asiatic monsoon is itself the result of a complex interaction between the distribution of land and water, topography and tropical and mid-latitude circulations. The simple model of Fig. 5.20 provides a good first approximation to the summer circulation pattern, with low

233

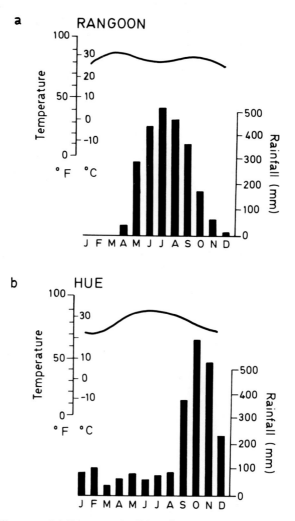

Fig. 5.19. Am climates. (a) Rangoon (16°N, 96°E) and (b) Hue (16°N, 108°E) (for location see Fig. 5.3).

pressure centres over the northern part of the Indian subcontinent and northern southeast Asia. Warm moist air is drawn into the thermally created low pressure areas of the continental interior where it rises, releasing both precipitation to create the wet season and latent heat to provide the energy necessary to continue the system. Over the oceans is a compensating descent of cold dry air. Once this pattern is established the on-shore winds bring the monsoon rains. The moisture laden winds are highly susceptible to orographic influences, so that the coastlands of India for example, which are backed by mountain ranges, receive the most rainfall. The interior, of course, also receives rain

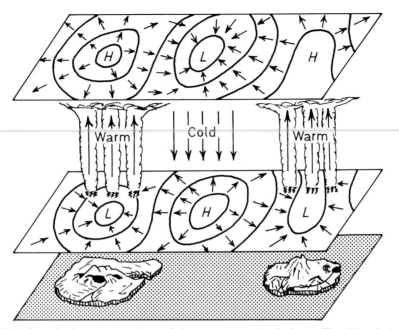

Fig. 5.20. Idealised representation of the monsoon circulations. The islands in the figure represent the tropical continents in the summer hemisphere. Solid lines represent isobars or geopotential height contours near 1000 hPa (lower plane) and 14 km or ~200 hPa (upper plane). Short solid arrows indicate the cross-isobaric flow. Vertical arrows indicate the sense of the vertical motions in the middle troposphere.

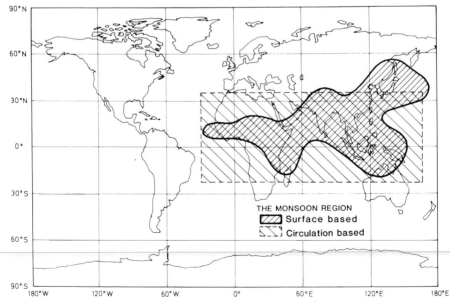

Fig. 5.21. The monsoon region: the areas of monsoon climate differ depending on whether the area is defined by surface-based observations or by considerations of the atmospheric circulation patterns.

235

since the whole area is one of rising air. The Himalayas effectively provide a barrier to the north, confining the circulation to the area south of them.

The detailed pattern of climates associated with this summer circulation is, however, much more complex. The Tibetan Plateau, at a **height exceeding 4000 m above sea level, provides a high level heat** source which appears to have a significant influence. Above the Plateau is a high pressure region, which is part of the zonal subtropical high pressure belt. As the equatorial trough associated with the ITCZ moves to about 25°N over India at the height of the summer, a strong north to south pressure gradient is established. This is reinforced by a north to south temperature gradient and together they lead to the development of an easterly jet stream at a height of about 150 hPa (15 km). This jet extends from the South China Sea across southeast Asia, central India and the Arabian Sea into the southeastern Sahara of Africa. The dynamic effects associated with this jet lead, for example, to an enhancement of precipitation to the north of the main axis in southeast Asia and to its south over the Arabian Sea and the Horn of Africa. The strength and position of this jet, therefore, has a profound influence on the surface weather and climate of the area.

With the onset of autumn, two related general changes occur. Firstly, the thermal contrasts between land and water decrease, weakening the circulation. Secondly, the westerlies of mid-latitudes begin to migrate southwards. Eventually part of this westerly airstream blows to the south of the Himalayas and completely disrupts the tropical circulation. The easterly jet is replaced by strong westerly winds aloft. At the surface a north to south pressure gradient extending from Siberia almost unbroken to the equator is established, with a high pressure centred over Siberia. The resulting airflow near the surface is from the north or northeast. This air combines with that resulting from subsidence below the westerlies to ensure that dry air covers the region, leading to the relatively cold conditions as noted for Hue. Over much of India, air has blown downslope off the Himalayas. Although this has been warmed adiabatically, it still gives relatively cold conditions.

This type of pattern persists through the spring. As summer approaches the conditions for the summer monsoon circulation pattern are slowly developed. The onset of the actual circulation is retarded, however, by the continued influence of the westerlies south of the mountain barrier. As soon as the main line of the westerlies moves sufficiently far north that the flow south of the Himalayas ceases, the summer monsoon circulation starts. The very rapid establishment of warm moist airflow from the south creates the phenomenon known as the 'burst' of the monsoon. Cool, dry conditions are replaced by warm humid ones, with copious rainfall, almost overnight.

### *Timing of monsoon rainfall*

The time of the monsoon burst is thus highly dependent on conditions in mid-latitudes. In general, the later the burst occurs the smaller is the total rainfall in the area. Since it does occur at a variable time, there is great variability in the annual rainfall amount (Fig. 5.22). Failure of the monsoon, or even its delay, significantly affects the agriculture of the region and thus the livelihood of millions of people. Accurate prediction of the onset of the monsoon is thus of vital importance. However, as indicated, this is a complex problem requiring an understanding not only of the circulation patterns of the tropics, but also consideration of mid-latitude circulations as well.

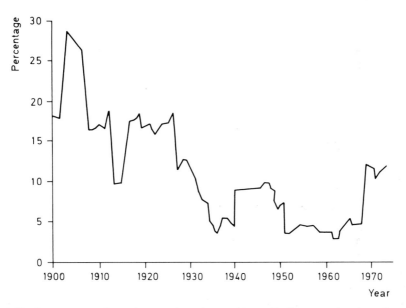

Fig. 5.22. Percentage of weather stations in northwest India reporting less than half normal annual rainfall in a given year. The curve shows 10-year running averages. (National Defense University, Washington, DC).

## 5.5 Mid-latitude weather

In mid-latitudes there is no single atmospheric feature that can be said to dominate the climate in the way that the Hadley cell does for the tropics. In the tropics any influence of topography or continentality serves mainly to modify the basic pattern. In mid-latitudes, however, while it is possible to think of the Rossby waves as being the most characteristic feature, they are themselves partly controlled by topography, continentality, cryospheric extent, air masses and depressions. Hence no simple explanation of climatic types and regions can be given for mid-latitudes. However, in this section we provide a brief description of mid-latitude weather and then, in the following section, synthesise it to provide a regional climatology.

Climatic descriptions in mid-latitudes seem to reinforce the notion of climate as a 'series of weather events' since such a description does indeed require understanding and knowledge of atmospheric activity on the synoptic scale usually considered the preserve of the meteorologist. To simplify matters, we shall temporarily regard Rossby waves and jet streams as 'steering mechanisms' and concentrate on the weather-producing features of depressions, anticyclones and air masses. To avoid an unnecessarily large number of parentheses we describe the circulation and movement of systems in the Northern Hemisphere. As explained in Chapter 4, the direction of circulations is reversed in the Southern Hemisphere.

### Depressions (cyclones)

A depression is a synoptic feature with cyclonic (counterclockwise) air circulation around a low pressure centre. Near the surface the air spirals inwards towards the centre, creating convergence, uplift and a tendency for cloud formation.

The classical model of depression development and characteristics was produced by the 'Bergen school' of meteorologists in the early 1920s. Although many refinements have been made, particularly in the light of upper air soundings and satellite observations, the main features of this model are still pertinent (Fig. 5.23). In the model the depression originates with a *front*: a region where two masses of air at different temperatures are in contact. A general definition of a front is thus a feature with a marked horizontal temperature gradient of restricted lateral extent but elongated for, usually, several hundreds of kilometres. Traditionally a depression was assumed to form on the polar front (see sect. 4.6), which is no different from any other except that it is usually very well marked and exceptionally elongated.

### Depression development

When a wave begins to develop on the front, an embryonic cyclonic circulation is initiated, with a distinct low pressure region developing at the crest of a wave. The front becomes divided into warm and cold portions. As depression development progresses pressure falls at the crest of the wave and the trailing cold front moves more rapidly than the leading warm front. The region of warm air between the fronts, the 'warm sector', becomes progressively smaller. Eventually the cold front overtakes the warm one and occlusion takes place. Thereafter the depression fills in. The polar front is frequently re-established somewhat farther south than it was previously. The whole sequence may take a week or so for completion, during which time the system may have moved two or three thousand kilometres in a general east-

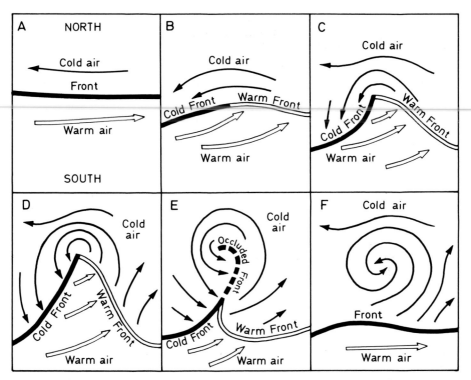

Fig. 5.23. Plan view of the six idealised stages in the development and final occlusion of a depression (or extratropical cyclone) along the polar front in the Northern Hemisphere. Stage (D) shows a well developed depression system and stage (E) shows the occlusion.

ward direction. In addition, new depressions may form on the trailing cold front of an old one. Thus there can be *depression families*, usually a sequence of three or four depressions occurring at any one time, each being younger than its predecessor and occurring progressively farther south.

There is a distinct sequence of weather conditions associated with the passage of a depression. Figure 5.24 illustrates in schematic form the surface weather map conditions of an idealised 'mature' depression, together with a cross section along line AB, showing conditions in the area with the most marked weather sequence. The warm front is a relatively gently sloping feature with a preponderance of horizontally developed cloud, while the cold front is much steeper and vertical cloud development is common. Satellite photographs show these cloud distributions very clearly (Fig. 5.25).

The actual sequence of events for a particular depression system will not follow the above model exactly. Indeed, an individual weather map is unlikely to show the features as clearly as is suggested here although

239

(a)    DEPRESSION

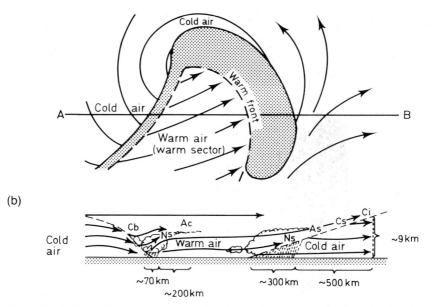

(b)

Fig. 5.24. Schematic representation of a depression system as seen: (a) in plan; and (b) in cross-section. The cross-section is taken along the line AB. The cloud types are: Cb – cumulonimbus; As – altostratus; Ac – altocumulus; Cs – cirrostratus; Ns – nimbostratus; Ci – cirrus.

many are easily recognisable (e.g. Fig. 5.26). However, as climatologists rather than meteorologists, we are mainly concerned with identifying the general characteristics of depressions, not the particular characteristics of any single one. Studies of numerous individual depressions in the Southern Hemisphere, especially in the southern Indian Ocean, suggest that our model may have to be modified somewhat if we are to characterise conditions in the southern mid-latitudes correctly.

## Anticyclones

In mid-latitudes we can often think of *anticyclones* (Fig. 5.27) as the opposite of depressions. Certainly in a general sense they are, being regions of high surface pressure and divergence, giving air spiralling clockwise (in the Northern Hemisphere) out from the centre. Descending air dominates near the centre. Thus they are regions of generally clear sky conditions although strong surface heating often leads to cumulus cloud formation locally.

There are several classes of anticyclone, depending on their size and mode of origin. The descending limbs of the Hadley cells are large, more or less permanent anticyclones. Similarly, the high pressure

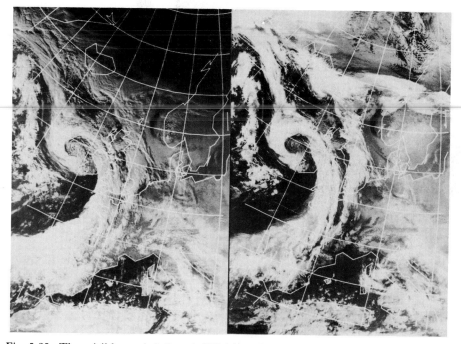

Fig. 5.25. The visible and infrared NOAA polar orbiting satellite images for 20 January 1982 taken at 1354 Z, clearly showing the partially occluded depression system adjacent to the UK. Compare these images with the surface chart in Fig. 5.26. (Courtesy: P. Baylis, Department of Electrical Engineering, University of Dundee).

region of wintertime interior Asia and the companion region over Arctic Canada are persistent seasonal phenomena. There are, however, a group of anticyclones which are small-scale synoptic features roughly comparable to depressions.

These *travelling anticyclones* occur in a number of ways. They occur in embryonic form between individual members of a depression family, giving a period of clear weather between depression passages. Commonly behind the last member of such a family there is a 'burst' of cold air of polar origin which establishes a high-pressure region. Similar scale anticyclones are also established as a result of the index cycle (Fig. 4.23). These usually take the form of *blocking anticyclones*. These features can persist for several weeks, travelling eastward only very slowly, if at all. Thus they can greatly influence regional weather conditions. It is in a blocking situation with the anticyclone over north-western Europe, for example, that Britain commonly experiences warm dry airflows from the south or southeast. This gives respite from the continuous stream of depressions, which are now diverted around the blocking feature and pass to the north of the country (Figs 5.28 and 5.29).

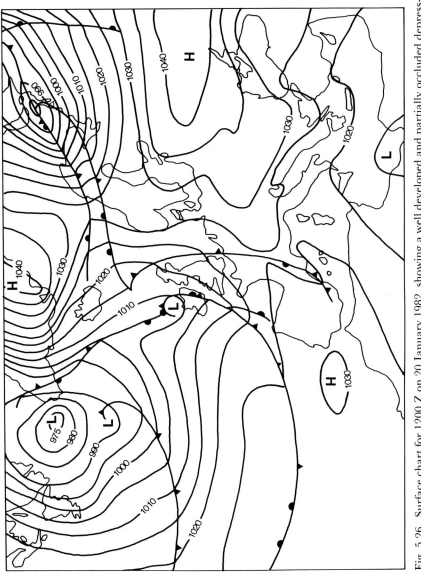

Fig. 5.26. Surface chart for 1200 Z on 20 January 1982, showing a well developed and partially occluded depression system approaching the UK with, ahead of it, a weaker depression system with a much wider warm sector.

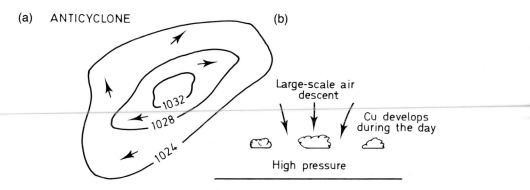

Fig. 5.27. Schematic representation of an anticyclone as seen: (a) in plan (units of pressure are hPa), and (b) in cross-section. Anticyclonic conditions can cause large-scale air pollution by restricting vertical mixing. Winter nights are often very cold as clear skies allow rapid cooling but high values of incident surface radiation can lead to cumuliform cloud development on summer days.

### Air mass source regions

Any region where calm, or relative calm, conditions persist for a few days can become a source region for an *air mass*. An air mass is a portion of the atmosphere where temperatures are approximately horizontally uniform and conditions are barotropic. Thus it is usually a region of clear skies and low wind speeds. Although theoretically any area on Earth could become a source region, in order for the characteristics to develop sufficiently to become a significant feature of the circulation, an area of the order of a million square kilometres or more is needed. Thus in practice there are only a few common source regions (Fig. 5.30). Most notable are the centres of the semi-permanent high pressure areas at the outer limits of the Hadley cell and, in winter, the thermally induced high pressure regions over the poleward continental margins. The other common source regions are the semi-permanent low pressure regions in the high mid-latitudes, especially those over the oceans. Since the air remains for several days over a source region, it takes on the thermal and moisture characteristics of the region. Its stability is also influenced by the underlying surface; air which comes into a cold source region will lose heat to the underlying surface and will tend to become a stable air mass, while air moving into a warm source region will tend to become hydrostatically unstable. The main characteristics of the common types of air masses are summarised in Table 5.2. The variation in stability indicated for the maritime tropical air mass arises because of the effects of ocean currents on the underlying temperature. We can think of the basic condition as being one of near neutral stability, but the warm waters of the west side of the ocean converts this to an unstable tendency, while cold currents on the east side lead to stability. It was fashionable at one time to provide

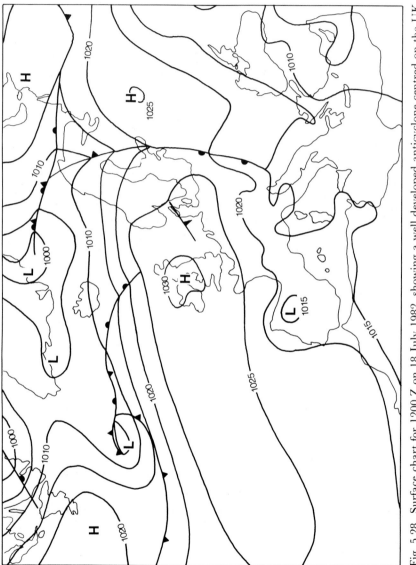

Fig. 5.28. Surface chart for 1200 Z on 18 July 1982, showing a well developed anticyclone centred on the UK which is forcing depression systems to travel far to the north of their usual path.

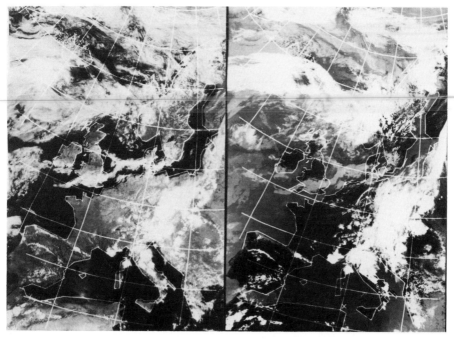

Fig. 5.29. Visible and infrared NOAA polar orbiting satellite images for 1348 Z on 18 July 1982. The position of the weak depression system being forced northward (cf. Fig. 5.28) is clearly seen. In addition, the use of combined information from two wavelength regions for identification of cloud height is clearly illustrated by the low lying stratus cloud overlying southern Ireland and southern England. This cloud band appears bright in the visible image but is relatively dark in the infrared image, suggesting that it is low level cloud with temperatures not very different from those of the surface. (Courtesy: P. Baylis, Department of Electrical Engineering, University of Dundee).

numerous subclasses of air mass types but the resulting variations were of degree, rather than nature. Hence the simple scheme used here, using six basic types is, for all practical purposes, adequate.

### Air mass movement and development

Changes in the pattern of the general circulation will cause these air masses to move from their source regions. In general, polar air masses will move over increasingly warm surfaces. When such movement takes place the air mass is symbolised as mPK or cPK, the K indicating that the air mass is colder than the surface over which it is moving. These air masses will be warmed from below and will decrease in stability as they move. Thus some cloud formation may become associated with them. However, they do tend to move as a coherent body of air, frequently taking on anticyclonic characteristics as they move into the main line of the westerlies. They can thus themselves influence the route of the Rossby waves. The tropical air masses, when they move,

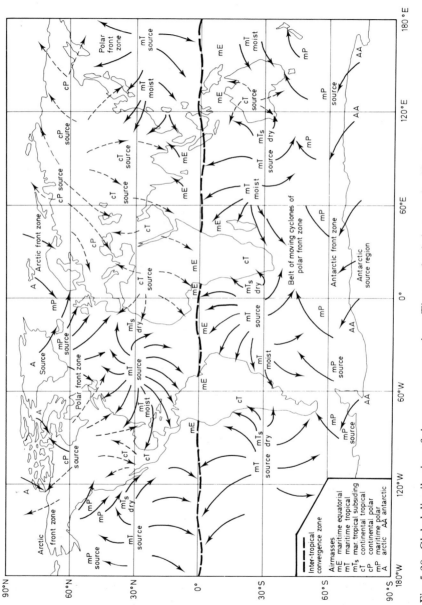

Fig. 5.30. Global distribution of air mass source regions. The standard air mass classification is given in Table 5.2.

**Table 5.2  Classification of air masses**

| Major group | Subgroup | Source regions | Properties at source |
|---|---|---|---|
| Polar (P) | Maritime Polar (mP) | Oceans poleward of approx. 50° | Cool, rather damp, unstable |
| | Continental Polar (cP) | Continents in vicinity of Arctic Circle; Antarctica | Cold and dry, very stable |
| | Arctic (A) or Antarctic (AA) | Polar regions | Cold, dry, stable |
| Tropical (T) | Maritime Tropical (mT) | Trade wind belt and subtropical oceans | Moist and warm, stability variable: stable on east side of oceans, rather unstable on west |
| | Continental Tropical (cT) | Low latitude deserts, chiefly Sahara and Australian deserts | Hot and very dry, unstable |
| | Maritime Equatorial (mE) | Equatorial oceans | Warm, moist, generally slightly stable |

will become more stable, since they are warmer than the surfaces over which they pass. The appropriate symbols are therefore cTW and mTW. These, however, are only general guides. Regional variations can easily occur. mT air that affects the southeastern United States, for example, has its source in the subtropical high pressure region of the North Atlantic Ocean. It is likely to be mTW in winter when the continent is cold, but mTK in summer when the continent is considerably warmer than the ocean.

Much mid-latitude weather forecasting was developed using air mass movements as the basic guide, an approach termed *air mass analysis*. Modern meteorology no longer uses this approach extensively; however, from a climatological viewpoint it provides a useful concept. Since air masses, anticyclones and depressions are closely connected, and each in turn influences, and is influenced by, the Rossby waves and the jet stream, movements in any one feature lead to movements in them all. Furthermore, we can think of the polar front and depressions as precipitation controlling features, and air masses and anticyclones as temperature controlling features. Thus air mass analysis combines most mid-latitude circulation features and provides a relatively simple means of characterising mid-latitude climates. Indeed, this approach has often been used as a framework for a genetic classification of climate.

# 5.6  Mid-latitude climate regions

The analysis of the climate regions of mid-latitudes can be approached by considering the frequency and seasonality of the influence of each of the mobile features on particular areas. In general, the Rossby waves, because of their great variability in position from day to day, allow penetration of polar air masses equatorward at certain times, and tropical air masses poleward at others. Since there is energy exchange with the surface during all of these movements, the end result or 'average' conditions is a more or less regular poleward temperature decrease across mid-latitudes. This would be expected from consider-ations of global energy exchanges alone. Further, since we have already demonstrated that the Rossby waves are likely to be closer to the equator in winter than in summer, it follows that polar air can more easily penetrate farther equatorward in winter than in summer, thus enhancing the seasonal contrasts anticipated from global energy considerations.

Since the precipitation bearing depressions are guided by the Rossby waves, we can also anticipate a seasonal precipitation distribution in keeping with the seasonal changes in wave position. Further, since depressions are guided from west to east, it is easy to speculate that they are likely to yield the most precipitation on the west coasts, with a gradual decrease in amounts inland. However, it is likely that the truly dry climates in mid-latitudes will be confined to continental interiors, since on east coasts moisture can be advected over the continents both as the result of air mass movements, particularly moisture-laden mT air, and monsoonal-like motions associated with continentality effects.

It is difficult to go beyond these general statements for most regions, however, because the details for each continent are very much influ-enced by the topography of the continent and, partly as a consequence, its continentality effect.

Some detailed statements applicable to all continents can be made, however, for a group of climates which might be called 'west coast climates', which in the Köppen system have the symbols Cs, Cf and Df as one moves successively poleward.

### West coast climates

The *Mediterranean* climate, Cs, is one of warm to hot dry summers and

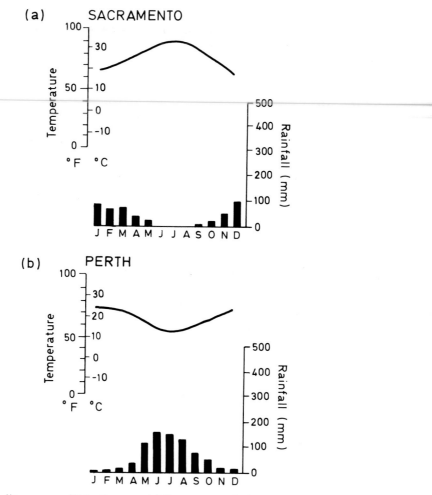

Fig. 5.31. Mediterranean (Cs) climates. (a) Sacramento (39°N, 22°W) and (b) Perth (32°S, 116°E) (for location see Fig. 5.3).

cool, wet winters (Fig. 5.31). During the summer the areas are influenced by the poleward limb of the Hadley cell, with clear, cloudless conditions. The Rossby waves, with their depressions, are poleward of the areas, and only occasionally does a trailing cold front bring summer rain to the regions. In winter, however, these depression tracks are more nearly overhead and bring precipitation to these areas. Their west coast position makes it relatively rare for cP air to move off the continent against and across the planetary waves, so that very low temperatures are rarely encountered. Cold spells are usually associated with mP air, which is considerably less cold than cP air and is likely to give cool, cloudy conditions.

Poleward of the Mediterranean climatic region are the Cf climates, frequently called *marine west coast* climates, which are influenced by depressions all year round and so have rain at all seasons (Fig. 5.32). On average, of course, they are cooler than their equatorward neighbours. This is most marked in the summer, where frequent cloudy periods ensure relatively low monthly temperatures. Again, however, winter cP is a relatively rare air mass, although not as uncommon as farther equatorward.

Proceeding poleward to the Df climate, which may be called the *cold west coast* climate, the major change is a lowering of temperatures

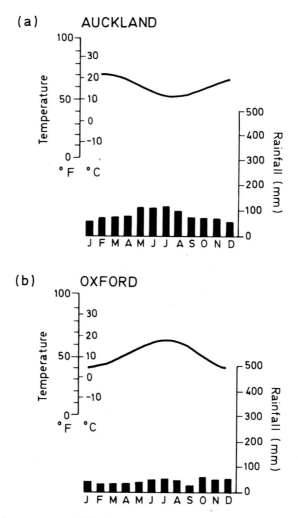

Fig. 5.32. Marine west coast (Cf) climates. (a) Auckland (37°S, 174°E) and (b) Oxford (52°N, 2°W) (for location see Fig. 5.3).

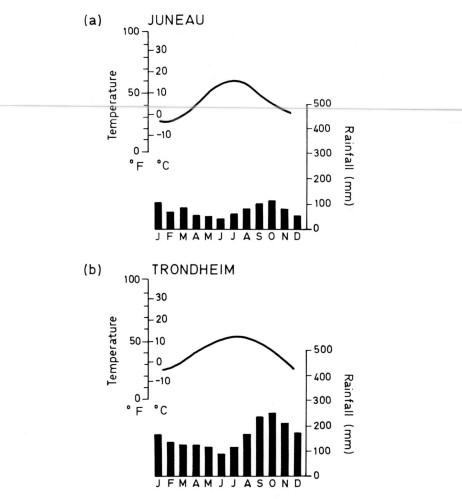

Fig. 5.33. Cold west coast (Df) climates. (a) Juneau (58°N, 134°W) and (b) Trondheim (63°N, 11°E) (for location see Fig. 5.3).

throughout the year and a resultant much shorter growing season (Fig. 5.33).

The summer minimum in precipitation so characteristic of the Mediterranean climate is generally carried through to the other west coast climates. The major reason is that the winter depressions tend to be much more vigorous than the summer ones. The trend is certainly very well developed in North America and New Zealand, but is less clear in Europe, being completely obscured at Oxford (Fig. 5.32(b) ). This difference is largely a result of the different topography of the continents and of the stations chosen. In Europe there is no mountain barrier oriented north to south and the west coast

251

climates penetrate a great distance inland. The maritime influence can be felt a thousand kilometres from the sea, the major changes inland being a slow decrease of precipitation amounts and an increase in temperature range as the continentality effects are enhanced. The Alps serve mainly to provide a distinct separation of the Mediterranean climates of the south from the more temperate ones to the north. In North America the situation is almost the reverse. The Rocky Mountains provide a barrier to the penetration of west coast climates inland, but there is a more gradual transition as one moves north along the west coast. Only in northern Europe is there a comparable mountain barrier, but even here the differences are great. The Rocky Mountains steer depressions into the Gulf of Alaska and tend to ensure that they remain there, whereas the mountains of Scandinavia force the depressions to detour to the north, but allow them to continue their eastward progress. The west coast of South America is on its equatorial side akin to North America, but as one moves poleward strongly resembles Scandinavia and has a similar climate. West coast climates also occur in restricted areas of southern Africa and Australasia.

### Interior desert climate

Away from the west coasts the difference in mountain barrier orientation between North America and Europe has a profound influence on their respective climate patterns. The west coast influence can easily penetrate inland in Europe and the *interior desert* climates are not reached for thousands of kilometres. In North America, however, the deserts occur immediately east of the mountain barrier.

### Dynamic and air mass climatology of North America

For North America the Rocky Mountains provide a real climatic barrier. In this case 'barrier' is a highly appropriate word, for the desert areas are in a gigantic 'rain shadow' of these mountains. Here rain shadow is used in both its traditional sense, and in the sense that the Rocky Mountains influence the position of the Rossby waves. Commonly these waves have a crest at, or close to, the northern Rocky Mountains. They thus commonly swing south over the western interior, reach a trough over the Mississippi River valley and then move off the continent in the northeastward direction (Fig. 5.34). Since there is no appreciable east–west oriented mountain barrier, these waves are capable of moving cP air deep into the interior and very far south. They can equally easily move mT air of Gulf of Mexico origin to the margins of the Arctic. Thus short period temperature fluctuations can be very marked.

Expressing these ideas in a complementary way, the polar front can vary in position over a very wide area in North America. The average winter and summer positions are shown in Fig. 5.34. Thus, as far as

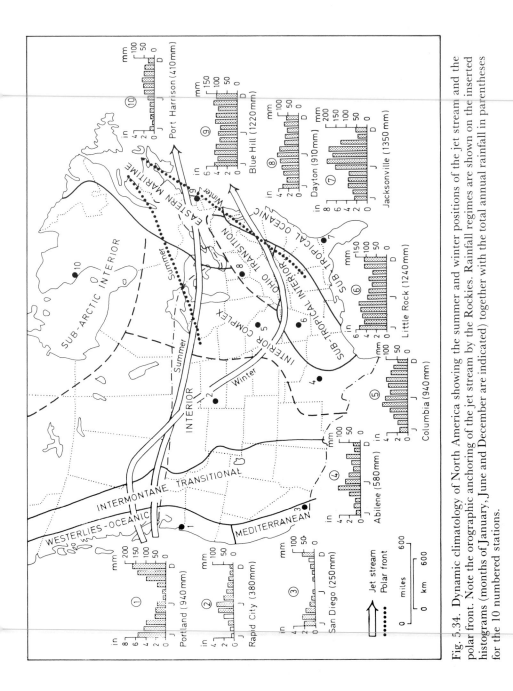

Fig. 5.34. Dynamic climatology of North America showing the summer and winter positions of the jet stream and the polar front. Note the orographic anchoring of the jet stream by the Rockies. Rainfall regimes are shown on the inserted histograms (months of January, June and December are indicated) together with the total annual rainfall in parentheses for the 10 numbered stations.

253

temperatures are concerned, the extreme southeast of the continent is dominated by mT air all the year round. As one moves northwestward the time of mT dominance is slowly decreased until somewhat north of the US/Canadian border cP air dominates throughout the year. This effect is combined with the continentality effect to produce not only lower mean annual temperatures in the interior, but also a much higher annual temperature range (Fig. 5.35).

The cP air mass is cold and dry. As it moves south it may become slightly unstable and produce some cloud, but rarely does it contain enough moisture to provide significant precipitation. The mT air mass, on the other hand, has its source region over the tropical Atlantic or

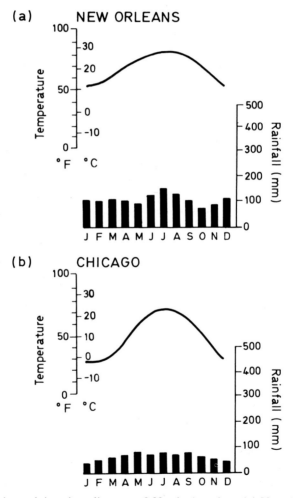

Fig. 5.35. Continental interior climates of North America. (a) New Orleans (30°N, 90°W), Da climate, and (b) Chicago (42°N, 88°W), Cfa climate (for location see Fig. 5.3).

the Gulf of Mexico. It is warm and humid and, especially in summer when its instability is increased as it moves over the land, gives rise to precipitation. The major moisture source for the North American continent east of the Rocky Mountains is thus Atlantic air. Without the action of depressions there would be a very marked decrease in precipitation from the east coast inland. One type of depression forms on the polar front west of the Mississippi. Frequently the warm sector consists of mT air which has had a long trajectory over land. Hence precipitation is rather sparse. However, once air from the Gulf of Mexico is incorporated directly, usually in areas east of the Mississippi, the vigour is enhanced and precipitation amounts increase. A second type of depression is the storm that manages to cross the mountain barrier from the west. Usually such a storm loses much of its moisture but maintains its energy while crossing. Consequently it can rapidly replenish itself as it moves over a moist surface, or can release its remaining moisture over the interior. These storms are thus the main source of precipitation in the interior of the northern United States and southern Canada.

### Eurasian interior climates

The east coast of Eurasia falls within the same climate types as does eastern North America. However, there are several differences in the causal mechanisms and in the resultant climatic details. Probably the major difference is again topographic. In place of the low level plains that characterise much of eastern North America there are the high plateaux of China. This region is also the monsoon region and certainly the effects of this regime, particularly the precipitation distribution, can be seen far away from the area usually regarded as the true monsoon area (Fig. 5.36(a)). Only in the far north is level, low lying land encountered and a more even precipitation regime developed (Fig. 5.36(b)). However, in both cases the large annual temperature regime is clearly demonstrated. This persists inland throughout eastern Asia.

### Southern continental interiors

The relatively small amount of land in the Southern Hemisphere precludes extensive development of continental interior and east coast climates. They are best developed in South America, where the topographic conditions are somewhat similar to those in North America. Thus there is an area of east coast climate, Köppen symbol Cf, comparable to that in the southeast and moist interior of the United States. Asuncion (Fig. 5.37(a)) displays the seasonal temperature variations and the distinct summer maximum in precipitation typical of such climates. South of this moist region is a much drier area. This corresponds in latitudinal and topographic setting to the

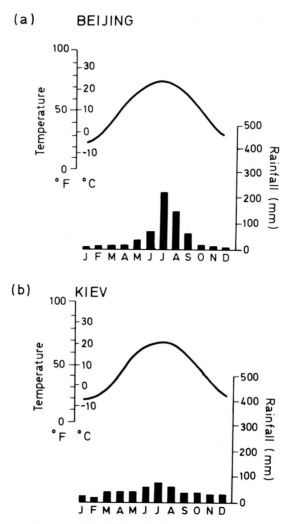

Fig. 5.36. Eurasian climates. (a) Beijing (40°N, 116°E), Da climate, and (b) Kiev (51°N, 31°E), Db climate (for location see Fig. 5.3).

southwestern deserts of the United States. However, unlike the northern continent, its extent is restricted so that only the 'cold' desert, symbolised by BWk, occurs (Fig. 5.37(b)).

Over the other southern continents the arid areas are hot, BWh, deserts. They primarily occur as a result of the descending limb of the Hadley cell and thus are a feature of the tropical circulations. The south and east margins of these continents, however, do have regions influenced by the westerlies, with the resultant moist east coast climates clearly indicated in Figs 5.31 and 5.32. These are all warm, C type climates. Only at the southern tip of South America do any of

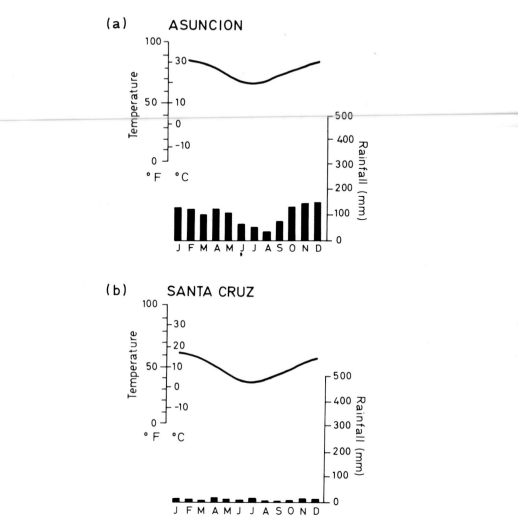

Fig. 5.37. Southern Hemisphere east coast interior climates. (a) Asuncion (25°S, 58°W), Cf climate and (b) Santa Cruz (50°S, 69°W), BWk climate (for location see Fig. 5.3).

these southern continents penetrate far enough poleward to experience a cold, Köppen type D, climate.

## 5.7   Polar climates

Polar climates are, of course, cold climates. In the Köppen system they are defined as those climates where the average temperature of the warmest month is not above 10 °C. Even with this constraint, however, there are variations in the climate from place to place. In particular,

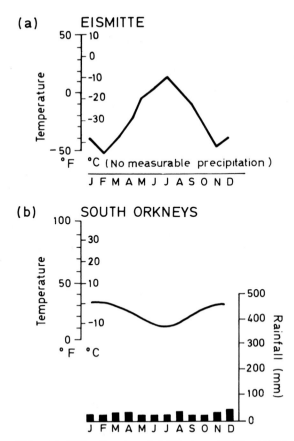

Fig. 5.38. Polar (EF and ET) climates. (a) Eismitte (71°N, 41°W) and (b) South Orkneys (61°S, 45°W) (for location see Fig. 5.3).

continentality effects occur (Fig. 5.38) and annual temperature ranges can be large in inland areas (Fig. 5.38(a)). Diurnal temperature changes are similar to those of mid-latitudes but they are not as marked, mainly because the solar zenith angle changes by a relatively small amount during the day. Indeed, right at the poles there is no change in zenith angle in any given 24-hour period. Instead there is a 'day' 6 months in length, followed by an equally long 'night'. Thus 'freezing days', days when the temperature never rises above 0 °C, are relatively uncommon in maritime areas even in winter although in inland areas in winter, periods of many days continuously below freezing are common.

In areas where the mean annual temperature is at or below −9 °C, *permafrost*, permanently frozen ground, can exist. The depth of permafrost must be, at least partially, a function of palaeoclimatic conditions since deep layers are insulated from current surface conditions and temperature changes take a long time to penetrate. In Canada, where over 50% of the land area is permanently frozen, the depth is about 2 m at the southern edge of the permafrost zone, and as great as 300 m in the north. A thin surface layer may thaw if air temperatures above freezing occur during summer. However, this promotes growth of some sparse vegetation, which acts as an insulating layer for the lower levels, which thus remain frozen.

### Aridity and windiness at high latitudes

Low precipitation (e.g. Fig. 5.38(a)) is the second major characteristic of polar climates which occur in the area of the polar cell. While this cell is relatively weak and is certainly not a major feature like the Hadley cell, the basic tendency is towards stable conditions and relatively little vertical motion. This stable tendency is enhanced near the surface in winter, when a radiatively caused inversion is common in the lowest 1 km of the atmosphere. Furthermore, the cold air is not capable of holding a great deal of moisture and so, however vigorous the release mechanisms, the resulting precipitation amounts must be small. Although precipitation amounts are low, conditions are often favourable for slow, gentle uplift, so that in summer low stratus cloud is common. Cloudiness may be around 40% in winter, but rise to over 70% in summer.

Since there is little energy available for evapotranspiration, most of the precipitation that does fall remains on the ground. Hence in summer, when the ground is unfrozen, the surface is likely to be wet. This effect is exaggerated in the permafrost region where infiltration is negligibly small even in the summer. There is an additional feature associated with moisture in the wintertime. When temperatures fall below $\sim -30\,°C$ atmospheric moisture freezes rapidly into tiny ice crystals in the air close to the ground, forming an *ice fog*. This can be a serious hazard for people and animals, since the ice crystals can block nasal passages and cause suffocation.

Wind conditions associated with the polar climates of the Antarctic continent are very different from those of the Arctic Basin. The interior of the latter is frequently a region of calms. The Arctic margins can also be calm when the polar cell dominates. At other times this marginal area is strongly influenced by mid-latitude conditions. Depressions are often steered for thousands of kilometres along the boundary between sea ice and open water (Fig. 5.39(a)) and can bring strong winds as well as precipitation to the Arctic. Whilst similar steering occurs along the margins of Antarctica, the high plateaux and mountains in the interior of this continent themselves generate strong winds. Cold dense air originating in the interior sweeps down the slopes creating almost continuous windy conditions. Port Martin (66°55′S, 141°24′E), believed to the windiest place on Earth, has an

(a)

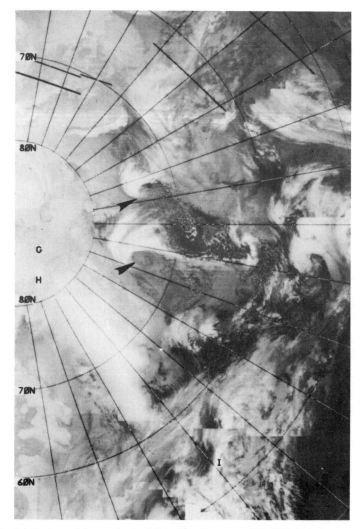

Fig. 5.39. (a) Defense Meteorological Satellite Program (DMSP) infrared image for 12 January 1979 showing two cyclones (arrowed) travelling close to the sea ice edge in the Greenland and Norwegian Seas; (b) DMSP image of Greenland and the North Pole (dark spot) for 20 May 1979 showing polynyi (dark areas) and cloud shadowing on the bright surface. (Courtesy: D. Robinson Lamant).

average annual wind speed of 64 km h$^{-1}$, while individual 24-hour averages may exceed 105 km h$^{-1}$.

### Difficulty of obtaining polar climatological data
The hostile nature of the polar climates means there is a rather sparse network of surface-based observation stations. Consequently our knowledge of the detailed surface climate variations in time and space

(b)

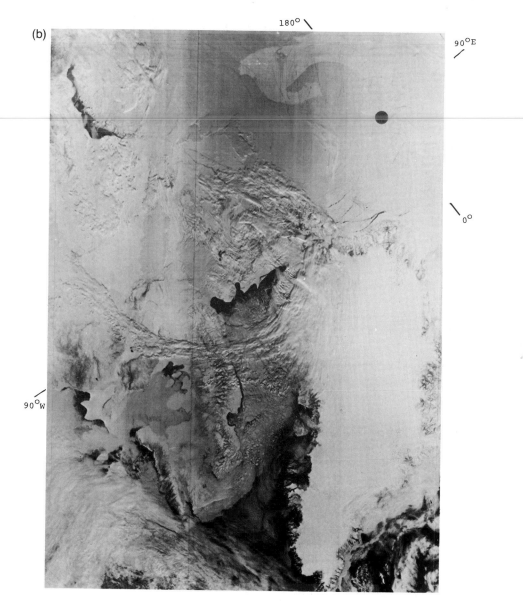

is incomplete. However, our knowledge is increasing through use of satellite observations. For example, energy exchange considerations indicate that a fundamental feature of global climate fluctuations is the variation in cryosphere extent. Until the advent of polar orbiting satellites, it was not possible to obtain a regional view. Now we can say, with some confidence, that there are some, albeit relatively small, long-term fluctuations in sea-ice extent. Seasonal fluctuations are such

261

that in the Northern Hemisphere the sea-ice extent grows from around $7.0 \times 10^6$ to $14.1 \times 10^6$ km$^2$ from summer to winter. In contrast, the seasonal change in the Southern Hemisphere is from $2.5 \times 10^6$ to $20.0 \times 10^6$ km$^2$, reflecting the different continental configurations of the two polar areas. It has also been observed that less than usual ice cover in the Barents and Kara Sea area is associated with more than usual ice in the Chukchi Sea. Similarly, the extent of ice off Alaska in August appears to be associated with the amount off Greenland in the previous June and July. Temporal variations of this type must be reflected in the surface climates of the polar regions. In addition, since depressions are steered along the margin between the sea ice and the open water (Fig. 5.39(a)), apparently because of the temperature discontinuity across the junction, sea-ice changes must also affect mid-latitude climates.

The presence of *polynyi*, open water areas, within the sea ice is also easier to study using satellite observations (Fig. 5.39(b)). These are the only places where the relatively warm waters below the sea ice can interact directly with the air. Hence they are of great importance in establishing the energy balance of the area and understanding the processes which create polar climates.

Despite our lack of understanding of polar climates, or even an adequate description of them, it is clear that the polar regions, through their interaction with mid-latitudes, have a profound effect on the climate of the whole globe. It is also clear that they can be modified relatively easily by anthropogenic influences. Atmospheric pollutants have been observed in the polar regions. Since these pollutants typically have a much lower albedo than the polar surface, marked changes in the surface energy balance can be anticipated. Also, irrigation schemes in the USSR have been proposed which would involve diversion of rivers which naturally drain into the Arctic Basin. This could increase Arctic Ocean salinity and hence lead to decreased sea-ice cover and an altered albedo. Any change that modifies the albedo could have climatic consequences far beyond the polar regions themselves.

## 5.8 Changes in regional climates

The climatic regions introduced in this chapter are not fixed, invariable, or inevitable. Certainly the climate of the whole planet undergoes constant natural fluctuations. These must lead to changes in the climate of all the component regions. Man-induced changes are possible on a local scale. These usually occur at scattered and isolated places within a region and do not appear to result in regional scale changes. Consequently it is usual to regard a particular regional climate as a fairly stable entity which has not been significantly influenced by human activity. However, as the scale of human activity increases, such regional modifications become increasingly likely as has just been suggested for the polar regions. Fortunately, we are beginning to understand more about the processes creating regional climates and thus we can suggest the potential impact of human activity. As examples of this newer type of information we can consider conditions in an equatorial climate and a continental interior one. For each the satellite observations yield monthly average values of net radiation and outgoing longwave radiation at the top of the atmosphere, together with the planetary albedo for the area (Fig. 5.40). Also included are monthly average cloudiness values, obtained from ground-based observations.

For the equatorial climate, illustrated here for São Gabriel, Brazil (Fig. 5.40(a)), the net radiation curve exhibits a double maximum, each corresponding to the time when the overhead Sun leads to a solar radiation maximum. The lower values in 'spring' appear to be a direct result of the increase in albedo caused by the greater cloud amount. Although the highest value in this season would normally occur in March, the extra increase in albedo in this month slightly depresses the value. This increase may itself be a result of changes in cloud conditions, the longwave radiation curve indicating a minimum emission in this month, consistent with the suggestion that there is a variation in cloud amount, and possibly cloud type, in this season. The difference in the magnitude of the two minima in net radiation, approximately 25 W m$^{-2}$, is less easy to explain. The outgoing longwave radiation in June only exceeds that in December by about 7 W m$^{-2}$. Part of the remainder can be accounted for by the slightly different albedos in the two months. However, it has been suggested that the difference in the distance between the Earth and the Sun in these two months may play a significant role, a factor which has received scant attention in the past. Overall, however, the annual variation in the fluxes is small, as would be expected in this equatorial region where surface temperatures vary little throughout the year. Nevertheless, it can be suggested for the year as a whole that cloud conditions play a vital role in creating and maintaining this regional climate.

Much more marked seasonal changes are apparent when a midlatitude continental interior climate is considered (Fig. 5.40(b)). In particular, the albedo curve has a distinct annual cycle. This can be related directly to the change in surface conditions from summer vegetation to winter snow. Clouds play a somewhat minor role in these albedo variations, although the decrease in cloudiness from December to February leads to an increase in albedo as more of the highly reflective surface is exposed to solar radiation. Thus for this region in this season the cloud effect is in the opposite sense from that in the equa-

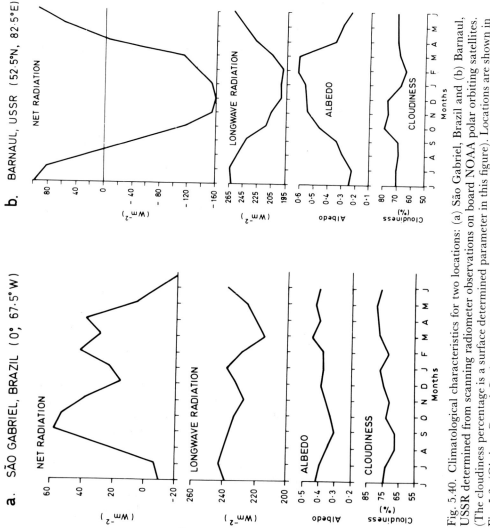

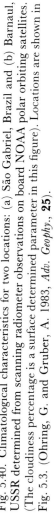

Fig. 5.40. Climatological characteristics for two locations: (a) São Gabriel, Brazil and (b) Barnaul, USSR determined from scanning radiometer observations on board NOAA polar orbiting satellites. (The cloudiness percentage is a surface determined parameter in this figure). Locations are shown in Fig. 5.3. (Ohring, G. and Gruber, A. 1983, *Adv. Geophy.*, **25**).

torial climates. The net radiation curve is a response to the great seasonal variation in solar radiation. The annual fluctuation is somewhat damped, however, because outgoing longwave radiation is at a minimum during winter, a direct response to the effect of surface temperature. Thus, again in general terms, the condition of the surface appears to be the vital factor to consider in the development of continental interior climates.

Information such as that just discussed can be used to suggest the potential impact of anthropogenic effects on regional climates. Such suggestions must be obtained by incorporating the postulated human effects into simulation models of the present climate. At present this must be undertaken by adapting global scale models to regional conditions by allowing them to emphasise those processes, such as cloud amount or surface conditions, deemed important for a particular region. Most attention has been focused on the role of surface changes, since it is in this area that mankind is beginning to make what can be regarded as regional scale changes. Particular concern has been expressed about two potential impacts: desertification and deforestation.

### Desertification

It has been suggested that changes in surface type in arid and semi-arid areas can reinforce drought conditions in these regions. The sparse vegetation natural to these areas can easily be removed as a result of relatively minor changes in the climate, or by Man-induced overgrazing. When the vegetation is removed and bare soil exposed there is a decrease in soil water storage, because of increased runoff, and an increase in albedo. These would tend to affect surface temperatures in opposite ways. With less moisture available the decreased latent heat flux would lead to an increase in surface temperature, whilst an increased albedo would produce lower temperatures. Model calculations indicate that the latter would dominate. It can then be hypothesised that the increased cooling would lead to large-scale subsidence. In this descending air, cloud and precipitation formation would be impossible and aridity would increase. The validity of this hypothesis cannot be tested by actual observations in the arid regions since the surface albedo changes little on an annual basis. However, energy flux information of the type obtained for the continental climate discussed above, where albedo does change significantly, can be used to develop a model of the potential effects. The result of one such model simulation is shown in Fig. 5.41. This is a global simulation, but concentrating on a surface albedo change for a group of semi-arid areas. It can be seen that an increase in surface albedo does seem to decrease rainfall. Use of a global simulation emphasises that all parts of the climate system are interlinked. Although this particular model includes many

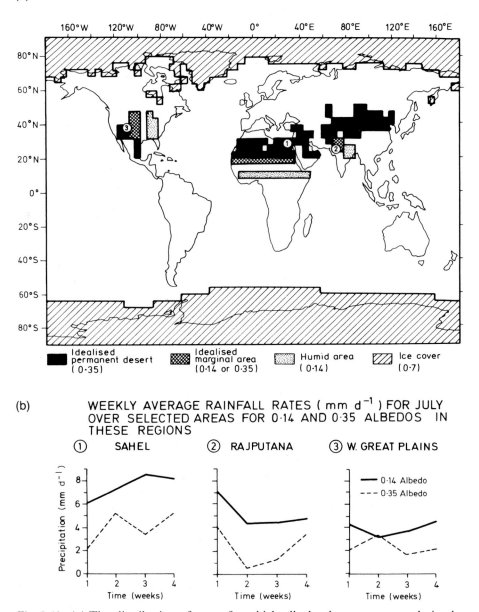

(a) GLOBAL ALBEDO DISTRIBUTION ASSUMED FOR EXPERIMENTS

Idealised permanent desert (0·35) | Idealised marginal area (0·14 or 0·35) | Humid area (0·14) | Ice cover (0·7)

(b) WEEKLY AVERAGE RAINFALL RATES ( mm d⁻¹ ) FOR JULY OVER SELECTED AREAS FOR 0·14 AND 0·35 ALBEDOS IN THESE REGIONS

① SAHEL  ② RAJPUTANA  ③ W. GREAT PLAINS

—— 0·14 Albedo
- - - 0·35 Albedo

Fig. 5.41. (a) The distribution of areas for which albedo changes were made in the climate model experiments designed to examine desertification. (Ice extent is also shown.) (b) The graphs show the result of increasing the surface albedo from 0.14 to 0.35 in three regions when free evaporation was permitted. The model integration results for the first 4 weeks are shown. In the case of the Sahel, for example, precipitation is reduced by ~50%. (Henderson-Sellers, A. and Wilson, M.F., 1983, *Rev. Geophys. Space Phys.*, **21**, 1743.).

simplifications, the results are illustrative of the type of surface-induced climatic effects which are possible.

## Deforestation

Whenever land is cleared for agricultural purposes there is a change in surface character. This change can be especially marked when forests are replaced by cropland. At present about 10% of the land area of the planet is under tillage while 30% is forested. However, the amount of forest land, particularly in the tropics, is rapidly being reduced and so the surface characteristics of large areas are being changed. One area which is undergoing deforestation is the Amazon Basin in Brazil. The data for São Gabriel presented above indicate that cloud amount, as well as surface conditions, control the climate. Thus, unlike the conditions leading to desertification, the important change is in the water characteristics rather than in albedo changes. It is well documented, for example, that many European forests have a potential evapotranspiration of around 850 mm yr$^{-1}$, whereas similarly located open ground has a value closer to 450 mm yr$^{-1}$ (see also Fig. 6.18). The values are certainly higher in tropical areas, but the differences are still very apparent. To investigate the effects of deforestation in Brazil, therefore, a global simulation model was developed which emphasised the change in moisture flux in tropical areas. In this model the tropical forest vegetation over an area of $5 \times 10^6$ km$^2$ of the Amazon Basin was converted to savanna grassland. Although this is a massive alteration, at the present rate of deforestation it could occur within 30–60 years. The model converted the vegetation almost instantaneously and then took about 5 model years to return to approximately stable climatic conditions. At the end it was found that precipitation and evaporation both decreased by around 10%. There was little change in surface temperature, probably because the decreased latent heat flux away from the surface was compensated by a slight increase in albedo. The moisture changes were a regional scale phenomenon, but no global scale changes were detected. However, the model did not take account of the increase in atmospheric carbon dioxide which would result from such a massive deforestation.

The examples considered here suggest that regional scale climate changes can occur and that they may not necessarily be associated with global changes. However, they do require modifications in surface type that cover large areas. Consequently, although minor fluctuations are probably occurring continuously from entirely natural causes, and although the possibility of significant Man-made changes exists, the impact of human activity is at present small. Certainly most human modifications are on the local scale and do not as yet have the potential to alter the present distribution of regional climates.

# Chapter 6
## Local Climates

6.1   Factors controlling local climates

6.2   The importance of surface type

6.3   The urban climate

   6.3.1   Airflow around obstacles and its implications for city planning

6.4   The influence of topography

   6.4.1   Local winds

6.5   The influence of larger-scale atmospheric features

6.6   Defining and measuring local climates

6.7   Inadvertent climate modification

   6.7.1   Air pollution

6.8   Deliberate climate modification

6.9   The human response to climate

# Chapter 6
# Local Climates

The smallest scale of climate variations can conveniently be termed local climates. The spatial range for these varies from a few cm$^2$ – the conditions around a growing plant – to a few km$^2$ – the climate of a city. Although this, as Fig. 1.5 indicates, spans the range of what is conventionally both micro- and meso-climate, they are treated together here since this is the scale where humans interact directly with the climate. It is the local climate which we experience every day that dictates what crops we grow, that determines our home heating bills and influences our city drainage system design. It is also on the local scale that we can deliberately modify the climate to help meet our need for comfort and also the scale upon which the major inadvertent modifications have already taken place.

The local climate depends for its general characteristics upon the regional climate and ultimately upon the global climate system. It is therefore useful to keep in mind constantly that the local climate of a particular place is a variation on the regional climate. Indeed the mechanisms acting to create a local climate are essentially the same as those creating the global climate. The major differences are those of emphasis. In particular, the character of the surface and how it varies spatially and interacts with the overlying atmosphere are the most vital considerations.

The character of the surface includes virtually everything we see when we look at a landscape. The type of surface, whether it be grass, forest, concrete or water; the nature and size of upstanding objects such as fences, trees, or tall buildings; the general topography of the area and its overall altitude, all influence the surface character and thus the local climate. Summarising these effects, we can say first that the surface characteristics and their variations are the major determinant of local differences in the energy balance. The main physical consequence is that there are spatial temperature variations. These are likely to lead to air density and pressure differences which can create local

271

winds when the regional atmospheric conditions are favourable. These regional conditions are also greatly influenced by the surface characteristics to produce local cloud and precipitation regimes. These not only influence the local water balance, but also the local energy balance, thus bringing us full circle.

To appreciate the nature and complexity of local climates, therefore, first in this chapter we shall consider in a comprehensive way the factors creating the local climate. Specific aspects of the surface characteristics will then be introduced. This will be followed by discussion of the role that the atmosphere plays. Finally, consideration will be given to the interaction of humans with the climate.

## 6.1   Factors controlling local climates

Local climates are the result of spatial variations of surface characteristics and the interaction of the surface with the overlying atmosphere. These effects can be approached through the energy and water balances discussed earlier (Chs. 2 and 3). First we shall consider the energy balance and its implications, since for this we can concentrate on the surface characteristics without great need to consider the effects of the overlying atmosphere. At the end of this section we shall introduce the water balance and examine how this is influenced by both the surface and the atmosphere.

The basic concept of the energy balance was introduced in Chapter 2. The governing equation is

$$Q^* = H + LE + G \qquad\qquad [6.1]$$

where $Q^*$ is the net radiation, $H$ and $LE$ are the sensible and latent heat fluxes to the atmosphere and $G$ is the heat flux into the underlying medium.

As discussed in Chapter 2, the net radiation $Q^*$ can be expanded to emphasise the surface

$$Q^* = (1-A)K\!\downarrow \; + \; L\!\downarrow \; - \; \epsilon\sigma\, T_{\mathrm{s}}^{\,4} \qquad\qquad [6.2]$$

where $K\!\downarrow$ and $L\!\downarrow$ are the incoming fluxes of solar and terrestrial radiation respectively and $T_{\mathrm{s}}$, $A$ and $\epsilon$ are the surface temperature, albedo and emissivity respectively. Variations in $A$ and $\epsilon$ with surface type have already been considered (Table 2.1).

In considering in more detail the nature of these energy fluxes, we can first discuss $H$ and $LE$ together. We have already considered heat conduction and convection on the macro-scale in sections 2.9 and 3.4. However, for local climates it is necessary to consider these effects on the micro-scale.

## Surface boundary layer

Near the surface the effects of friction on the airflow pattern are marked. Adjacent to the surface is the *laminar boundary layer,* a layer never more than a few millimetres thick, where there is no turbulent mixing and all heat transfer is by conduction. Above this is the *turbulent boundary layer,* where the air is mixed as a result of both hydrostatic and dynamic instability. The thickness of this layer depends on the nature of the underlying surface, but can be visualised as extending, with various modifications, throughout the friction layer of the atmosphere up to the point where the geostrophic approximation becomes valid (sect. 4.3). For most of this chapter, however, we shall be concerned only with the lowest few tens of metres of this turbulent boundary layer.

In the laminar boundary layer the rate of energy transfer is dictated by molecular processes. Thus the sensible and latent heat fluxes are given by

$$H = -\rho \, c_{\mathrm{p}} \, K_{\mathrm{h}} \, (\partial T/\partial z), \;\; LE = -\rho \, L \, d \, (\partial q/\partial z) \qquad [6.3]$$

where $q$ is the specific humidity, $K_{\mathrm{h}}$ and $d$ are, respectively, the thermal diffusivity of air ($0.16$ to $0.24 \ \mathrm{cm}^2 \, \mathrm{s}^{-1}$) and the diffusivity of water vapour in air ($0.20$ to $0.29 \ \mathrm{cm}^2 \, \mathrm{s}^{-1}$). In conditions of rapid surface heating or evaporation, the vertical gradients necessary to maintain fluxes can become very great. It is not unusual, for example, to find a temperature gradient of $30 \, °\mathrm{C} \ \mathrm{mm}^{-1}$ in this very thin layer in the afternoon over a desert surface.

It should be noted that in these equations, and the ones that follow immediately, the transport of any gaseous entity is proportional to its gradient. This can be usefully applied, for example, to the flux of carbon dioxide to and from the surface of the Earth. Equations of similar form also relate the way in which frictional stress or drag controls the flow of momentum from the atmosphere to the surface. This is rarely noticeable over a land surface, but is the effect that creates waves on water surfaces. Such equations can also be used to characterise the transport of particulate matter into the atmosphere, a topic to be dealt with in section 6.7.

## Wind profile in the boundary layer

In order to consider further the fluxes in various atmospheric and surface conditions in more detail, it is necessary first to consider the variation of wind speed with height, the *wind profile.* A typical set of wind profiles is presented in Fig. 6.1 in which it should be noted that the vertical axis (height) has a logarithmic scale. When conditions are near neutral (0600 and 1900 hours) the wind speed increases almost logarithmically with height (drawn as a straight line on these axes).

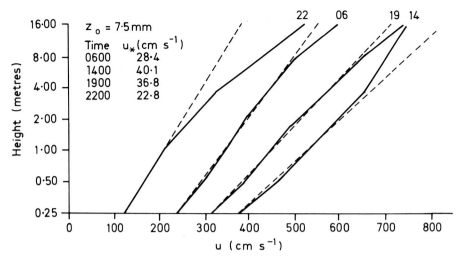

Fig. 6.1. The average variation of wind speed with height and time of day in the layer near the ground. The observations (solid lines) were taken over short grass at O'Neill, Nebraska during the summer. The dashed line is a best fit straight line using the three lowest observed values.

The general equation for these neutral curves can be shown to be

$$u = \frac{u^*}{k} \ln (z/z_0) \qquad\qquad [6.4]$$

where $u^*$ is the friction velocity, $z_0$ is the roughness length and $k$ is a constant, the *von Kármán constant*, with a value approximately equal to 0.4. Note that in Fig. 6.1 $k/u^*$ is the slope of the line and that $u^*$ depends on the overall wind speed. Although this equation holds in all stability conditions at heights below about 1 m, above this height stability influences the results. In stable conditions (e.g. 2200 hours local time) the wind speed increases more rapidly with height than in neutral conditions, largely because there is no hydrostatic instability to aid in turbulent mixing. In unstable conditions (e.g. 1400 hours) this hydrostatic instability combines with any dynamic instability present to enhance mixing and thus smooth out the profile.

The *roughness length* is the height above the surface at which the wind speed would go to zero if the turbulent layer extended completely to the ground and if there were no upstanding elements to obstruct the windflow. The values for a particular surface are independent of wind speed or atmospheric stability, provided, of course, the surface elements do not bend in the wind. Typical values are given in Table 6.1.

## Table 6.1 Roughness lengths ($z_0$) for various surfaces

| Type of surface | Height of stand (cm) | $z_0$(cm) |
|---|---|---|
| Fir forest | 555 | 283 |
| Citrus orchard | 335 | 198 |
| Large city (Tokyo) | | 165 |
| Corn | 300 | |
| $u_{5.2}^a = 35$ cm s$^{-1}$ | | 127 |
| $u_{5.2} = 198$ cm s$^{-1}$ | | 71.5 |
| Corn | 220 | |
| $u_{4.0} = 29$ cm s$^{-1}$ | | 84.5 |
| $u_{4.0} = 212$ cm s$^{-1}$ | | 74.2 |
| Wheat | 60 | |
| $u_{1.7} = 190$ cm s$^{-1}$ | | 23.3 |
| $u_{1.7} = 384$ cm s$^{-1}$ | | 22.0 |
| Grass | 60–70 | |
| $u_{2.0} = 148$ cm s$^{-1}$ | | 15.4 |
| $u_{2.0} = 343$ cm s$^{-1}$ | | 11.4 |
| $u_{2.0} = 622$ cm s$^{-1}$ | | 8.0 |
| Alfalfa brome | 15.2 | |
| $u_{2.2} = 260$ cm s$^{-1}$ | | 2.72 |
| $u_{2.2} = 625$ cm s$^{-1}$ | | 2.45 |
| Grass | 5–6 | 0.75 |
| | 4 | 0.14 |
| | 2–3 | 0.32 |
| Smooth desert | | 0.03 |
| Dry lake bed | | 0.003 |
| Tarmac | | 0.002 |
| Smooth mudflats | | 0.001 |

[a] The subscript gives the height (in metres) above the ground at which the wind speed u is measured.

### Fluxes in the turbulent boundary layer

Within the turbulent boundary layer the energy transfer rate is controlled by a process analogous to that for the laminar layer, but with the molecular processes replaced by turbulent ones which tend to be more efficient and thus more rapid. The amount of turbulent mixing is largely controlled by the wind speed and its rate of change with height (the wind profile). Several approaches can be used to characterise this turbulence. For our purposes a full appreciation of the techniques of deriving them, and their respective strengths and weaknesses, is not required. Instead we can quote one treatment, often called in micrometeorology the aerodynamic method, as an example. In this the fluxes are calculated by replacing the (molecular) thermal diffusivity in equation [6.3] by the turbulent diffusion coefficient given by $ku*z$. These equations can then be integrated over a height $z_1$ to $z_2$. This gives, by use of the logarithmic wind profile (equation [6.4]), values for the fluxes $H$, $LE$ as

$$H = -\rho \, c_p \, k^2 \, \frac{\Delta T \, \Delta u}{(\ln z_2/z_1)^2} \quad ; \quad LE = -\rho \, L \, k^2 \, \frac{\Delta q \, \Delta u}{(\ln z_2/z_1)^2} \qquad [6.5]$$

where $\Delta u$, $\Delta T$ and $\Delta q$ are the differences in wind speed, temperature and specific humidity between heights $z_2$ and $z_1$ respectively. Since the wind speed at any height depends on $z_0$, values of the fluxes in equation [6.5] depend on the surface roughness and thus on the type of surface.

An analogy has often been drawn between this bulk aerodynamic approach and Ohm's law (which says that the electrical resistance is equal to the potential divided by the current). Replacing the potential by the concentration and the current by the flux leads to a climatic analogue of the electrical resistance, called the surface resistance. This permits relatively easy consideration of the separate roles played by plants and soil in producing the moisture flux. Thus, for example, the bulk stomatal resistance of vegetation can be calculated and used as part of the above method to determine the role of transpiration in the latent heat flux.

### Importance of surface physics
The flow of heat, $G$, into the underlying medium is the final component of the energy balance. The contrast between land, which allows only conduction, and water, which allows both conduction and convection, has been treated in Chapters 2 and 4. For local climates we are essentially concerned only with land surfaces. In these conditions $G$ is given by

$$G = -K \frac{\Delta T}{\Delta z} \qquad [6.6]$$

where $K$ is the thermal conductivity of the medium. The thermal conductivity and the related conductive capacity vary between surfaces (Table 2.2) and this leads to variations in the rate of heat flow, the range of surface temperatures and the depth of penetration of the heat.

276

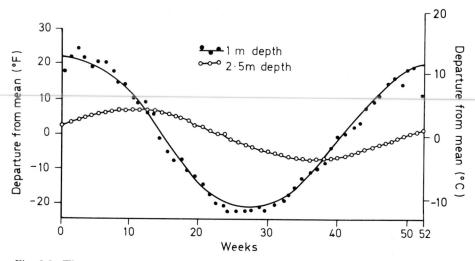

Fig. 6.2. The annual temperature wave at soil depths of 1 m and 2.5 m measured at Griffith, Australia with fitted sine curves. Temperature departures from the mean value are plotted as a function of weeks in the year.

Typical annual temperature variations at various depths within a clay soil are shown in Fig. 6.2. These results, a decrease in amplitude of the temperature wave with depth and an increasing delay in the time of the maximum with depth, are typical of all surfaces. Generalising the results, the lower the conductive capacity the greater the near surface temperature change, while the greater the thermal conductivity, the greater the depth of penetration of the change.

The foregoing analysis has deliberately emphasised the role of surface variations in influencing the surface energy budget, rather than a detailed discussion of the fluxes themselves. The latter approach is important in defining conditions exactly in a particular location and has been widely used in, for example, determining evaporation rates for cropped surfaces. However, it is sufficient for our present purposes to reiterate that equations [6.1]–[6.6] control the energy budget of the surface and thus the surface temperatures. Although clouds will influence the radiation regime, and precipitation will largely control the availability of moisture for the latent heat flux, the major controls of surface temperature variations from place to place in a local area are the characteristics of the surfaces themselves.

The other elements of the local climate depend to a greater extent on the atmospheric processes above the surface rather than the surface itself. However, we can conclude this section by mentioning these elements and noting their intimate connection with the surface energy balance. Spatial variations in temperature, caused by surface type differences, can lead to the creation of local winds. In calm or near calm conditions the air in contact with the surface exchanges energy with it and tends to approach the same temperature as the surface. Thus density differences will occur. In rolling terrain these differences may lead to hydrostatic readjustments, manifested as local winds. Density differences are also highly likely to lead to pressure differences, creating a pressure gradient and again leading to local winds. Thus when the regional scale winds are light, topographic effects can lead to the development of near surface winds whose characteristics are controlled more by the surface than by the regional wind.

The local water balance of an area is dependent on both the surface and the overlying atmosphere. The rate of evaporation, of course, is dependent on the surface type as is the amount of moisture that is retained in the underlying medium. Topography and vegetation influence the amount of runoff. Precipitation production can also be influenced by the surface, since local energy balance variations can produce hot spots to initiate convection, whilst the topography can create orographic clouds and precipitation. Much of the precipitation for a local area, however, is fundamentally dependent on the regional scale situation, with local factors being mainly a secondary effect modifying the amounts received.

## 6.2 The importance of surface type

The effects of contrasting surface types on the components of the energy balance can be seen clearly in the annual variations of the budget components in Fig. 6.3. The moist conditions at Hamburg lead to a flux of $LE$ that almost always exceeds $H$. The dry Arizona surface has a very low $LE$ and almost all of the net radiation absorbed by the surface is removed by the sensible heat flux. Over the water of Lake Mead high evaporation rates occur throughout the year. In individual months there is a large energy exchange between the surface and the underlying water, but the annual average value of $G$ is close to zero.

The influence of surface type, and its resultant effect on temperature, can be explored in more detail using diurnal information. Fig. 6.4 illustrates the diurnal variations in the components of the energy balance for three contrasting surfaces. All observations were made under clear skies (as compared with Fig. 2.29 which shows the effect upon the radiation terms of cloudy skies) in light to moderate wind conditions. The contrast in $LE$ between the grass surfaces and the bare desert soil is immediately apparent and not surprising. The high $LE$ at the Arizona site is the result of a high humidity gradient between the irrigated grass and the dry atmosphere. The humidity gradient at Hancock, Wisconsin is much smaller. Note that the sensible heat flux in Arizona is directed towards the ground, indicating that the energy needed for evaporation is being drawn from the warm air as sensible

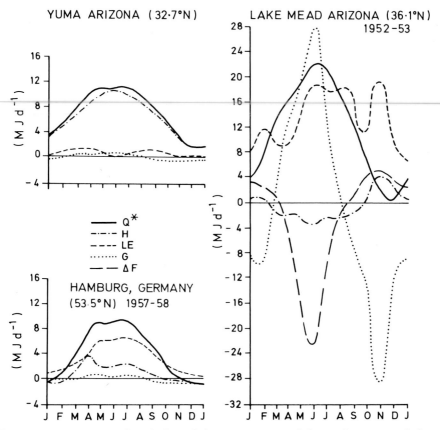

Fig. 6.3. Average annual variation of the components of the surface energy balance at Yuma, Arizona; Lake Mead, Arizona and Hamburg, FRG. $Q^*$ is net radiation, $H$ and $LE$ are sensible and latent heat fluxes, $G$ is the heat flow into the surface and $\triangle F$ is the net sub-surface flux of heat out of a water column.

heat as well as from the radiation fluxes. The diurnal temperature changes right at the surface, calculated from the energy fluxes, clearly indicate the role of the surface in creating local temperatures.

### Oasis effect: an example of the effect of change of surface type

The conditions at Tempe (Fig. 6.4(b)) are an example of the *oasis effect*, common whenever wind blows off one surface on to one with different moisture characteristics. This effect is most marked when strongly contrasting surfaces are juxtaposed such as, for example, when there is an area of groundwater discharge in a desert (Fig. 6.5). We can assume that the wind approaches the oasis having blown for some distance over a hot dry desert. The air is therefore in equilibrium with the desert surface, being itself warm and dry. Upon reaching the oasis edge evaporation rapidly increases, with sensible heat being extracted

279

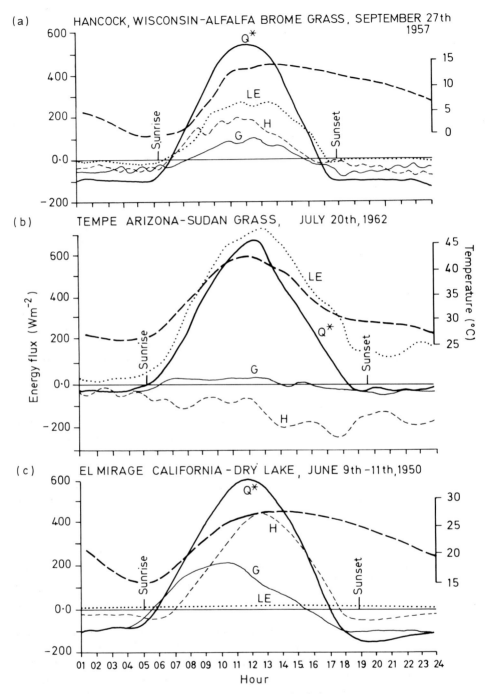

Fig. 6.4. Diurnal variation of the surface energy budget components over grass at (a) Hancock, Wisconsin and (b) Tempe, Arizona; and (c) over soil at El Mirage, California. The dashed curve shows the variation in surface temperature.

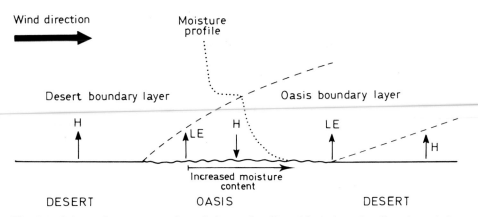

Fig. 6.5. Schematic representation of the oasis effect. Note that the direction of the sensible heat flux, *H*, changes. This is because sensible heat is removed from the air to achieve maximum evaporation from the oasis.

from the air to supplement the radiant energy to establish the high rates. The temperature thus decreases somewhat as the air begins to come into equilibrium with the new surface. Indeed, a new *internal boundary layer* is created, increasing in depth as more of the oasis is crossed and adjustment continues. This adjustment is a continuous process, however, since the humidity gradient is decreasing as more water vapour is evaporated into the air. Eventually, provided the oasis is large enough, complete adjustment will be achieved and the air will be in equilibrium with the new surface. Thereafter the energy balance components and the temperatures will not change downwind. Few situations are as stark as that of an oasis, but the adjustment process and the development of internal boundary layers will occur whenever the surface type changes. This may involve wind blowing over a city surface into an urban park, blowing from irrigated to unirrigated land, or simply from a wet to a dry surface.

### Surface roughness changes

Whenever the surface type changes in this way, almost always there is a change in surface roughness. This leads to changes in the wind profile and thus a change in the fluxes. Again an internal boundary layer will be developed while the airflow comes into adjustment with the new surface. For example, when the wind blows from an orchard into an open field there will be a rapid decrease in the roughness length (Fig. 6.6). The near-surface wind speed will increase rapidly with height, further enhancing the rate of the sensible and latent heat transfers in the lowest layers. The opposite effect will occur when, for example, air approaches an irrigated crop area.

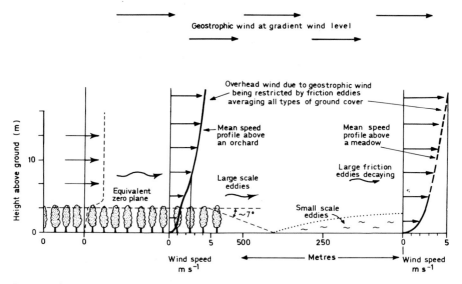

Fig. 6.6. Schematic representation of wind passing from an extensive orchard to an open field. The effect of the orchard dominates for a short distance near the surface before an internal boundary layer, dominated by small-scale eddies, is established. Thereafter an increase in wind speed occurs at all levels in the friction layer.

Fig. 6.7. Cypress trees used as wind breaks near Les Beaux, Provence, France. (Courtesy: A.M. Harvey).

282

## Shelter belts

The reduction in wind speed resulting from an increase in surface roughness can be used in the construction of *shelter belts*. In agricultural areas subject to high winds there is a danger that wind will cause bending and breaking of stalks and the shaking loose of grain or that livestock will be exposed to damaging conditions. Shelter belts can be used to provide some protection in those regions where high winds usually blow from a preferred direction. The most effective type of shelter belt is a belt of trees immediately upwind of the area to be protected. Figure 6.7 shows a well developed array of cypress tree shelter belts which protect delicate crops in the south of France. Such belts provide a semi-permeable barrier to windflow (Fig. 6.8). The wind profile begins to adjust to the new surface, but, since the obstacle is thin, it never reaches complete adjustment. Instead it begins to return to the upwind profile. The effect, however, is that some low level wind passes through the gaps between the trunks with markedly reduced speed. At higher levels the more densely packed crowns provide a more substantial barrier and force the air to flow over them. Any subsequent wake that is created by this wall of tree crowns is concentrated above and just downwind of the belt, being prevented from reaching the ground by the presence of the low level airflow. Thus near-surface wind speeds are reduced until the wind profile readjusts, which is at a distance equivalent to several tens of tree heights downwind.

## Snow fences

A similar phenomenon is produced by *snow fences* which are intended to prevent snow accumulation in particularly sensitive areas, such as roads. A slatted fence is installed some distance away from the road on its upwind side. As the airstream containing blowing snow encounters the barrier, its speed is reduced in much the same way as that of the shelter belt. This speed reduction reduces the snow-carrying ability of the wind and the snow is deposited just downwind of the fence. Beyond that region the wind speed will again increase, but now its snow load has been reduced and it sweeps across a clear roadway without impeding visibility or traction.

Shelter belts and snow fences work successfully mainly because they operate as surfaces of different roughness rather than acting as true obstacles to the wind. They change the wind direction by only minor amounts, whereas true obstacles will lead to significant changes in wind speed and direction. In the next section we shall include the effects of Man-made obstacles as components of the local climate and in section 6.4 we shall introduce topography as an obstacle to windflow.

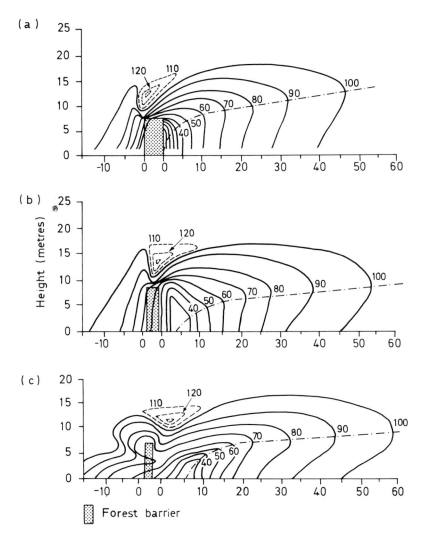

Fig. 6.8. Percentage of undisturbed wind speed at different heights experienced as a result of (a) thick, (b) moderately wooded and (c) thin forest belt as measured at Kirov, USSR. The abscissae are multiples of the forest belt height. There is a great reduction in speed immediately downwind of the thick belt, but the effects of the thinner belts persist for a greater distance downwind. (Munn, 1966).

## 6.3 The urban climate

Probably the most marked contrast in surface type occurs between a city and its rural surroundings. Differences in surface materials, drainage characteristics, sources of heat, configuration of surfaces and pollutant loading act to change all aspects of the local climate of a city. The major changes are summarised in Table 6.2.

**Table 6.2  Comparison of urban and rural climatic parameters**

| Element | Parameter | Urban compared with rural (− less; + more) |
|---|---|---|
| Incoming radiation | On horizontal surface | −15% |
| | Ultraviolet | −30% (winter); −5% (summer) |
| Temperature | Annual mean | +0.7 °C |
| | Winter maximum | +1.5 °C |
| | Length of freeze-free season | +2 to 3 weeks (possible) |
| Wind speed | Annual mean | −20 to −30% |
| | Extreme gusts | −10 to −20% |
| | Frequency of calms | +5 to +20% |
| Relative humidity | Annual mean | −6% |
| | Seasonal mean | −2% (winter); −8% (summer) |
| Cloudiness | Cloud frequency and amount | +5 to +10% |
| | Fogs | +100% (winter); +30% (summer) |
| Precipitation | Amounts | +5 to +10% |
| | Days (with less than 5 mm) | +10% |
| | Snow days | −14% |

## Urban albedo and urban 'canyons'

All components of the radiation budget at a city surface are influenced by the presence of the city. Incoming solar radiation is depleted by pollution in the urban atmosphere. It is also decreased because of the additional cloud amount usually associated with a city. The solar radiation reaching the ground encounters a surface with a different albedo from rural areas. The albedos of the urban materials are not significantly different from those of many rural surfaces, but the configuration is vastly different. To a good approximation a rural surface is predominantly horizontal, allowing only a single reflection of radiation before it is returned through the atmosphere. In contrast, a city surface is a mixture of vertical and horizontal elements, creating urban canyons; this leads to multiple reflections between streets and buildings, with some absorption in each case, before any radiation escapes to the atmosphere. Hence the city albedo depends on the

285

surface configuration and can vary rapidly over short distances. The overall effect, however, is an urban albedo lower than that for a rural area. Although this enhances shortwave absorption, the effect is partially offset by the smaller amount of radiation actually reaching an urban surface.

Incoming longwave energy is often increased as a result of the enhanced atmospheric pollution and cloud amount. The radiation emitted by the surface is, of course, a function of the surface temperature. However, it is also a function of surface geometry. Within urban canyons there is a much greater chance of absorption, so that less outgoing longwave radiation escapes from the city than from a rural area with the same surface temperature. Thus there is an increase in net longwave absorption in the city, enhancing surface temperatures. This enhancement is modified by the alteration in the other surface energy fluxes. Both the latent and sensible heat fluxes are influenced by the increase in surface roughness as one moves from the relatively smooth countryside through the suburbs into the city centre. The increase in turbulence associated with this tends to increase the magnitude of these fluxes.

### Modifications to the surface energy budget in urban areas

The major differences in the latent heat flux between rural and urban areas, however, occur because of the differences in the surface moisture content. For many regions on Earth a rural surface is a moist surface and the latent heat flux is a significant contributor to the surface energy balance. Most city surfaces, however, are designed to ensure rapid removal of rainwater, with little free water retention on the surface. Consequently only during or immediately after a rainstorm does a city have a significant latent heat flux. At other times this flux is confined to energy emanating from vegetation within the city, which usually covers only a small fraction of the total surface. Thus for much of the time the city acts in the opposite way to an oasis, tending to have flux characteristics akin to those of Yuma (Fig. 6.3) and El Mirage (Fig. 6.4).

The energy exchanges are thus restricted to the sensible and ground heat fluxes. These two, while acting in the same way for city and country, must remove a greater amount of energy in the city in order to attain an equilibrium temperature. This implies that the temperature gradient must be increased, which in turn requires a higher surface temperature. Since the conductive capacity for city materials is generally lower than for moist soil, a greater subsurface temperature gradient is needed to maintain the same ground heat flux. Furthermore, the air above a city is frequently directly heated by exhausts from automobiles, industrial processes and the ejection of heat from buildings. This also requires the temperature of the city surface to

## Table 6.3  Hypothesised causes of the urban heat island

1. Increased counter radiation ($L\downarrow$) due to absorption of outgoing long-wave radiation and re-emission by polluted urban atmosphere.

2. Decreased net longwave radiation loss ($L^* = L\downarrow - L\uparrow$) from canyons (tall buildings and narrow sidewalks) due to a reduction in their sky view factor by buildings.

3. Greater shortwave radiation absorption ($K^* = K\downarrow - K\uparrow$) due to the effect of canyon geometry on the albedo.

4. Greater daytime heat storage due to the thermal properties of urban materials and its nocturnal release.

5. Anthropogenic heat from building sides.

6. Decreased evaporation due to the removal of vegetation and the surface 'waterproofing' of the city.

7. Decreased loss of sensible heat due to the reduction of wind speed in the canopy.

increase in order to maintain the temperature gradient necessary to remove heat from the surface by the sensible heat flux.

### Urban heat island

The net result of the energy exchanges discussed above and summarised in Table 6.3, is to create an *urban heat island*, making the city a few degrees warmer than its surroundings. The intensity of the heat island is largely a function of building density and the amount of incorporated vegetation, as would be expected from energy considerations. There is a general relationship between city size and heat island intensity:

$$T(\text{urban} - \text{rural}) = P \log p \qquad [6.7]$$

where $p$ is the population. The constant of proportionality, $P$, appears to differ between European and North American cities, probably reflecting their different urban characteristics. Nevertheless, cities of all sizes display an urban heat island (Fig. 6.9).

In human terms, the negative aspects of this increase in urban temperature include aggravation of heat stress during prolonged heat waves. On the other hand, in cold climates, increased temperatures are a benefit since they increase comfort and tend to reduce power demand. Some cities, such as Stuttgart, Germany, have deliberately incorporated vegetation into their city planning to mitigate the heat island effect. Others, such as Dayton, Ohio, USA have experimented with 'green parking lots', car parks whose surface is an intermixture of tarmac and grass, with the same objective. Such strategies, unless

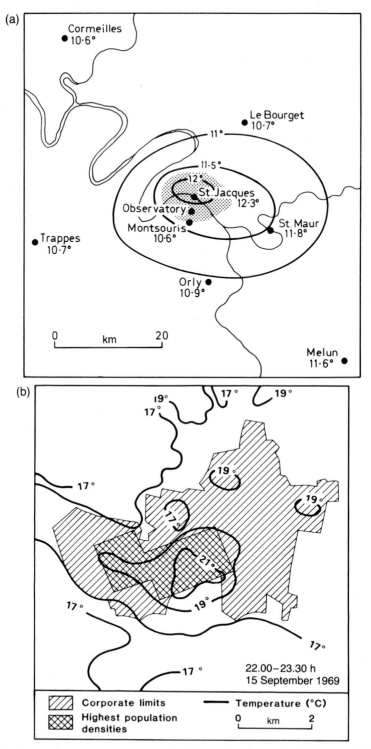

Fig. 6.9. Urban heat island maps at (a) Paris, France and (b) Chapel Hill, North Carolina, USA.

288

adopted city-wide, have only a local effect, since we must consider a city as being composed of a series of interlinked microclimates. Certainly in calm conditions this is obvious since the temperature variations within a city can be clearly sensed by any urban dweller. For example, during a summer afternoon in Columbia, Maryland, USA the temperature over a lawn was found to be 31 °C whereas a nearby car park had a temperature of 44 °C.

The urban heat island itself influences other elements of the climate. By providing a hot spot, convection can be enhanced, leading to an increase in cloudiness over and immediately downwind of the city. In addition, the increased roughness usually associated with a city may cause it to act as a barrier to the regional windflow, forcing the air to rise, further increasing the possibility of increased cloud amount. In some cases these effects are suspected of being the agents responsible for increased precipitation downwind of the city (Fig. 6.10). However, the role of the regional climate in such enhancement must be stressed. Regional airflows must contain the necessary moisture and must be susceptible to the vertical motions induced by the city. Estimates of precipitation enhancement by the city are difficult to obtain since it

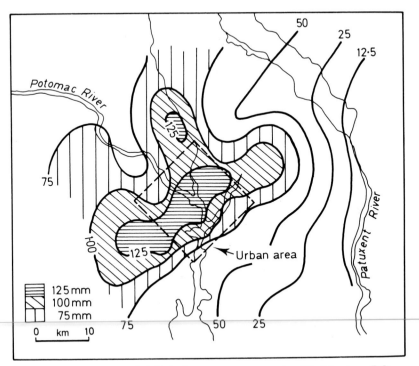

Fig. 6.10. Distribution of rainfall during a storm over the Washington, DC area on 9 July 1970 showing the precipitation enhancement due to the presence of the conurbation. The wind was from the northeast.

is difficult to extract the urban effects from the usual temporal and spatial precipitation variations. Most of our information, therefore, comes from cities in areas of relatively homogeneous terrain with a well marked prevailing wind direction. The influence of cities in other situations is not as clearly demonstrable.

### Characteristics of urban air circulation

In calm or near calm conditions the city may generate its own circulation (Fig. 6.11). This can be particularly marked at night, when warm air rising from the city surface is replaced by cool rural air. If the rising air encounters an elevated inversion, a not uncommon situation, it is forced to spread laterally. As it spreads it is cooled radiatively, its density increases and it descends over the rural areas. It is then 'sucked back' into the city to complete the circulation. It is a microscale version of the Hadley circulation, but truly thermally driven. An urban dome is thus created, which can often be seen when pollution sources are present in the city. This pollution is retained within the urban circulation and concentrations increase throughout the night. Pollution concentration will not decrease until the increase in incoming solar radiation changes the regime as it breaks the overlying inversion in the morning.

In windy conditions the city acts as a much rougher surface than does the rural area, leading to a decrease in wind speed within the city (Fig. 6.12). This leads to the establishment of an urban boundary layer

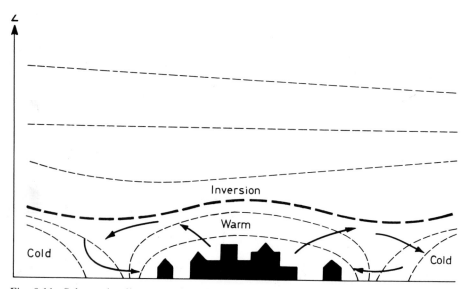

Fig. 6.11. Schematic diagram of typical urban circulation patterns when regional winds are light. The dashed lines are isotherms. The temperature inversion rises over the 'urban dome'.

290

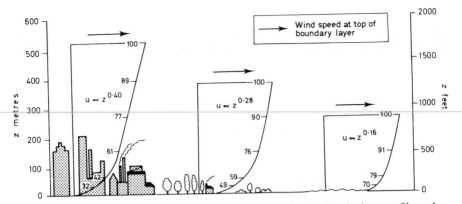

Fig. 6.12. Diagrammatic representation of horizontal wind velocity profiles above urban, rural and sea surfaces. The numbers are percentages of the wind speed at the top of the boundary layer.

in much the same way as was discussed above for rural surfaces. However, the roughness elements are typically much bigger than for rural areas and tend to act as obstacles to the windflow. The result is an increase in turbulence within the city and the development of a distinct internal boundary layer which forces the regional wind to rise as it flows over the city. The result is often called the *urban plume* (Fig. 6.13). At the upper edge of such a plume, where the regional winds are forced to rise, enhanced cloud formation commonly occurs and sometimes increased precipitation (Fig. 6.10).

Since the surface elements that create the roughness act as obstacles to the windflow, they change its direction and create great spatial variability in speed. Hence the overall statement that wind speed in a city is less than that outside can be very misleading. A city will have spots which, for many wind directions, remain virtually calm, while nearby streets may channel the wind and produce high speeds. The

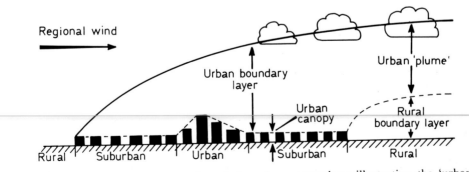

Fig. 6.13. Schematic representation of the urban atmosphere illustrating the 'urban plume'.

calm spots may foster high pollution levels. High wind speeds may make city streets difficult to walk along and may produce tremendous stresses on buildings which act as obstructions at the end of the channelling streets. Identification of these potential trouble spots prior to the construction of new buildings depends on a knowledge of the airflow around obstacles and is of great concern to city planners.

### 6.3.1 Airflow around obstacles and its implications for city planning

*Airflow around obstacles*

Generalisation of the characteristics of airflow around obstacles is difficult because of the dominating influence of the geometry of the obstacle. However, some insights can be gained by considering the results from observations, both in wind tunnels and the ambient atmosphere, of simple obstacles. When air encounters an obstacle (Fig. 6.14) the flow separates and accelerates. In turbulent flow this will increase the velocity with which eddies impact the upwind face of the building, increasing the stress on that building. Once past the obstacle, the nature of the airflow depends greatly on its speed. In light steady winds the air will decelerate and return to its upwind conditions (Fig. 6.14(a)). At higher wind speeds airflow separation will occur and

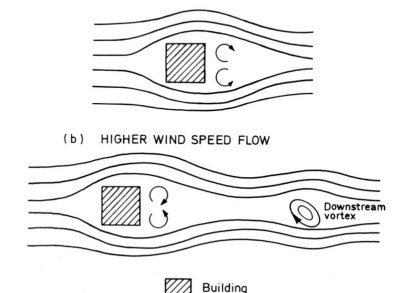

Fig. 6.14. (a) Schematic representation of airflow around an obstacle. At the higher wind speed (b) downstream vortices can be created.

a wake vortex will develop. This will not be a continuous phenomenon, but will serve to intensify the turbulence inherent in the wind, creating highly turbulent and variable conditions in the building wake. Frequently vortex shedding will occur at more or less regular intervals, these vortices travelling downstream, rotating alternately clockwise and anticlockwise, at a speed less than that of the mean wind (Fig. 6.14(b)). This stream of vortices is called a *Karman vortex street*.

With a three-dimensional building the vertical motions become important (Fig. 6.15). As air approaches the building face, the air near the edges will flow around the building, but the air near the centre must rise over the obstacle. A surface of separation, commencing at the upwind edge of the roof, is developed. Below this surface the wind patterns are the result of air movements induced both by the overlying flow and air incorporated from the wake vortex. Thus the wind direction immediately adjacent to the roof may be in the opposite sense to that of the major windflow.

The presence of turbulence in the wind leads to rapid variations in

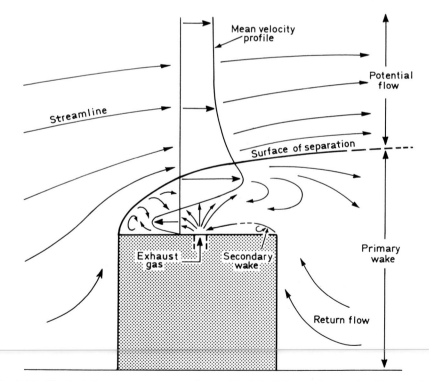

Fig. 6.15. Typical flow pattern around a cubical building orientated with one face normal to the wind. As the wind direction immediately adjacent to the roof is opposite to the mean flow exhaust gases can become trapped, recirculated and even enter windows or air-conditioning intakes lower down the building.

the strength and movement of the wind and its associated vortices. In particular, rapidly varying stresses on the building can be created, which may, in association with the upwind stresses, lead to collapse of the building. Further, when the wake vortex leaves the building there is a distinct pressure drop. In a well-sealed building this may set up a pressure difference across the building wall, which, not infrequently, causes the blowing out of windows.

### Air channelling by complex building configurations

The common situation in most cities, of course, is for a collection of complex buildings which have complex interactions with the wind field. Studies of various 'typical' situations have been undertaken. Two common conditions are illustrated in Fig. 6.16. In Fig. 6.16(a) air is funnelled between two converging buildings, causing increased wind speed as a result of the *Venturi effect*. This only becomes a significant problem, however, when the buildings have certain geometries. The second effect (Fig. 6.16(b)), the creation of transverse currents when buildings are aligned approximately perpendicularly to the wind direction, is the result of pressure differences created between upwind and downwind faces of the buildings.

These general indications of airflow patterns provide a starting point for an analysis of the likely effect of a new building, or the removal of an old one. However, if detailed consideration for a particular situation is needed, wind tunnel studies using a physical scale model of the situation are required.

(a)  VENTURI EFFECT                    (b)   TRANSVERSE CURRENTS

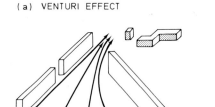

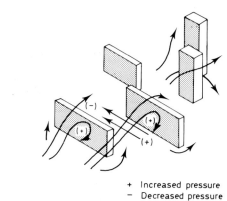

+  Increased pressure
−  Decreased pressure

Fig. 6.16. Airflow patterns around buildings illustrating flow resulting in (a) the Venturi effect and (b) transverse currents. In (a) wind speed increases as air is forced to flow through a narrow opening. In (b) an induced pressure gradient results in an unexpected cross flow of air.

## Induced wind, turbulence and urban planning

There are several reasons why city planners are concerned with the effects of winds in an urban area. First is consideration of pedestrian comfort and safety. The effect of a steady wind can be compared to the effect of walking uphill (Table 6.4). When turbulence is incorporated, experiments have suggested that wind speeds above 5 m s$^{-1}$ are annoying, above 10 m s$^{-1}$ disagreeable, and above 20 m s$^{-1}$ they can be dangerous. A second consideration is the ability of a building to withstand the wind stresses placed upon it. This safety aspect involves not only any new building, but also the possible impact of the new building on older buildings adjacent to it. A final consideration is the possibility of the creation of pockets of calm within the city in positions where they may lead to the development of high pollution levels. As a result of these potential problems several cities are requiring that the impact of new buildings be assessed before building permits are issued. Some of the most formal regulations are applied at Calgary, Alberta, Canada. That city requires a model simulation study for all new developments, to ensure that the development meets the guidelines specified in Table 6.5. Other cities, such as Ottawa, Canada and Dayton, Ohio, USA have created models of their downtown areas for use in wind tunnel studies and have undertaken detailed investigations of their local climates, as an aid to urban development. These cities, while emphasising wind considerations, are concerned with all aspects of the local urban climate and the microclimates within it.

## Table 6.4  Hill slopes and wind speeds which demand the same muscular power

| Hill slope | 1/20 | 1/10 | 1/7 | 1/5 | 1/4 | 1/3 |
|---|---|---|---|---|---|---|
| Wind speed (m s$^{-1}$) | 9 | 13 | 15.5 | 18.5 | 21 | 24 |

## Table 6.5  Guidelines for acceptable levels of urban wind modification

1. Wind speed should, at all locations, be less than double the wind speed approaching the city.

2. In locations where pedestrians will be active, e.g. sidewalks, the wind speed should be equal to, or less than, 1.5 times the approaching wind speed.

3. In locations where pedestrians will be lounging, e.g. parks, sidewalks, etc., the wind speed should equal the wind speed approaching the city.

4. Wind speed should be above 0.4 times the wind speed approaching the city to prevent high pollution levels from occurring in stagnant air regions.

# 6.4   The influence of topography

## *Variation of received solar radiation on slopes*

So far we have tacitly assumed that the landscape we are considering is flat. This, of course, is rarely the case, and topography influences the local climate as much as does the surface character. Any landscape can be considered to be composed of a series of slopes of varying angle to both the horizontal and the north–south axis. Each 'element' therefore receives a different amount of incoming solar radiation, dictated by the astronomical relationships discussed in Chapter 2. Examples of the amount of radiation received on slopes at various angles and azimuths in clear conditions are given in Fig. 6.17. The great differences between slopes, both in amount and in temporal distribution, are readily apparent. It is difficult to summarise the results, but we can note that for the Northern Hemisphere the maximum radiation is received on south facing slopes, while slopes inclined at an angle approximately equal to the latitude have maximum annual total radiation. The actual amount received on a particular slope during a particular time period will, of course, depend on the atmospheric conditions above the slope. These may themselves be influenced by topography; for example, when orographically formed clouds considered in Chapter 3 (sect. 3.3) are common.

## *Drainage and evaporation on slopes*

The variable distribution of solar radiation is the major driving force for variations in the rest of the energy balance components, thus leading to variations in the local climate in the ways we considered earlier. In particular the rate at which water drains from slopes under the action of gravity depends on the angle of the slope and the vegetation and soil conditions on it. Generally the more gentle the slope, the thicker the soil and the denser the vegetation, and hence the slower the drainage rate. A slow drainage rate allows greater retention of moisture which is available for evaporation. A typical example of the effect of changing the vegetation of a slope by clearing the land is given in Fig. 6.18.

A major consequence of energy and water balance differences between slopes is that the rate of temperature decrease up a slope depends on the angle and orientation of the slope. Some of these features can be simulated in local scale climate models as described in section 7.3 of Chapter 7. Although the basic rate of change will be the environmental lapse rate, this will be modified by the local energy balance characteristics of the slope surface. In general, in the Northern Hemisphere, a south-facing slope is likely to be warmer than a north-

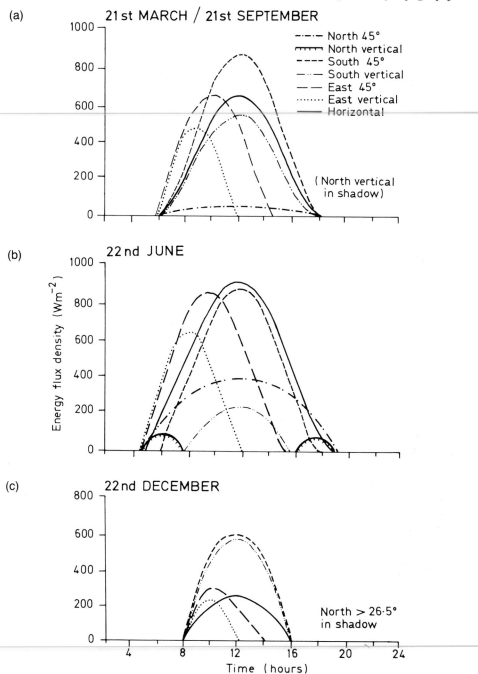

Fig. 6.17. The diurnal variation of direct beam solar irradiance upon surfaces with different angles of slope and aspect at latitude 40 °N for: (a) the equinoxes – 21 March and 21 September; (b) summer solstice – 22 June; and (c) winter solstice – 22 December.

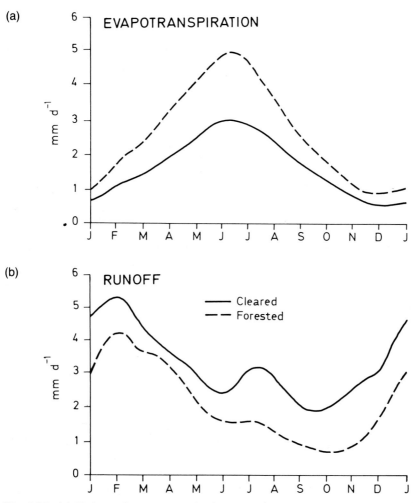

Fig. 6.18. (a) Estimated evapotranspiration; and (b) estimated (cleared) and observed (forested) runoff for cleared and forested catchments near Coweeta, North Carolina.

facing one. The persistence of this is frequently reflected in the vegetation distribution up the slope, notably the higher tree line on the south-facing slope (Fig. 6.19).

### *Wind flows in non-planar topography*
Local differences in the energy balance of slopes can also lead to the creation of local winds by a valley. Considering the valley as consisting of three slope elements, valley floor, valley sides and mountain ridge, there will be different temperature regimes for each element. In near calm regional wind conditions, these temperature differences will lead to the development of a valley wind system. On clear nights longwave

Fig. 6.19. Photographs of valley treelines in the Rocky Mountain National Park in Colorado, USA. (Courtesy: A.M. Harvey).

radiation loss from the mountain ridge will lead to considerable cooling of that surface and its overlying air. Cooling of the valley sides and floor will be much less marked because of radiation exchanges between the two valley walls and the floor. Consequently the cooler, denser air at the top will sink to the floor, moving down as a slow, generally smooth flow. This is the *katabatic wind* (Fig. 6.20(a)). Although the speed of the flow depends on the angle of slope and the roughness of

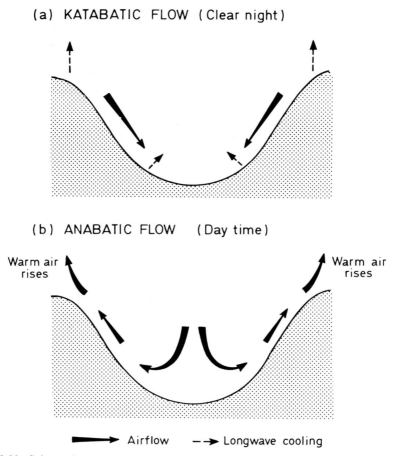

(a) KATABATIC FLOW (Clear night)

(b) ANABATIC FLOW (Day time)

Warm air rises

Warm air rises

➡ Airflow   ⇢ Longwave cooling

Fig. 6.20. Schematic representation of (a) katabatic flow on a clear night and (b) anabatic flow during the day.

the surface, it is found to be approximately proportional to the square root of the temperature difference between the top and bottom of the valley, the speed increasing approximately linearly with the distance from the top. However, the speed of the katabatic wind rarely exceeds a few m s$^{-1}$. The temperature difference between the top and the bottom, $\triangle T$, depends on the net longwave radiation loss, $L^*$:

$$\triangle T \propto L^{*2/3} \tag{6.8}$$

The cold air that flows down the valley sides will collect at the valley bottom. It will then flow out of the valley unless there is an obstruction to the flow. Any obstruction will act as a 'dam' and lead to the build-up of a cold air pocket. Such a feature is known as a *frost hollow*. The obstruction could be a variety of things, either natural, such as a constriction in the valley or a glacial ridge, or artificial, such as a road or railway embankment. Frost hollows are by no means confined to valley situations. Any topographic depression is capable of producing such a feature when there is a net loss of longwave radiation in near calm conditions. The Gstettneralm sinkhole, near Lunz, Austria, is one of the coldest places in central Europe. It is not unusual for the base of the sinkhole, approximately 150 m below the general ground level, to record temperatures 20 °C below the surrounding area (Fig. 6.21). On occasional nights temperatures below −50 °C have been observed.

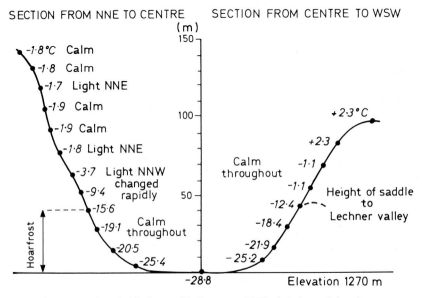

Fig. 6.21. Gstettneralm sinkhole on 21 January 1930 showing night temperatures in two sections.

The distribution of low night-time temperature associated with cold air drainage can have important consequences for agriculture. The effect of this can clearly be seen in the distribution of vineyards in many wine-producing areas. High quality wines require that the grapes be exposed to some climatic stress, necessitating their location in regions susceptible to frost. However, frost itself is detrimental to the vine. Whenever possible, therefore, the vines will be planted on the hillsides, avoiding the colder hilltops and valley bottoms (Fig. 6.22). This is thus a form of frost protection. Other forms, to be discussed in section 6.8, must be used in areas where it is desirable to use all available land, despite the risk of frost in low lying areas. Agriculturalists will be aware of the risk of frost and can frequently take steps to combat it, provided forecasts are available a few hours prior to the

Fig. 6.22. Hillslope vineyards in the Appennine Mountains in Italy.

onset of the frost. Satellite information has been used in Florida, USA to produce maps of the regions where frost hollow effects are likely to occur and to produce the required forecast information (Fig. 6.23).

In conditions favourable to the development of the katabatic wind (Fig. 6.20(a)), the air in the centre of the valley above the valley floor

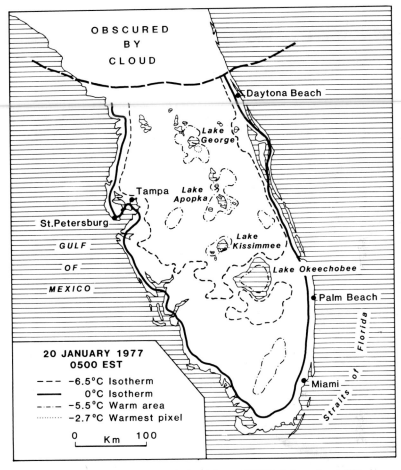

Fig. 6.23. Florida temperature map derived from GOES infrared satellite imagery. It is hoped to be able to use remotely-sensed information of this type to permit rapid identification of the onset of subzero temperatures where sensitive plants are in jeopardy. (Chen, E., Development of nocturnal GOES infrared data as a source of climate information, NOAA, 1982).

is likely to be considerably warmer than the surface air, and thus an inversion is created. This may do no more than confine the cold air to low levels, giving a low lying valley fog if the temperature falls below the dew point. However, if a low level pollution source exists within the valley system, the whole circulation pattern may serve to trap the pollution and lead to increasing pollution concentrations while the circulation persists.

The *anabatic wind* is the daytime equivalent of the katabatic wind we have been considering. Again, this is most often developed when regional winds are light and conditions are cloudless. Now the valley tops receive the maximum incoming solar radiation, and the maximum heating, in the system. The air just above the valley top thus becomes warmer than the air at the same level over the valley itself. This warm air rises. Air then flows up the valley sides, as the anabatic wind, to replace this rising air, and air from the centre of the valley descends to replace this (Fig. 6.20(b)). This may lead to a dynamically caused inversion, with its implication for pollution levels. However, this daytime circulation is rarely as well developed as its nighttime counterpart. Not only are daytime regional winds usually stronger than those at night, but also the general instability in this regional airflow is likely to swamp the circulation induced by the valley and create the dominant flow patterns. Nevertheless, an anabatic type circulation can be developed early in the morning as the nocturnal inversion in the valley is broken. Results from simulation models (e.g. Fig. 6.24) suggest that this circulation is initially an intermittent feature, only becoming firmly established some two hours after sunrise.

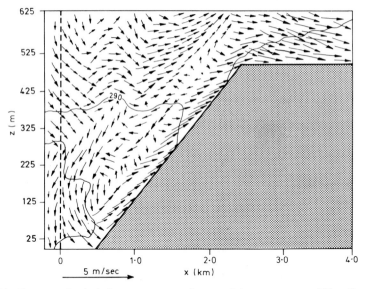

Fig. 6.24. Computed wind flow pattern and potential temperature (K) adjacent to a valley side (shaded) 210 minutes after the onset of radiational heating. By this time the anabatic circulation is well established.

With stronger regional winds, hills and valleys become obstacles to the wind flow. Valleys, for example, can generate flows at right angles to the regional wind in much the same way as the obstacles considered in Fig. 6.16(b). The downdraught created in the lee of a valley slope can develop a rotor (Fig. 6.25). This can have detrimental effects both for pollution in the valley and for aircraft flying through it.

Isolated hills which act as barriers to the wind have flows that are

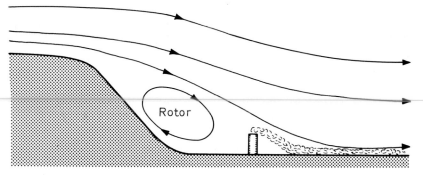

Fig. 6.25. Schematic representation of the impact of valley downdraught, caused by descent in the lee of an orographic obstacle, on an emitted plume.

very similar to those indicated in Fig. 6.15. The major differences are those of scale. In addition, topographic features tend to be rather more rounded than the stark shape of most urban obstacles. The resultant flow therefore tends to be somewhat smoother, although severe turbulence is still likely to be generated. Mountain ranges also act as barriers although the scale of the barrier here tends to produce the sort of thermodynamic effects that have been considered in connection with orographic effects. They therefore tend to produce their own regional climates, as was suggested in Chapter 5, with the local climates being formed within them by the processes being considered in this chapter.

The winds we have considered so far have been essentially microscale, the forces needed for their creation being derived mainly from the surface characteristics. However, there is a set of local winds which can be regarded as meso-scale, which result from the interaction of the atmosphere and the surface. These include the group of winds with local names which are beloved by the compilers of crossword puzzles.

### 6.4.1 Local winds

The small-scale winds we have considered in previous sections have been driven by surface conditions. The local winds we consider here depend upon the interaction of the surface with the overlying atmosphere.

#### Sea breeze

The simplest, most widespread and most persistent of local winds is the sea breeze circulation (Fig. 6.26). This is a small-scale analogue of the monsoon circulation considered in Chapter 5 (sect. 5.4) and results from the differential heating of land and sea along a coastline in near calm conditions. The more rapid heating of the land during the daytime (Fig. 6.26(a)) results in the development of a temperature gradient across the coast. This leads to ascent over the land and

305

descent over the sea and hence a pressure gradient which initiates an airflow from sea to land, the *sea breeze*. At the same time there is a compensating return flow aloft. The flow develops through the day and by the middle of the afternoon may extend several tens of kilometres inland. The uplift over the land may lead to cumulus cloud formation (Fig. 6.26(b)), but this rarely is sufficiently developed to give increased precipitation. At night the situation is reversed and the flow is from the colder land to the warmer sea, as a *land breeze* (Fig. 6.26(c)). This is generally less well formed than the sea breeze.

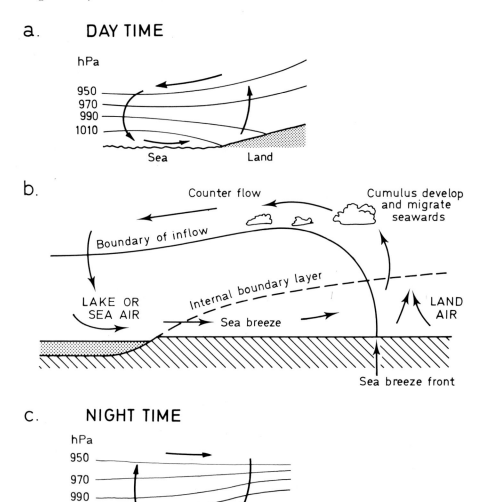

Fig. 6.26. Schematic representation of (a) and (b) day-time and (c) night-time land/sea breezes.

A group of local winds are associated with the descent of air from a mountain range or high plateau (Table 6.6). The descending air is adiabatically warmed and arrives at the lower elevations as a hot, dry wind. The Föhn of Europe and the Chinook of North America are examples. Such winds will occur only when the regional winds are in the correct direction to facilitate the downslope flow. The descending air will emulate the downwash flow of Figs 6.15 and 6.25, so that the wind is likely to be very gusty. Further, it will affect only a small area to the lee of the mountains unless the synoptic situation is such that it will aid in the removal of the cold air ahead of the Föhn. Then the wind can penetrate several hundred kilometres away from the base of the mountains. Although this penetration could occur at any season of the year, these winds tend to be most strongly developed in spring, when depression systems moving over the mountains provide ideal conditions.

A second group of winds (also included in Table 6.6) is associated almost entirely with changes in the synoptic situation, but changes that occur frequently enough, and produce weather changes noticeable enough, to be impressed on the local consciousness. The Mistral is an example. In winter cold air collects in the valleys and on the plateaux of the Alps. This air will remain stagnant, with continued cooling, until it is forced to move, usually when a depression passes through the Mediterranean. It descends into the Rhone Valley and, even though

## Table 6.6  Some local winds of the world

| Name | Location | Characteristics | Season |
|---|---|---|---|
| Bora (Latin, *boreas*, north) | Adriatic Coast | Cold, gusty north-easterly wind. Frequency at Trieste, 360 days in 10 yr. Mean wind speed 14 m s$^{-1}$, summer 10 m s$^{-1}$ | |
| Chinook (from Chinook Indian Territory) | Eastern slopes of Rockies | A warm wind that may, at times, result in sudden and drastic rise in temperature. May attain 15 or 20 °C in spring with relative humidity of 10% | Most violent in winter (may reach 100 km h$^{-1}$) |

307

| Name | Location | Characteristics | Season |
|------|----------|-----------------|--------|
| Etesian (Greek, *etesiai*, annual) | Eastern Mediterranean | Cool, dry, north-easterly wind that recurs annually | Summer and early autumn |
| Föhn (German, possibly from Latin, *favonium* = growth, i.e. favouring wind) | Alpine lands | Similar to Chinook. Characterised by warmth and dryness | Most frequent in early spring |
| Haboob (Arabic) | Southern margins of Sahara (Sudan) | Hot, damp wind often containing sand. Of relatively short duration (3 h), average frequency of 24 per year | Early summer |
| Harmattan (Arabic) | West Africa | Hot, dry wind characteristically dust laden. | All year but most effective in low-Sun season |
| Khamsin (Arabic, Khamsin = 50) | North Africa and Arabia | Hot, dry south-easterly wind. Regularly blows for a 50-day period. Temperatures often 40–50 °C. Same wind with adiabatic modifications include Ghibli (Libya), Sirocco (Mediterranean), Leveche (Spain) | Late winter, early spring |
| Levanter (from Levant, eastern Mediterranean) | Western Mediterranean | Strong, easterly wind, often felt in Straits of Gibraltar and Spain. Damp, moist, sometimes giving foggy weather for perhaps 2 days | Autumn, early winter to late winter, spring |

| Name | Location | Characteristics | Season |
|---|---|---|---|
| Mistral (maestrale of Italy = master wind) | Rhone valley south of Valence | Strong, cold wind channelled down Rhone Valley. May reach 28 m s$^{-1}$ in north. Can cause sudden chilling in coastal regions. (Note also the Bise, an equivalent cold north wind in other parts of France) | Most frequent in winter |
| Norther | Texas, Gulf of Mexico to West Caribbean | Cold, strong, northerly wind whose rapid onset may suddenly & drastically lower temperatures (also Tehuantepecer of Central America) | Winter |
| Pampero | Pampas of South America | Southern Hemisphere equivalent of the Norther | Winter |
| Zonda | Argentina | A warm, dry wind, on lee of the Andes. Can attain 33 m s$^{-1}$. Comparable to Chinook and Föhn. In dry weather carries much dust | Winter |

somewhat adiabatically warmed, is experienced as a cold wind moving to the Mediterranean shore. A similar situation occurs in the production of the Santa Anna of southern California. Here, however, the source is the hot dry deserts of the interior, giving a hot dry wind over the Californian coast.

Seasonal changes in the general circulation pattern create the third group of local winds (Table 6.6). Here an example is the Harmattan. As the subtropical high pressure region drifts southward in winter, the humid tropical air over west Africa is replaced by dry, dusty air of Saharan origin. The return northward of this high pressure belt in spring produces the Sirocco. This blows from the Sahara across the Mediterranean basin. It arrives at the European coast as a hot wind and, although it may have picked up a little moisture over the sea, it is still dry.

It should be noted that this grouping is not mutually exclusive. Local winds can be created by a combination of factors. Certainly in the Mediterranean basin and surrounding areas, an area of diverse terrain that is influenced by both synoptic features and the seasonal shifts of the general circulation, numerous local winds are developed, with local causes and local names (Fig. 6.27).

# 6.5 The influence of larger-scale atmospheric features

Local surface type and topography have a continuous, pervasive influence on local climates. Although, as we have seen, they may lead to atmospheric phenomena, the driving force that we have emphasised is the surface itself. However, large-scale atmospheric motions frequently have specifically local expressions. This will be the focus of this section. The atmospheric events we shall consider are of two types: firstly the true atmospheric meso-scale features, such as thunderstorms and tornadoes, and secondly, the local expressions of large-scale atmospheric features, such as freezing rain or dust storms.

### Convective storms

The major meso-scale feature affecting local climates is the convective storm. Details of the basic nature and development of such storms were given in Chapter 3 (sect. 3.9). Convective storms in general can be classified into four types: those formed as a consequence of convergence, those associated with frontal uplift, those resulting from orographic uplift and true air mass storms. All can bring significant rainfall to a particular area. The consequences of convergence, particularly in the tropics, have already been considered. Orographic uplift, and the precipitation associated with it, have also been noted. Both

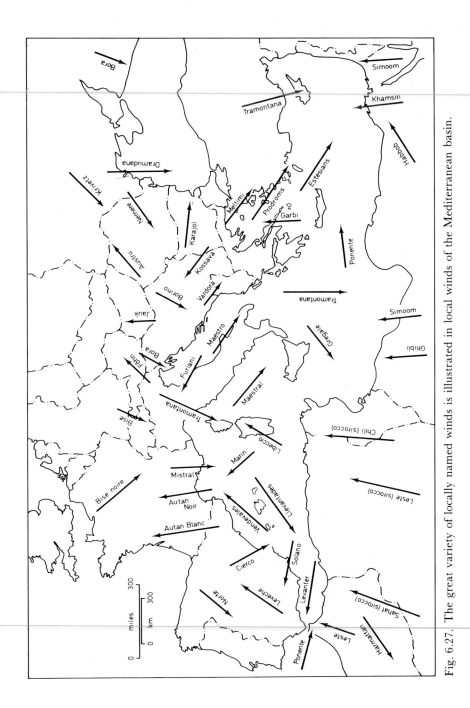

Fig. 6.27. The great variety of locally named winds is illustrated in local winds of the Mediterranean basin.

311

phenomena are likely to give consistent if sporadic precipitation in the local areas affected.

True air mass storms, originating seemingly at random within an air mass, are generated by the micro- or meso-scale features of the air-flow or the underlying surface. They are often the major source of precipitation to mid-latitude areas during periods when depressions are not in the vicinity. These storms, which rarely consist of more than three or four individual cells, are highly localised and fairly short lived. Hence they are a somewhat unreliable source of precipitation for a particular location. Storms associated with frontal uplift, particularly those that develop into squall line storms, tend to be longer lived, more intense and, of course, only affect areas where depressions are likely to occur. Both air mass and squall line storms give significant precipitation amounts in mid-latitudes in summer (Fig. 6.28). The long-term averages of this figure mask the marked variation in spatial distribution of precipitation from a particular storm. A typical example, obtained using a variety of observational techniques during a detailed investigation of the urban climate of St Louis, Missouri, USA and vicinity, is shown in Fig. 6.29.

Hail, as noted in section 3.7, is a frequent consequence of convective storms. This can be very destructive of crops and property. However, the amount of destruction depends greatly on the severity of the storm. Thus a climatological assessment of hail needs to take into account not only the frequency but also the intensity of the event. Such a climatology is indicated in Fig. 6.30 which shows a regionalisation of the United States based on hail frequency and intensity. This can be compared directly with the monetary losses associated with hail for the

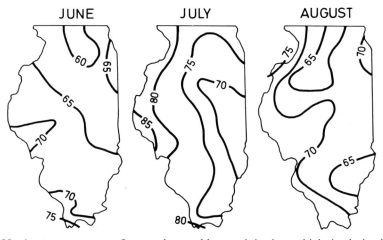

Fig. 6.28. Average percent of normal monthly precipitation which is derived from thunderstorms during the months of June, July and August for the state of Illinois, USA (Changnon, S.A. and Huff, F.A. 1980, *Bulletin* 64, Illinois State Water Survey).

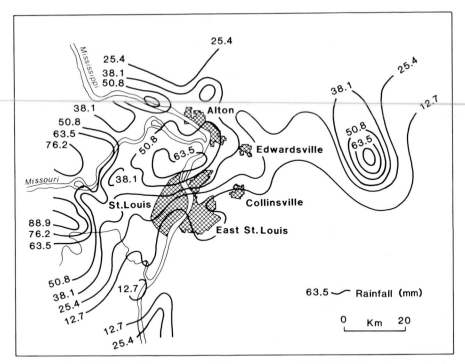

Fig. 6.29. Rainfall (mm) from a squall line storm on 23 July 1973 as it passed over St Louis, Missouri, USA (see Fig. 6.28 for source data).

various regions (Table 6.7). It is apparent, for example, that region 7 has approximately twice as many hail days as region 12, but that the latter are more destructive.

When we turn to the local expression of the larger scale, synoptic features, we are concerned with the frequency with which those features occur at a particular location. As with hail, we are usually primarily interested in individual events which have some societal or economic impact. As an example, we may use the case of freezing rain which has a direct impact on the icing of power lines. For freezing rain to occur a particular temperature distribution between the precipi- tating cloud and the ground is needed (Fig. 3.20). This situation is commonly associated with the passage of a warm front, but by no means all such fronts, even in the cold period of the year, lead to ice storms. While a general climatology of the likelihood of the event in a given year can be produced (Fig. 6.31), local surface effects, especially those controlling the temperature structure close to the ground, will be the final determinant of the event at a particular lo- cation. Similar arguments can be made for features such as fog frequency, dust storm frequency, or any other phenomenon which depends on both the conditions close to the ground and in the atmosphere.

313

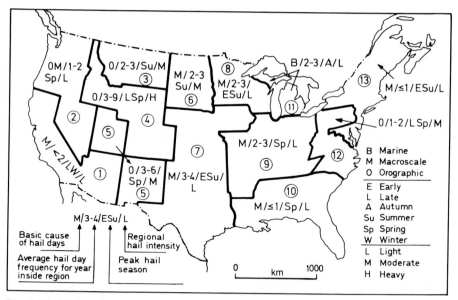

Fig. 6.30. Hail regions of the USA classified, as shown on the key, by: (i) basic cause of hail, (ii) average hail day frequency per year for the region; (iii) peak hail season and (iv) the regional hail intensity rating (see Figs 6.28 and 6.29 for source data).

## Table 6.7  Annual losses to crops and property due to hail, by hail regions

| Hail region | 1975 values in $1000s | | |
| --- | --- | --- | --- |
| | Crops | Property | Total |
| 1 | 18,520 | – | 18,520 |
| 2 | 8,906 | – | 8,906 |
| 3 | 27,204 | 520 | 27,724 |
| 4 | 47,273 | 690 | 47,963 |
| 5 | 8,771 | 80 | 8,851 |
| 6 | 81,552 | 130 | 81,682 |
| 7 | 319,881 | 15,140 | 335,021 |
| 8 | 61,960 | 1,640 | 63,600 |
| 9 | 104,436 | 14,140 | 118,576 |
| 10 | 32,979 | 5,270 | 38,249 |
| 11 | 4,006 | 1,470 | 5,536 |
| 12 | 44,317 | 670 | 44,987 |
| 13 | 13,646 | 250 | 13,896 |
| Unallocated | – | 35,000 | 35,000 |
| United States | 773,511 | 75,000 | 848,511 |

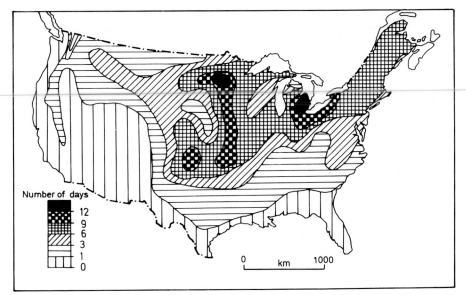

Fig. 6.31. Mean number of days with freezing precipitation for the USA. Data are averaged over the period 1939–48.

## 6.6 Defining and measuring local climates

The preceding sections have discussed the processes acting to create local climates and have emphasised their spatial variability. In many cases where climatological information is to be applied to particular problems, the local scale is the scale with which the climatologist is concerned. Indeed, many of the problems we have presented earlier have their most practical applications locally. Although in section 2.4.1 we discussed solar energy for the state of North Carolina as a whole, it is at a particular location that a solar energy system would be installed. Similarly, we noted in Chapter 3 (sect. 3.9.2) the problem of relating airport measurements of snowfall to values in downtown city streets. Answers to such problems demand a knowledge of the local climates.

Rarely will observations of the climatic parameters of interest be available at the site where we need them. Indeed, the network of observation sites for climatic parameters is not very dense when considered on a local scale. Further, the satellite observations which give greater areal coverage are not usually detailed enough for the requirements. Hence there are two courses open to a climatologist needing information for a site without observations. A measurement programme could be initiated, or records for existing sites could be converted to apply to the problem site.

## Measurement programmes for micro- to meso-scale climatology

The advisability of initiating a measurement programme depends on the nature of the problem and the time and financial resources available. Discussion of the types of instrumentation available is beyond the scope of this book. It is sufficient to say that very detailed micrometeorological results can be obtained, but that they require considerable expertise and finance. Less detailed methods are available, of course, with a concomitant reduction in cost and expertise. In many cases such measurements must be undertaken. Nuclear power plants in the United States must maintain meteorological observations to monitor the likely effects of any emissions. There is no possibility in a situation such as this for using existing observation stations away from the plant. For some other problems, however, it may be inappropriate to initiate measurements. If, for example, a crop yield model (see sect. 7.3.1) is to be produced, a long period of record is needed in order to establish the correlations between the yield and the climatic parameters. It is impractical to wait for decades for the results. Further, the results will depend on a great number of factors, not all of which can be measured, so that the meteorological data required need not be of exceptionally high accuracy. Data from a nearby site, with a long period of record, may therefore provide the best information.

## Transferability of local-scale climatological data

In transferring data from one site to another, all aspects of the local climates of the two sites must be taken into consideration. Thus the various effects introduced above must be incorporated. For some parameters quantitative models are available to assist in the transfer. Unfortunately such models are the exception and most transfers will have to be accomplished using educated guesses based on the basic principles of the controls of local climates.

We can illustrate some of the ideas involved in such data transfers using temperature as an example. We assume that the point of interest has several temperature observing stations within a few kilometres, but there are no observations at the site. For some purposes it is acceptable to use the average value, weighted in some way if needed, from the surrounding points. Weighting schemes, of which several are available, generally serve to emphasise the values of the nearest points and pay less attention to those farther away. This type of approach has been common in developing the basic grid-point temperature data required for input into climate models, as will be discussed in Chapter 7.

If we adopt this simple approach in an area of varying topography, several corrections may need to be made. First are variations caused by altitude. The correction to be applied will depend on the nature of the data requirements. If long-term average conditions are of interest, it may be simplest to assume that the average environmental lapse rate applies. However, if average conditions at a specific time of day are needed, consideration must be given to the atmospheric stability conditions. The temperature variation with altitude for the observation stations available should be investigated to give some insight into this. The problem becomes especially acute if we have a situation where drainage winds and frost hollows may be important. Unless there is a valley bottom station in a similar situation, the temperatures in the valley can only be vaguely estimated using estimated or observed values of longwave radiation loss and equations such as equation [6.8].

If the circumstances allow, it is possible to make some kind of check on these estimates by undertaking a short-term measurement programme at the site of interest. Standard statistical methods can then be used to check the validity of the assumptions.

So far we have considered temperature measurements without great concern as to what, exactly, is being measured. We have implicitly assumed that our temperature measurements have been of the standard type introduced in Chapter 2 (sect. 2.9). If we are concerned with a different level above the surface, or with the surface itself, another set of corrections must be considered. Here energy balance considerations are paramount. The nature of the temperature profile, dictated largely by the energy balance, will determine our corrections. For example, concern with overnight frost on the ground will necessitate a different correction than will a concern with ice accumulation on a broadcasting antenna, while both will be different from the correction needed for finding the summer afternoon heat load on a building.

These general considerations, of course, only give the outlines of possible approaches to particular problems. Each climatic element has its own set of techniques. This summary has served to emphasise the difficulty of specifying in a quantitative, rather than qualitative, way the spatial variations of local climates. Models are being developed to aid in this quantification, but still it is very difficult to generate a model that is sufficiently general to be of wide applicability to the almost infinite variety of local climates that occur over the Earth.

## 6.7   Inadvertent climate modification

The scale of local climate is, by our definition, the scale at which mankind operates. On a local scale humans can modify the climate deliberately or inadvertently, and can take steps to utilise it as a resource or avoid its adverse impacts. In the next three sections we explore each of these in turn, starting with mankind's inadvertent modification of climate.

Consideration has already been given implicitly to inadvertent climate modification. Any change in the surface characteristics by a human agency leads to a climate change. Examples would include the creation of a city, or the formation of a frost hollow by injudicious placement of an embankment. These are essentially local-scale phenomena, and indeed it is on this scale that the major modifications have taken place. However, consistent local modifications, implying that similar modifications take place at a large number of sites, could combine to produce coherent regional, or even global, alterations in climatically important parameters as suggested in section 5.8 of Chapter 5.

### *Effects of increasing use of energy*

The increase in surface temperature as a result of increased urbanisation is still a local phenomenon. Indeed, equation [6.7] suggests that the urban heat island effect is likely to lead to only small temperature increments as large cities expand. Probably more important is the input of waste heat into the atmosphere. This effect can be approximated by considering the worldwide use of energy, most of which eventually ends up as waste heat. If we use the 'worst case' conditions from various projections of total population and energy usage at some time in the future when 'equilibrium conditions' are reached, population will increase from $4 \times 10^9$ to $20 \times 10^9$ and per capita energy usage from 2 kW to 40 kW. This gives a total energy usage of approximately $8 \times 10^5$ GW, which is approximately 1% of the solar energy reaching the Earth. While this is a small fraction and of itself may not be a cause for great concern, the energy usage will not be evenly distributed. Although 58% of the world's population live in Africa, Asia, Australia and South America, they currently only account for 13% of the global energy usage. Hence, at least at present, most of the extra heat input to the atmosphere will come from Europe and North America, in the Northern Hemisphere mid-latitude belt. This imbalance is, thus, a cause for concern.

The potential climatic effects of carbon dioxide are worldwide in nature. Its effect upon climate can also be examined, along with that due to waste heat, with climate models. This will be the focus of Chapter 7. Air pollution, however, is as yet mainly a local, or in some instances a regional, problem.

### 6.7.1 Air pollution

Air pollution is usually regarded as consisting of any and all particulate and liquid matter (other than cloud droplets) in the atmosphere together with various noxious gases that are emitted by certain industrial processes. This definition, however, includes matter of natural origin, such as desert dust, pollen and sea spray. While natural material can occur in high concentrations locally, and be a 'pollution problem', it can be assumed that such material is a normal part of the climate and that the present climate system has adjusted to its presence. Indeed, the importance of such material in cloud and precipitation formation has already been considered in Chapter 3 (sect. 3.3). We are concerned here with pollution of anthropogenic origin.

There are many *sources* for pollution: industrial, agricultural, domestic and transport. Intensive cultivation of arable land and deforestation both enhance the natural emplacement of material in the atmosphere. This is a low level source of wide areal extent. In contrast, most industrial sources are relatively high level point sources, although they may be grouped together to give the effect of an areal source. In our discussion we shall assume that we are dealing with an elevated point source of industrial pollution, although extensions can be made for ground level and area sources. A schematic diagram indicating the major factors associated with air pollution is given as Fig. 6.32.

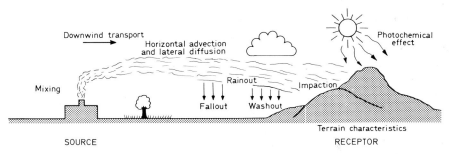

Fig. 6.32. Schematic representation of the transport and removal of air pollutants.

The point where the pollution enters the atmosphere can be characterised by the actual *stack height* or, more often, by the *effective stack height* (Fig. 6.33(a)). The effective stack height depends on the height of the stack itself and the exit velocity of the material. If the material is a gas, the effective height will also depend on its buoyancy and the atmospheric stability, since the gas coming from the stack acts in many ways like a parcel of air displaced from the surface. Thus, under normal conditions, the effective stack height will be at least the height of the stack itself, and usually considerably higher. Once at this height, however, the plume will begin to level off and be influenced almost solely by the atmospheric conditions. Of great importance is the relationship between the effective stack height and the local stability conditions (Figs. 6.33(b) and 6.34). Any of these conditions may persist for a long period of time, or may change diurnally in the way indicated in Fig. 3.14. However, the marked effects of inversions are clearly seen, emphasising that it is desirable to ensure that the effective stack height is above the local surface inversion in order to minimise ground level concentrations.

319

(a)

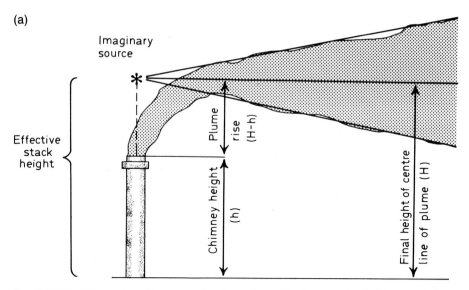

Fig. 6.33(a). Schematic diagram showing the effective stack height, which is compounded from the actual chimney height and the buoyant plume rise.

(b)

Fig. 6.33(b). Buoyant plumes from the cooling towers of an electricity generating station penetrating a low-level early morning inversion. (Courtesy: A. Abbott).

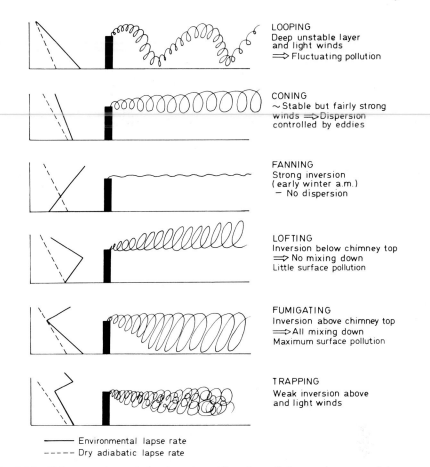

LOOPING
Deep unstable layer
and light winds
⟹ Fluctuating pollution

CONING
~ Stable but fairly strong
winds ⟹ Dispersion
controlled by eddies

FANNING
Strong inversion
(early winter a.m.)
– No dispersion

LOFTING
Inversion below chimney top
⟹ No mixing down
Little surface pollution

FUMIGATING
Inversion above chimney top
⟹ All mixing down
Maximum surface pollution

TRAPPING
Weak inversion above
and light winds

——— Environmental lapse rate
– – – – Dry adiabatic lapse rate

Fig. 6.34. Chimney plume behaviour as a function of the environmental lapse rate (solid line) and wind speed. Note that dispersion is also a function of turbulence which is often locally induced (e.g. Figs. 6.14 and 6.25).

### Chimney plume dispersion

As the plume moves horizontally away from the stack under the influence of the regional wind, turbulence and diffusion within it will cause it to expand. These are pseudo-random processes. Considerable research effort is currently being expended on determination of the correct formulations which will allow realistic predictions of pollution concentrations. We need be concerned here only with the results that have a bearing on climatological effects. Most evaluations start with the *Gaussian Plume Model*, which may be written in various forms for various boundary conditions. For illustrative purposes we present it for a continuous stack emission (i.e. a point source) of strength $S$. In this case the mean concentration, $C$, is given, at a point in space $(x, y, z)$ where the origin of the axes is situated at the stack top, by

321

$$C = \frac{S}{4\pi rk} \left[ \exp\left\{ -\frac{u}{4Kx}(y^2+(z-H)^2) \right\} + \exp\left\{ -\frac{u}{4Kx}(y^2+(z+H)^2) \right\} \right]$$

[6.9]

where $H$ is the effective chimney height, $u$ the mean wind speed, $r^2 = x^2 + y^2 + (z - H)^2$ and $K$ is a diffusion coefficient. In this example we assume that this coefficient is the same in all three directions and does not vary with elevation. In many cases this is an unrealistic assumption.

In the case of particulate emissions it is also necessary to include the rate at which the particles fall under gravity (Fig. 6.32). It is possible to derive expressions for the maximum height reached by these particles, $z_{max}$, their residence time in the atmosphere, $t$, and the distance of travel, $x_{max}$. These are shown in Fig. 6.35 for a range of particle terminal velocities. In the figure $m$ is the fraction of the initially injected particles that remain airborne. Those particles that are removed from the plume by gravitational settling, a process simply known as *fallout*, arrive at the surface as *dry deposition*. The particles may also be removed by impaction as part of the plume moves over the ground surface and the particles are collected on objects such as plant leaves. The values given in Fig. 6.35 must remain as general estimates which will vary with stability and wind speed. Nevertheless, they indicate clearly that the residence time, travel distance and height of the largest particles are several orders of magnitude smaller than for the smallest particles.

These values are also only appropriate for horizontally uniform terrain. The complexity of any models developed to characterise pollution concentrations increases tremendously as soon as topographic effects are introduced. In fact, no general models are yet available. The approach that is usually adopted is to estimate values for a particular area using a short-term set of measurements and considerations of the mechanisms creating the local climate, as suggested in the previous section.

### Removal of atmospheric pollutants

In addition to the fallout of particles as a result of gravitational forces and impaction, the particulate pollution may also be removed by precipitation. Particles may act as condensation nuclei in the precipitation formation process and reach the ground by *rainout*. The particles may also be swept up by falling drops, incorporated into them, and fall as *washout* (Fig. 6.32). The efficiency of this type of removal depends primarily on the types of processes discussed in Chapter 3 (sects. 3.3 and 3.6), but generally the average residence time of a particle decreases as the rainfall rate increases.

So far we have mainly been concerned with solid particulates.

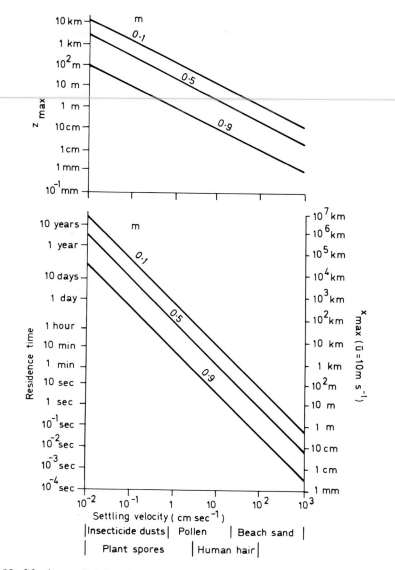

Fig. 6.35. Maximum height of rise, $z_{max}$, residence time and distance of travel, $x_{max}$, for various fractions, $m$, of particles of given settling velocity (cm s$^{-1}$) released from rest into an atmosphere with a (constant) mean wind speed of 10 m s$^{-1}$ and eddy diffusivity of $10^4$ cm$^2$ s$^{-1}$.

Gaseous matter can be treated in the same way, but with the additional consideration of chemical transformations and *photochemical effects*. These may lead to the gas-to-particle conversions noted earlier, or produce new gases. Gases are particularly important climatologically because, being effectively small and very light 'particles', they can rise to great heights, have long residence times and travel great distances. Gaseous emissions have the potential to affect the radiation budget over large areas and therefore to modify the regional and global climate. They may also have a detrimental effect both on the natural components of the Earth's atmosphere and on the precipitation that falls from it.

### Urban smogs

The major detrimental feature caused by gaseous air pollution at a local scale is the photochemical formation of *peroxyacetyl nitrate* or *PAN*. In the atmosphere the nitric oxide emitted by industrial processes is oxidised to nitrogen dioxide. In the presence of adequate ultraviolet radiation, this nitrogen dioxide dissociates to nitric oxide and free oxygen, the latter reacting with molecular oxygen to form ozone. Further reactions, several of which require ultraviolet radiation, occur between the nitrogen oxides, oxygen species and the hydrocarbons (originating from unburnt fuel). The result is a wide range of organic substances including PAN, which together form *photochemical smogs*. Since such smogs require both bright sunshine and gaseous air pollutants for their formation, they tend to be confined to large conurbations equatorward of about $50°$ (for example, Los Angeles and Sydney). However, many mid-latitude cities have experienced short-term pollution incidents when blocking anticyclonic conditions in the summer have been responsible for increased levels of insolation.

### Acid rain

Some gases react with rain to form acid solutions. The creation of carbonic acid ($H_2CO_3$) by the solution of carbon dioxide is universal and produces natural precipitation that is slightly acid. Similarly, nitric acid is commonly formed during thunderstorm activity and leads entirely naturally to acidic precipitation. However, the major concern of *acid rain* is with the acidity produced as a result of the various oxides of sulphur and nitrogen that are placed in the atmosphere as effluent from industrial processes, especially from electricity generating plants and from traffic. Sulphur dioxide, once oxidised, may be dissolved to give $H_2SO_4$ (sulphuric acid). The acidity is often enhanced by the presence of $HNO_3$ formed as the result of large emissions of various of the oxides of nitrogen ($NO_x$). The consequence of this is that the acidity of the precipitation is increased. Acidity is measured in terms of the pH, the concentration of free hydrogen ions, with neutral solutions having a value of 7.0, with acidity increasing as the pH decreases. Relatively few unambiguous measurements of the pH of natural precipitation are available, but it is usually estimated to be around 5.6. Recent measurements have suggested that many locations in the USA and northwest Europe now have rainfall pH values as low as 3.0.

Since the gases that are the cause of acid precipitation can travel great distances, acid rain tends to be the result of local processes but with regional consequences. Although 100% of the material may be deposited within a few kilometres of the site during a rainstorm, generally only approximately 10–20% of the deposition occurs within 50 km of the source. The fate of the remaining material depends greatly on the atmospheric conditions. For example, the high level plume from the INCO nickel smelter at Sudbury, Ontario, Canada has been tracked as far as Florida, a distance well over 2000 km. The stack at Sudbury was designed for widespread dispersion. At 381 m it is the highest in the world, and until very recently this plant was outputting up to 3% of the world's atmospheric emissions of sulphur. However, as noted, widespread dispersion only occurs in favourable atmospheric conditions. Although few situations are as stark as those at Sudbury, climatological conditions indicate that most pollutants will travel, on average, in a preferred direction. Hence the major industrial regions of the world are suspected of creating acid rain in nearby countries. Effluents from the UK and West Germany have been tracked to Scandinavia, leading to surface waters with low pH values (Fig. 6.36). Similarly, the impact of emissions from the industrial midwest of the USA is thought to occur in eastern Canada.

The most prominent effect of acid rain is alteration of the aquatic environment; for example, aluminium is leached by the acid rain from the soils surrounding a water body and is concentrated in the lake. Excessive aluminium cannot be tolerated by many fish species, so that a fish kill may occur with direct consequences for the economy and the

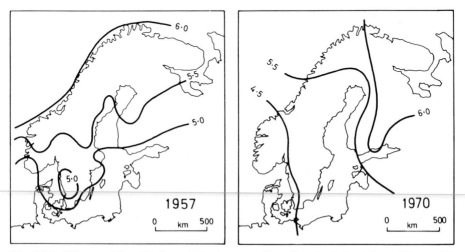

Fig. 6.36. The trend in pH of precipitation over Scandinavia derived from 12-monthly averages for the years 1957 and 1970. It can be seen that there is a regional decrease in pH.

ecology of the region. Trees are also detrimentally affected by lowered pH of rainfall.

# 6.8   Deliberate climate modification

Almost all attempts to modify the climate deliberately have been on a local scale. The large energies involved in atmospheric circulations at the larger scales precludes human modification with the current energy generation technologies. This does not mean that local scale changes could not be used as 'triggers' for regional ones. Several possible approaches have been suggested, such as coating selected areas of the Arctic ice with material of low albedo to promote melting and clearance of the shipping lanes. However, no-one has yet seriously suggested that mankind embark on a deliberate 'global experiment' of this nature. The complexity of the interactions and feedbacks within the climate system, which are poorly understood, would make it a very risky undertaking. Most intentional modifications have been on the local scale and are relatively simple applications of the ideas that have been developed throughout this book.

### *Crop protection climatology*
A knowledge of the effects of the surface energy balance, for example, has been used to develop frost protection techniques for agriculture. The objective is to prevent the temperature of valuable crops falling below freezing. Several methods are available when the short-term weather forecast indicates the likelihood of a ground frost, the type we have identified as a radiation frost, on a clear night. One method is to use smoke generating heaters within the area to be protected. These act in two ways: they warm the air directly and effectively, since the heat is unlikely to be transported to higher levels in these inversion conditions. However, the amount of heat needed to prevent the frost is too great to allow success by heating alone. It is therefore supplemented by the smoke given off by the heaters. This smoke rises somewhat, but tends then to spread laterally because of the inversion. The smoke thus creates an elevated surface for radiation exchanges above the crop being protected. The top of the smoke layer is radiatively cooled, while the bottom acts in the same way as a cloud deck for the actual surface and minimises the radiation loss from it. In these radiative frost conditions frost protection can also be obtained by utilising fans or propellers (Fig. 6.37). These are mounted, vertically or horizontally, above the crop, stirring the air and bringing to the surface the warmer air from aloft which replaces the cold surface air. The continual stirring prevents the continued temperature drop at the surface.

Both these methods are restricted to radiative cooling conditions and

Fig. 6.37. A frost protection 'wind machine' constructed from an aeroplane propellor and driven with an automobile engine. In winter this machine is used to mix warmer air downwards to prevent frost damage to the apple trees. (Courtesy: Greg Johnson, North Carolina State University).

are not effective when the cold air extends for a great depth within the atmosphere, giving a freeze rather than a ground frost. Alternative methods can be used for low growing crops, such as strawberries, which can tolerate temperatures one or two degrees below freezing but not lower temperatures. The crop is sprayed with water prior to the onset of the freeze, so that the fruit is completely surrounded with water. As temperatures fall the water freezes and releases latent heat. Provided the fruit is completely enclosed, some of this heat enters it, raising its temperature slightly. Thereafter the encircling ice acts as an insulator impeding heat loss from the fruit and minimising the internal temperature drop.

With a crop of high enough value, or small enough areal extent, or the crop of a backyard gardener, it is of course possible to cover the crop physically. This will provide good protection from radiative frosts and some protection from those associated with deep cold layers. The covering material and method may range from simple styrofoam cups placed over the susceptible parts of individual plants to permanent glasshouses.

For the agriculturalist the decision whether or not to invest in frost protection equipment is largely a climatological one. Historical records of the time and frequency of frosts can be used to assess the probability of frost affecting the crops to be planted, and thus the likely economic benefit of the investment. The decision is similar to that associated with the purchase of snow removal equipment cited in Chapter 3 (sect. 3.9.2). The only extension that needs to be mentioned is the possibility of installing complete protection, for example, by using glass-houses. This approach is used widely in the Lee Valley of northeast London. This potentially rich agricultural area is susceptible to cold air drainage from the surrounding slopes, encouraging heat loss not only by radiative exchanges but also by turbulent transport. As is usually the case, the glasshouses primarily act to provide protection from the wind and only peripherally from the radiative heat loss. The cost of installing, maintaining, and heating them is more than compensated by the revenue generated from the sale of early season, or even out of season produce.

### Local-scale climatic 'engineering'

Frost is also of concern to the transportation industry. Railroad switches (railway points) in areas prone to frost and icing are frequently electrically heated, while some sensitive or intensely used roadways have electrical heaters built into their roadbed. This is most commonly found on bridges, which lack the thermal inertia of the land-based road and are thus more sensitive to temperature changes.

The examples given so far have been highly localised and specific. Two very widely applied climate modifications are irrigation and housing. Both are so familiar that we rarely think of them as climate modifications, but it is useful simply to point out here that they represent two very successful and long established systems of climate modification. Further, as we come to understand more about climate, it is possible to continue to refine our irrigation and building practices so that they become more effective and efficient modifiers of climate.

### Cloud seeding

If it seems to be stretching definitions somewhat to call irrigation and buildings modifications of climate, the same cannot be said of cloud seeding, used for both precipitation augmentation and cloud dissipation. Indeed this has come to epitomise weather and climate modification. The basis for such modifications is artifical stimulation of the

Bergeron–Findeisen process discussed in Chapter 3 (sect. 3.6). Cloud seeding can only be undertaken when a cold cloud is already present, and so serves as a means of precipitation augmentation, not 'rainmaking' in an area with no naturally occurring chance of rain. Due to this restriction it is extremely difficult to demonstrate rigorously that a specific cloud seeding project has led to increased rainfall. Nevertheless, the balance of evidence suggests that the technique is useful in areas where a small increase in precipitation can lead to significant economic gain. Hence, for example, it is widely practised in the somewhat marginal agricultural areas of the American Midwest and in South Africa. It is also used to augment winter snowfall in the American Rockies, helping to enhance the amount of water in that frozen reservoir which will become available in the subsequent growing season.

Attempts are being made to use cloud seeding techniques to modify the microstructure of clouds as an aid to hail suppression. If a large number of small hail stones can be produced to replace a small number of large ones, there is likely to be a decrease in the damage the hail causes (Table 6.7). Although theoretically cloud seeding could produce this effect, there is not at present any reliable and reproducible observational evidence that this has been successful. A similar conclusion can be drawn from the attempts to modify hurricanes, either by decreasing their intensity or altering their direction, by cloud seeding.

Cloud and fog dissipation schemes use a very similar technique but here the aim is to overseed. Solid carbon dioxide, with a temperature around $-78\,°C$, dropped into a cloud from an overflying aircraft locally supercools the cloud and produces ice crystals by spontaneous nucleation. A 'dry ice' pellet 10 mm in diameter falling through a cloud at $-10\,°C$ produces about $10^{11}$ crystals before evaporating. This effectively overseeds the cloud, giving a very large number of small crystals which rapidly evaporate into the unsaturated air and thus the cloud dissipates. Silver iodide introduced near the top of a cloud behaves like ice crystals and again overseeding can cause dissipation. This technique is often used for the dispersal of cold fog at airports.

Thus deliberate modifications that are known to be successful are restricted to changes in near surface conditions, mainly using the principles associated with modification of the energy and water balances. There is some suggestion of successful modification of certain types of cloud but little evidence to suggest that we have successfully modified them when they are in their most destructive phases.

## 6.9   The human response to climate

Human responses to atmospheric phenomena are many and varied, and occur on all scales from the personal to the global. However, most responses to climate are not direct; but are almost always a response to the impact of climate on human activities.

Perhaps the major exception to this is in the field of human comfort. Human beings generally become acclimatised to the 'normal' atmospheric conditions of the area where they live and have built-in resilience for some variation about this value. Wide deviations from the normal can cause severe problems. Unusually hot conditions, for example, can cause severe heat stress. Most societies become adapted to their normal range of high temperatures, partly through physiological adaption, but mainly through cultural adaption by the use of appropriate building styles, life styles or air conditioning. Thus people in some areas of the world can easily deal with temperatures that cause problems in other regions. For example, much of southern and central England experienced a severe heat wave in June and July 1976. Mortality rates for people over 65, the group most susceptible to heat stress, reflected the weekly mean maximum temperatures in early July (Fig. 6.38). It is interesting to speculate whether these deaths so depleted the group of those susceptible to high temperatures that there was no similar jump in mortality figures in late August.

## Concept of effective temperature and human comfort
Almost all of these temperatures for the English summer of 1976 fell

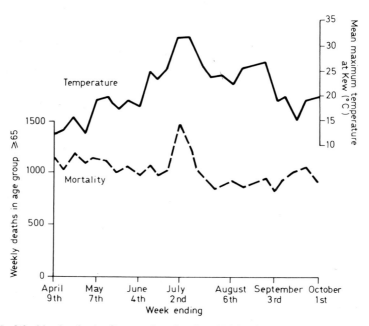

Fig. 6.38. Weekly deaths in Greater London in 1976 in the age group ≥ 65 years and mean maximum temperature as observed at Kew, London for the corresponding week.

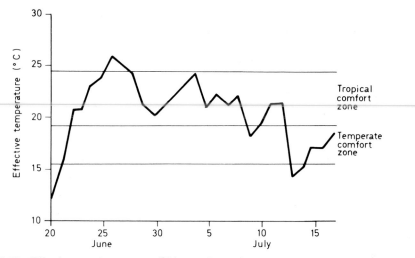

Fig. 6.39. Effective temperatures (°C) at the London Weather Centre at 1500 Z from 20 June to 17 July 1976.

within what has been defined as the 'tropical comfort zone' (Fig. 6.39), implying that peoples adapted to a tropical environment would have, in general, experienced no difficulty. The temperatures in this figure are given in terms of the *effective temperature*, defined as the temperature of still (average wind velocity 0.2 m s$^{-1}$), saturated air which has the same general effect upon comfort as the atmosphere under investigation. This index is just one of many which attempts to relate atmospheric variables to human comfort. It is not surprising that several indices have been developed, since the sensation of comfort depends on temperature, humidity, wind speed and the radiation environment in addition to the amount of clothing and the activity being undertaken by the body. Furthermore, there is no accepted means of measuring 'comfort', a sensation which may vary from person to person. Hence most indices have to be developed by exposing a group of people to various environmental situations and assessing their responses in terms of a generalised reaction.

The most commonly quoted index is the *temperature–humidity index* (THI) (Table 6.8), given by

$$\text{THI} = 0.4 \, (T_a + T_w) + 4.8 \qquad [6.10]$$

where $T_a$ and $T_w$, the dry and wet bulb temperatures, are in degrees C. THI is approximately equal to the effective temperature in near calm conditions.

The presence of wind decreases the apparent temperature experienced by the human body. This is beneficial at high temperatures, but

331

**Table 6.8  Apparent temperatures as a function of relative humidity, *RH***

| Air temp (°C) | Apparent temperature (°C) | Heat stress category | Average temperature |
|---|---|---|---|
| 60 | 52 | Caution | 27–32 |
| 57 | 49 53 | Extreme caution | 32–41 |
| 54 | 47 50 55 | Danger | 41–54 |
| 52 | 44 47 51 55 61 | Extreme danger | >54 |
| 49 | 42 44 47 51 54 59 64 | | |
| 46 | 39 42 44 46 49 53 57 62 66 | | |
| 43 | 37 39 41 42 44 47 51 54 58 62 66 | | |
| 41 | 35 36 38 39 41 43 45 48 51 54 57 61 65 | | |
| 38 | 33 34 35 36 37 38 40 42 43 46 49 52 56 59 62 | | |
| 35 | 31 31 32 33 34 34 36 37 38 40 42 43 46 48 51 54 58 | | |
| 32 | 28 29 29 30 31 31 32 33 34 35 36 37 38 39 41 43 45 47 50 | | |
| 29 | 26 26 27 27 28 28 29 29 30 31 31 32 32 33 34 35 36 37 39 41 42 | | |
| 27 | 23 23 24 24 25 25 26 26 26 27 27 27 28 28 29 30 31 31 32 32 33 | | |
| 24 | 21 21 21 22 22 22 23 23 23 23 24 24 24 24 25 25 26 26 26 26 27 | | |
| 21 | 18 18 18 18 19 19 19 19 20 20 21 21 21 21 21 21 22 22 22 22 22 | | |

| *RH* | 0  5 10 15 20 25 30 35 40 45 50 55 60 65 70 75 80 85 90 95 100 |

Equivalent and apparent temperatures are not real, measurable values. They are subjective impressions of temperature, exaggerated either because of wind or relative humidity. There are other formulae which give different results.

dangerous at low ones. Thus wind replaces humidity as the variable of concern at low temperatures. The most common formulation for the apparent temperature at low values is the *wind chill equivalent temperature* (Table 6.9).

Indicated in both Tables 6.8 and 6.9 are suggested responses to the atmosphere, responses designed not only to maintain human comfort but also physical well being. Obviously in regions where extremes occur frequently the society will have had to adapt to them, but the human response for outdoor activity, whether it be wearing suitable clothing or taking in adequate liquids or resting more frequently than usual, is a direct response to the climatic conditions.

Apart from this direct response in terms of comfort, the human response to climate on the local scale is one of responding to the impacts of climate. On this scale mankind can avoid or deliberately modify adverse climates and utilise favourable ones. Examples of each have already been given. Hill tops and valley bottoms are avoided by

**Table 6.9   Equivalent temperatures including the effects of wind chill**

| Estimated wind speed (m s⁻¹) | Actual thermometer reading (°C) | | | | | | | | | | | |
|---|---|---|---|---|---|---|---|---|---|---|---|---|
|  | 10 | 4 | −1 | −7 | −12 | −18 | −23 | −29 | −34 | −40 | −46 | −51 |
|  | Equivalent temperature (°C) | | | | | | | | | | | |
| 0 | 10 | 4 | −1 | −7 | −12 | −18 | −23 | −29 | −34 | −40 | −46 | −51 |
| 2.0 | 9 | 3 | −3 | −9 | −14 | −21 | −26 | −32 | −38 | −44 | −49 | −56 |
| 4.5 | 4 | −2 | −9 | −16 | −23 | −29 | −36 | −43 | −50 | −57 | −64 | −71 |
| 7.0 | 2 | −6 | −13 | −21 | −28 | −38 | −43 | −50 | −58 | −65 | −73 | −80 |
| 9.0 | 0 | −8 | −16 | −23 | −32 | −39 | −47 | −55 | −63 | −71 | −79 | −87 |
| 11.0 | −1 | −9 | −18 | −26 | −34 | −42 | −51 | −59 | −67 | −76 | −83 | −92 |
| 13.5 | −2 | −11 | −19 | −28 | −36 | −44 | −53 | −62 | −70 | −78 | −87 | −96 |
| 15.5 | −3 | −12 | −20 | −29 | −37 | −45 | −55 | −63 | −72 | −81 | −89 | −98 |
| 18.0 | −3 | −12 | −21 | −29 | −38 | −47 | −56 | −65 | −73 | −82 | −91 | −100 |

| >18 – little additional effect | Little danger for properly clothed people | **Increasing danger** | Great danger |
|---|---|---|---|
|  |  |  | Danger from freezing of exposed flesh |

cultivators of vineyards because of the possibility of frost. Solar energy systems can be utilised profitably in areas with large amounts of incoming solar radiation. Supercooled fogs can be dispersed to ensure that airports remain open. These examples indicate three areas of a local economy where climate has an impact. Others have been presented in earlier chapters indicating that almost all sectors of the economy are influenced, in one way or another, and to one degree or another, by climate. At the local scale the effects are usually obvious and can fairly easily be attributed to climatic causes.

*Cultural and economic impacts of climatic variations*
On the regional or national scale similar impacts are felt. For example, Table 6.10 lists some recent climatological hazards suffered in Australia. Here, of course, there is no possibility of climate modification, so the human response must be one of avoidance or utilisation. Most areas have, over a long period of time, 'adjusted' to the climate in, for example, their choice of agricultural practices or housing styles. Hence there is a pervasive utilisation of climate as a resource; although

**Table 6.10  Some recent climatological hazards in Australia**

| Feature (year) | Location | Effects (values are estimates) |
|---|---|---|
| **(1969–70)** | | |
| Drought | Widespread | 25% slump in net farm income |
| TC Glynis[a] | Dampier, WA | Breakwater damage |
| TC Ingrid | WA | Flooding, losses $1 million |
| TC Ada | Queensland central coast | 13 deaths, damage $12 million |
| Tornado | Bulahdelah, NSW | 1.5 million trees destroyed or damaged |
| **(1970–71)** | | |
| Lightning | NSW | Three deaths |
| TC Eva | Broome, WA | Extensive damage |
| TC Dora | Queensland south coast | Damage $1 million |
| TC Sheila | Roebourne, Dampier, WA | 220 km h$^{-1}$ wind gusts |
| Strong winds | Albany, WA | Damage $30,000 |
| Thunderstorms | Perth, WA | 200 houses damaged |
| Tornado | Adelaide, SA | Two houses unroofed |
| Strong winds | Port Macquarie, NSW | Damage $400,000 |
| Hail | Hobart, Tasmania | Damage $400,000 to apple crop |
| Frosts | Murray-Mallee, SA Mallee, Wimmera, Victoria | Extensive damage to cereal and vegetable crops |
| **(1971–72)** | | |
| TC Althea | Townsville, Queensland | Three deaths, damage >$50 million |
| TC Daisy | Bundaberg-Maryborough, Queensland | Two deaths, beach erosion, flood damage |
| Thunderstorm | Melbourne, Victoria | Record 1 h rain: 78.5 mm, wettest month ever: 238 mm, estimated release 100,000 tonne water over 1 km$^2$ of city |
| Tornadoes, thunderstorms | Southeast Queensland | Many houses damaged/destroyed, hail up to 100 mm diameter |

| Feature (year) | Location | Effects (values are estimates) |
|---|---|---|
| (1973–74) TC Una | Townsville-Rockhampton, Queensland | Flooding in coastal streams |
| TC Wanda | Brisbane-Ipswich, Queensland | Floods: 10,000 homes damaged/destroyed, 9000 people evacuated, damage $100 million |
| TC Zoe | Southeast Queensland/north-east NSW | Gales, flooding in coastal streams |
| Extratropical storm | Queensland/NSW coasts | Two large vessels driven ashore, six deaths, several yachts wrecked, extensive coastal erosion |
| Thunderstorm, tornado | Brisbane, Queensland | Damage >$2 million |
| (1975–76) Thunderstorm | Armidale, NSW | Damage $1 million including 500 hail-damaged vehicles |
| TC David | Yeppoon, Queensland | Houses unroofed, some demolished, flood damage $6 million |
| TC Joan | Port Hedland, WA | 85% town damaged, 100s buildings unroofed, major flooding, damage >$20 million |
| Thunderstorm | Toowoomba, Queensland | Damage $12–15 million |
| Tornado | Perth, WA | Damage $150,000 to state forest |
| (1976–77) TC Ted | Mornington Is./Burketown, Queensland | 1050 people lost homes, damage $8 million |
| TC Otto | Ingham-Tully, Queensland | Serious damage to sugar cane and property |

[a] *where*:
TC – tropical cyclone
WA – Western Australia
NSW – New South Wales
SA – South Australia

there is still wide scope for improvement in this area of applied climatology. More accurate specification of irrigation needs, or of the potential climate-related energy demand of an area, for example, uses the climate resource more efficiently and benefits the area economically. However, it is the adverse effects of climate that receive the most attention. Avoidance may not be possible, but adequate appreciation of, and adequate planning for, hazards may lessen their impact. The nature of the climatic event causing the impact, and the societal consequences, are many and varied. Figure 6.40 shows the regions of the United States having anomalous climatic conditions in a single month, while Table 6.11 outlines some of the possible impacts of these events.

The impact of climatic events on a regional scale usually have numerous consequences, some of them perhaps surprising. The summer of 1976 in Britain was exceptionally dry, qualifying as a drought with an estimated return period of around 100–200 years (Fig. 6.41). The lack of rainfall certainly retarded crop growth, but had

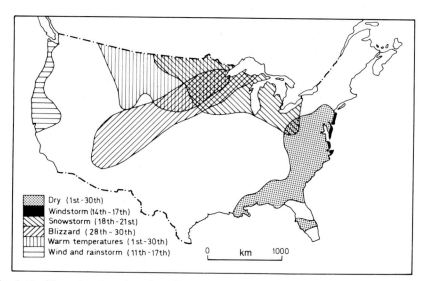

Fig. 6.40. Map showing the climatic events occurring during November 1981 in the USA as listed in Table 6.11.

an even greater effect on crop-destroying pests. The result was a harvest which was not too much below normal (Table 6.12). The unusually dry soil caused problems with the foundations of many buildings, a problem to the owners, but a boon to the construction trade. Similarly, resort areas in Britain reported an exceptionally profitable year. So, most climatic events have many and various impacts on the national economy, some sectors being favoured, some not.

**Table 6.11  Major weather events (USA) – November 1981**

| Event (date) | States significantly affected | Description |
|---|---|---|
| Wind and rainstorm (11th–17th) | California, Oregon, Washington | Beach erosion and coastal flooding resulted from a series of intense Pacific storms. Strong gale force winds generated swells of 5 to 6 m off the coasts of Washington, Oregon, northern California. Seasonal high tides on the western seaboard were intensified by the high southwest winds. Wave heights of 6–10 m were reported along the coast.<br><br>Storm warnings were issued for the coastline and bays while travellers' advisories were issued for inland areas. Gusts of 30–40 m s$^{-1}$ were common and speeds of up to 50 m s$^{-1}$ were recorded.<br><br>Although rainfall totals exceeded 25 cm in California no major flood damage was reported. |
| Windstorm (14th–17th) | Delaware, Maryland, New Jersey, North Carolina, Virginia | Strong northeasterly winds combined with high tides resulted in damages estimated close to $1 million. The hardest hit shoreline stretched from the Outer Banks of North Carolina to the beaches of New Jersey. During the four day period gale force winds and gusts of 35 m s$^{-1}$ were reported. Rainfall totals were generally 2.5 to 5.0 cm. No major flooding occurred. |
| Snowstorm (18th–21st) | Iowa, Michigan, Minnesota, North Dakota, Ohio, South Dakota, Wisconsin | A storm centred over Nebraska brought snow accumulation as high as 27.5 cm to Minneapolis and 35 cm to Houghton Lake, Minnesota, where a record for the greatest 24-hour snow fall was set. The storm had already left upwards of 15 cm in North Dakota. After immobilising Minnesota, Wisconsin and Michigan, the storm continued on through Ohio, Pennsylvania and West Virginia, leaving 22.5 cm in Union City, Pennsylvania. |

| Event (date) | States significantly affected | Description |
|---|---|---|
| Blizzard (28th–30th) | Arizona, Colorado, Iowa, Michigan, Minnesota, New Mexico, Nebraska, South Dakota, Utah | Blizzard conditions covered a path from Flagstaff, Arizona, through Minnesota. Snow accumulations ranged from 12.5 to 37.5 cm. Winds of up to 25 m s⁻¹ reduced visibility and created drifts of over a metre, closing roads and stopping ploughs. The storm continued through the month's end. |
| Dry (1st–30th) | Alabama, Delaware, Florida, Georgia, Louisiana, Mississippi, Maryland, New Jersey, North Carolina, Pennsylvania, South Carolina, Virginia, West Virginia | Concern over low water supplies, at start of the usually dry winter season, was increasing in the southeast portion of the country. Below normal rainfall again in November had aggravated an already dangerous situation. Areas in the Carolinas had received about 25–50% of normal rainfall for the period 31 August–30 November. Similarly, low percentage existed in the western panhandle of Florida.

Further north, through the middle Atlantic states, monthly precipitation totals were also below normal, both for November and the last quarter. Water supplies for agriculture, travel and drinking water were still too low to relax water restrictions in force since the previous year. |
| Warm temperature (1st–30th) | Minnesota, Montana, North Dakota, South Dakota, Wyoming, Utah | In contrast to October, monthly mean temperatures across the country averaged from near normal to 7 °C above normal. During the week of the 9th–15th temperatures exceeded 8 °C above normal. Record temperatures, some of the highest ever recorded this late in the season, were reported. A benefit of the warm temperatures included a large saving of energy dollars. Unfortunately the temperatures also contributed to incidents of dense fog. |

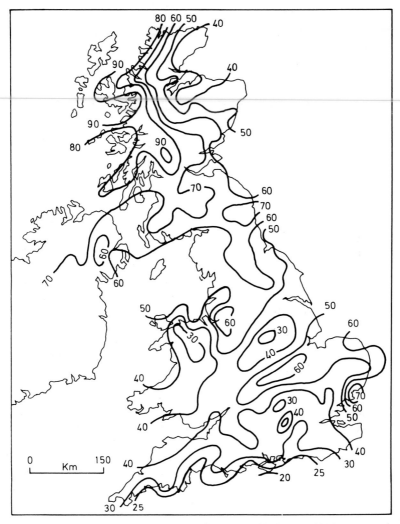

Fig. 6.41. Total rainfall in the period 1 April 1976 to 31 August 1976 expressed as the percentage of the 1916–50 average rainfall over the UK for the months April to August inclusive.

On the international scale climate events that have the most impact tend to be adverse effects which have a fairly long duration. The drought in the Sahel region and the Horn of Africa can be taken as a case in point. Through much of the 1970s and early 1980s rainfall was below average (Fig. 5.13). The resultant devastation of agriculture led to severe consequences in all segments of the society in the nations concerned, whether those segments were themselves directly affected by the climate or not. A disaster of this magnitude demands a response not

339

### Table 6.12 Annual estimates of average (UK) crop yields in tonnes per hectare

|  | 1974 | 1975 | 1976 | Percentage decrease in 1976 from 1974 yields |
|---|---|---|---|---|
| **England and Wales** | | | | |
| Wheat | 4.94 | 4.30 | 3.85 | 22 |
| Barley | 3.95 | 3.40 | 3.46 | 12 |
| Oats | 3.88 | 3.45 | 3.42 | 12 |
| Potatoes: early | 18.8 | 14.1 | 16.3 | 13 |
| maincrop | 33.9 | 22.1 | 20.4 | 40 |
| **Scotland** | | | | |
| Wheat | 5.75 | 5.58 | 4.98 | 13 |
| Barley | 4.99 | 4.79 | 4.10 | 18 |
| Oats | 3.66 | 3.45 | 3.58 | 2 |
| Potatoes: early | 24.1 | 19.3 | 19.4 | 20 |
| maincrop | 30.9 | 26.6 | 26.2 | 15 |

only from the nations directly concerned, but from all nations and the international community. In that sense, the impact of climate is a global impact.

Other truly global climatic events have yet to be detected. However, the threat posed by climate changes which may be induced by the increase in carbon dioxide in the atmosphere could lead to considerable impact throughout the world. When, how, or even whether, mankind can respond to such impacts is an open question which we consider in Chapter 7.

# Chapter 7
# The Future – Climate Change, Climate Models, Climate Impacts

7.1    Mechanisms of climatic change

   7.1.1   External causes of climatic change

   7.1.2   Internal causes of climatic change

7.2    Global-scale climate models

   7.2.1   Energy balance climate models

   7.2.2   One-dimensional radiative–convective climate models

   7.2.3   Two-dimensional climate models

   7.2.4   General circulation climate models

   7.2.5   Climatic feedback effects

7.3    Local climate models

   7.3.1   Agricultural production and climate

   7.3.2   Water resources

7.4    Changes and cycles in the climate system

   7.4.1   Geological record of climate

   7.4.2   Shorter time scale climatic changes

   7.4.3   Historical climatic change

7.5    Climatic changes and their impact on Man

7.6    Future climates – the probable impact of carbon dioxide

7.7    Epilogue

# Chapter 7
## The Future – Climate Change, Climate Models, Climate Impacts

All our discussion so far has stressed understanding and description of the nature and controls of the present climate. As important as such an understanding is, modern climatology is beginning to be able to go a step further and consider possible future climates. This is achieved through the use of climate models, simplified physical and mathematical descriptions of the controls on climate, which allow us to investigate the causal mechanisms of climatic change. These models are based on our understanding of the present climate. An examination of past climates is enabling us to identify those causal mechanisms of change which were of importance in the past, thus suggesting the major effects to consider when looking into the future.

Since these considerations indicate that climatic change in some form is inevitable, it is also important to investigate whether any postulated future changes will have a significant impact on society. Again, a study of impacts in the past gives us some insight into possible future impacts. However, a full analysis involves much more than climatology. Hence climatologists can only sketch in a few guidelines, providing the framework upon which society can operate to respond to any new climate that develops.

The discussions in this chapter must take a somewhat different form from those presented previously. Much of the material in earlier chapters consisted of well established and well documented facts, although, as we indicated, many areas of uncertainty remain. In the present chapter, however, there are very many uncertainties and relatively few well documented facts. The fields of climate modelling, past climate specification and climate impact analysis are areas of rapid development. Few definitive answers are available as yet. Consequently here we indicate the main lines of research and speculate on some possible answers that may emerge.

# 7.1 Mechanisms of climatic change

Climate changes can result from the action of any of the processes affecting the climate system. However, changes that affect the whole Earth, or major portions thereof, for at least several years, are likely to arise from a relatively small number of causes. These changes can be split into *external*, in which the agent of cause is outside the climate system, and *internal*, where the initial alteration is within the system itself (Fig. 1.3 in Ch. 1). Isolating a particular cause for a particular change, however, is extremely difficult because the interlinked nature of the system ensures that there are feedbacks, so that a change in one component leads to a change in most, if not all, other components.

This feedback effect can be illustrated by considering climate changes on the longest possible time scale, that of the evolution of the Earth. It is generally agreed that solar luminosity has increased by between 20% and 40% during the $4.5 \times 10^9$ years of the Earth's existence. Low incoming solar radiation suggests that surface temperatures should be low, certainly below the freezing point of water. Geological evidence suggests that this was not the case, leading to the 'weak Sun – enhanced early surface temperature' paradox. The explanation appears to be that higher concentrations of carbon dioxide in the primaeval atmosphere enhanced the 'greenhouse effect' and created higher temperatures. With the advent of biological activity the amount of oxygen in the atmosphere increased while the $CO_2$ concentration decreased. This, combined with associated surface changes, ensured that temperatures remained stable as the Sun's output increased.

## 7.1.1 External causes of climatic change

Potential external causes of climatic change include changes in the luminosity of the Sun and in the astronomical relationship between the Earth and the Sun and changes in the character of the Earth's surface as a result of continental drift and mountain building forces. In addition, changes in the polarity of the Earth's magnetic field may influence the upper atmosphere and thus the whole climate. Of these, possible changes resulting from astronomical changes and luminosity changes in sunspots have received the most attention.

The astronomical theory of climate variations, also called the *Milankovitch theory*, is an attempt to relate climatic variations to the changing orbit of the Earth around the Sun. There are several different ways in which the orbital configuration can affect the received radiation and thus possibly the climate. They are (Fig. 7.1):

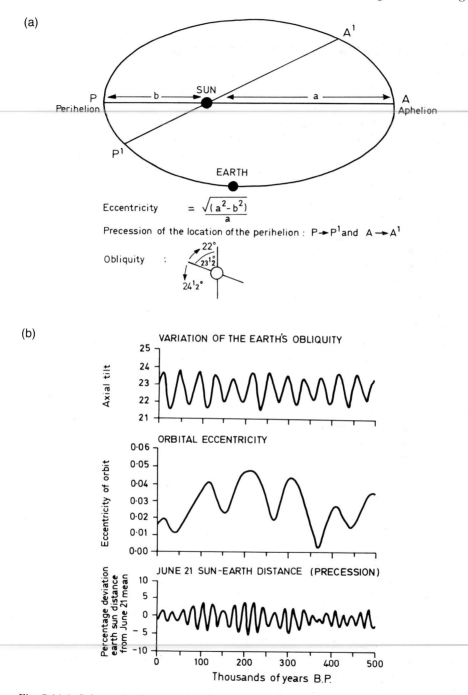

Fig. 7.1(a) Schematic diagram showing the variations in the three orbital components: eccentricity, precession of the location of the perihelion and obliquity. (b) Variations in these three components as a function of time. (Broecker, W.S. and Van Donk, J. 1970, *Rev. Geophys.* **8**, p. 188).

1. *Eccentricity variations*: The Earth's orbit becomes more circular and then more elliptical in a pseudo-cyclic way, completing the cycle in about 110,000 years. The mean annual incident flux varies as

$$\text{flux} \propto (1 - e^2)^{\frac{1}{2}} \qquad\qquad [7.1]$$

where the eccentricity of the orbit is $e$. For a larger value of $e$ there is a smaller incident annual flux. The current value of e is 0.018. In the last 5 million years it has varied from 0.000483 to 0.060791. These variations would result in changes in the incident flux of from $+0.014\%$ to $-0.17\%$.

2. *Obliquity*: The obliquity, the tilt of the Earth's axis, is the angle between the Earth's axis and the plane of the ecliptic (the plane in which the bodies of the solar system lie). This tilt varies from about 22° to 24.5°, with a period of about 40,000 years. The current value is 23.5°. Seasonal variations depend upon the obliquity (Fig. 2.7). If the obliquity is large so is the range of seasonality. However, the total received radiation is not altered, but a greater seasonal variation in received flux is accompanied by a smaller meridional gradient in the annual radiation.

3. *Precession of the location of perihelion*: The orbit of the Earth is an ellipse around the Sun, which lies at one of the foci. The point in the orbit when the planet passes closest to the Sun is called the perihelion ('close to Sun') point. Due to gravitational interaction with the other planets, primarily Jupiter, this point moves in space so that the ellipse is moved around in space. This precession will cause a precession in the time of the equinoxes. These changes occur in such a way that two periodicities are apparent: 23,000 years and 18,800 years. This change, like (2), does not alter the total received radiation but affects the distribution of the energy in both time and space.

## Confirmation of the Milankovitch hypothesis?

For long-term temperature data, spectral analysis, which permits the identification of cycles, has shown the existence of cycles with periods of $\sim 20 \times 10^3$, $\sim 40 \times 10^3$ and $\sim 100 \times 10^3$ years (Fig. 7.2). These correspond rather closely with the Milankovitch cycles. However, the strongest signal in the observational data is in the longest time period, with a cycle of 100,000 years. This would be the result of eccentricity variations in the Earth's orbit, which produce the smallest insolation changes. Hence the effect of the Milankovitch cycles is far from clear. While they offer an interesting 'explanation' for long-term, cyclic climatic changes, simulation models incorporating these external changes cannot yet produce temperature changes close enough to the observations to be conclusive. Almost certainly these external changes

346

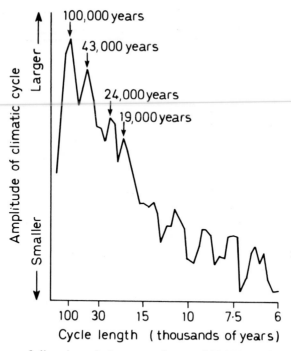

Fig. 7.2. Spectrum of climatic variations over the past 500,000 years. The graph shows the importance of the climatic cycles of 100,000 years (eccentricity); 43,000 years (obliquity) and 24,000 and 19,000 years (precession of the location of the perihelion). The curve is constructed from an isotopic record of two Indian Ocean cores.

trigger large feedback effects in the climate system which are yet to be fully understood.

### Sunspots

Variations in the climate during historical times have been linked with the sunspot cycle, the second possible external cause of solar-produced climatic change. This cycle occurs with a 22-year periodicity, the 'Hale' double sunspot cycle (Fig. 7.3(a)). The overall amplitude of the cycles seems to increase slowly and then fall rapidly with a period of 80–100 years. There also appears to be a quasi-cyclic fluctuation of the order of 180 years.

There have been correlations of sunspots with very short-term changes in the solar 'constant' and each of these apparent periodicities has been associated with suggested climate periodicities. In particular, the Little Ice Age (see sect. 7.4.3) has been linked to the '*Maunder minimum*' in sunspots (Fig. 7.3(b)). Between 1650 and 1730 there was both a sudden and unaccountable reduction, almost to zero, in the sunspot activity of the Sun, and low surface temperatures on Earth.

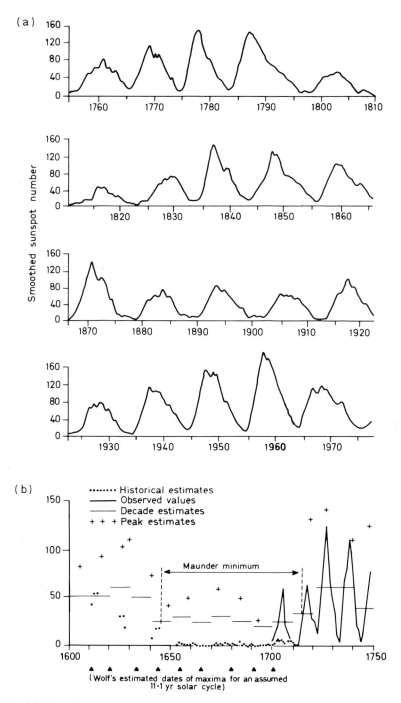

Fig. 7.3. (a) Variations in the annual mean sunspot number showing the 11-year sunspot number cycle which, when the magnetic polarity of the sunspots is noted, is found to be half the true 22-year cycle. (b) Estimated annual mean sunspot numbers in the period AD 1610 to 1750. The absence of sunspots during the Maunder minimum is clearly seen.

However, the actual period of the Little Ice Age itself seems to vary according to the geographical area from which data are taken. Certainly as yet no mechanistic link has been shown to relate sunspot numbers directly to the mean surface conditions on Earth.

## 7.1.2  Internal causes of climatic change

Any change in any component internal to the climate system will lead to a climatic perturbation. Most current concern centres on anthropogenically created effects and their consequences, which could operate on the relatively short time scales necessary to create noticeable changes within the next century. The only natural effect likely to be important on this time scale is volcanic activity.

### *Volcanoes*

Volcanoes provide a huge local source of particulates, heat and water vapour for the atmosphere, while volcanic ejecta, in the form of lava flows, alter the surface character. Vulcanism can certainly produce measurable temperature anomalies of at least a few tenths of a degree (Fig. 7.4(a)). However, an eruption and the resultant effect on the atmosphere is short-lived compared to the time needed to influence the heat storage of the oceans. Hence the temperature anomaly is unlikely to persist or lead, through feedback effects, to significant long-term climatic changes.

The energy and type of the volcanic eruption largely determine its climatic effect. Most eruptions inject aerosols into the troposphere at heights between 5 and 8 km. These particles are soon removed either directly by gravitational fallout or by rainout (sect. 6.7.1) and their climatic effect is small. Violent eruptions can hurl debris into the upper troposphere or even into the lower stratosphere (15–25 km) (e.g. Mount Agung in 1963 and El Chichón in 1982). These are much less common but can lead to more extensive climatic effects. At these heights there is a smaller chance of immediate removal by direct interaction with atmospheric water. Furthermore, the particles have a long residence time in the stratosphere, of the order of a year for particles of radii 2–5 $\mu$m but as long as 12 years for smaller particles of radii 0.5–1.0 $\mu$m. Thus they can become widely distributed by the stratospheric circulation. For instance, after the Krakatoa eruption in 1883 red skies, blue moons and fantastically colourful sunsets were observed in many places for several months.

Immediately following an eruption the stratosphere is dominated by particles which scatter radiation of wavelengths up to 10 $\mu$m very efficiently, being about ten times better scatterers than the normal stratospheric particles. They also absorb visible radiation. The contribution of the particles to the 'clear sky' optical thickness, $\tau$, can rise to 0.1 (20 times the normal value) after large eruptions. After about

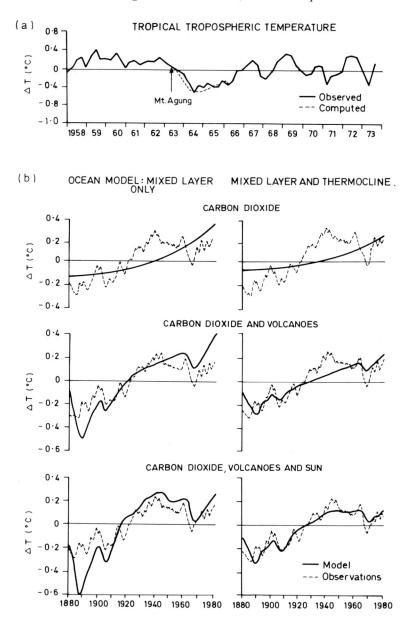

Fig. 7.4. (a) Observed tropospheric temperatures for the region between 30 °N and 30 °S and model-computed temperatures for the period after the eruption of Mount Agung in 1963. Both computed and observed temperatures exhibit a significant decrease as a result of the volcanic eruption. (b) Global temperature trends obtained from a one-dimensional radiative–convective (1-D RC) climate model taking changes in atmospheric $CO_2$, volcanic activity and solar luminosity into account. The results are in good agreement with the observed surface temperatures.

6 months there is an increase in sulphate production, which further increases the visible scattering and slightly increases the absorption in the infrared. These changes will affect the atmospheric heating rates. To attempt to assess the resultant impact on the climate it is necessary to calculate how the ratio of absorption to backscatter for both short-wave and longwave radiation alters relative to the normal situation. This seems to be partly dependent upon surface albedos. If the surface albedo is very high, warming can occur if the eruption causes approximate equality between absorption and backscatter because the radiation which is now being absorbed was previously being lost to space. Over a surface with a lower albedo, however, most of the incident radiation was being absorbed anyway so that the only way to produce a net heating is if the absorption is altered to be greater than the backscatter. It is more likely in this case that backscatter from the volcanic particles will cause a local cooling.

Model simulations suggest that radiation effects will, overall, produce a global cooling when large-scale volcanic eruptions occur. Figure 7.4(b) shows a typical simulation attempting to reproduce recent temperature changes, taking into account changes in solar luminosity and atmospheric $CO_2$ in addition to volcanoes. The volcanic effect was established using a 'dust veil index' developed from a catalogue of volcanic eruptions. The results clearly support the idea that volcanoes have a cooling effect. In all analyses, however, a complication arises because volcanic particles can serve as condensation nuclei, influencing the type and amount of cloudiness and thus possibly further altering the climate.

### Tropospheric aerosols

The effects of tropospheric aerosols on the radiative fluxes in the atmosphere are compared with those of gases and clouds in Table 7.1. The particle effect is generally small. Furthermore, there is no direct evidence of increasing particulate loading in the atmosphere and many measurements suggest that there has been no change in the atmospheric loading except following large volcanic eruptions. This suggests that such particles can be neglected as a possible climatic change mechanism. However, some recent measurements do seem to indicate that the electrical conductivity of the Northern Hemisphere troposphere has been steadily increasing over the last 70 years. One reason proposed for this is an increase in the concentration of aerosols in the size range 0.02 to 0.2 $\mu$m which, in turn, are likely to be attributable to human activities rather than natural sources. This change in particle size distribution may lead to variations in both radiative exchanges and cloud amount.

351

## Table 7.1 Interaction between atmospheric constituents and electromagnetic radiation in short and long wavelength regions

| Agent | Effect | Magnitude of effect | |
|---|---|---|---|
| | | Shortwave | Longwave |
| Gases ($CO_2$, $H_2O$) | Absorption | Moderate | Large |
| | Scattering | Small | Negligible |
| Aerosols (particles) | Absorption | Small | Small (large for Man-made) |
| | Scattering | Moderate | Small |
| Clouds | Absorption | Small | Large (varies with thickness) |
| | Scattering | Large | Large, but dominated by absorption |

### *Clouds*

Variations in cloud type and amount can result from many processes. One possible factor is the abundance of cloud condensation nuclei. Injections of excess aerosols, both natural and anthropogenic, into the lower troposphere tend to be removed fairly quickly. While this probably means that there is an increase in total condensation and rainfall, the effects are local and over a short time period. Similarly the input into the stratosphere is unlikely to affect cloudiness because the water vapour is generally unable to penetrate into the stratosphere. However, injections into the upper troposphere may have far-reaching effects. At heights of 10–18 km it is likely that some particles, whether volcanic debris or of human origin, could increase the formation of ice clouds. An increase in cirrus cloud cover has recently been described as one of the possibly critical climatic perturbing agents (Fig. 7.5). This is because an increase in cirrus cloud amount will increase absorption of the infrared radiation and therefore lead to a general climatic warming.

### *Stratospheric ozone*

A major detrimental effect on the Earth's atmosphere that is believed to arise from gaseous pollution emission is the perturbation of the ozone layer in the lower stratosphere. The prime concern is with chlorofluoromethanes, commonly called 'freons', which are used as refrigerants and spray-can propellants. Thus effectively there is an extensive areal source in the Northern Hemisphere. In the troposphere freons

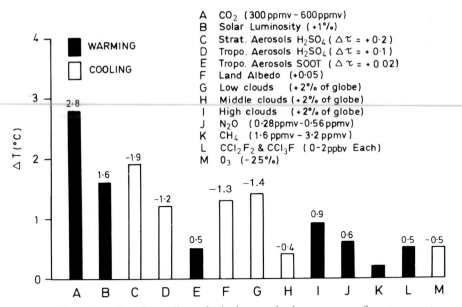

A  CO$_2$ (300 ppmv – 600 ppmv)
B  Solar Luminosity ( + 1°/o )
C  Strat. Aerosols H$_2$SO$_4$ ( $\Delta \tau$ = + 0·2 )
D  Tropo. Aerosols H$_2$SO$_4$ ( $\Delta \tau$ = + 0·1 )
E  Tropo. Aerosols SOOT ( $\Delta \tau$ = + 0 02 )
F  Land Albedo ( +0·05 )
G  Low clouds   ( +2°/o of globe)
H  Middle clouds ( +2°/o of globe)
I  High clouds  ( +2°/o of globe)
J  N$_2$O  ( 0·28 ppmv – 0·56 ppmv)
K  CH$_4$  ( 1·6 ppmv – 3·2 ppmv )
L  CCl$_2$F$_2$ & CCl$_3$F  ( 0–2 ppbv Each)
M  O$_3$  (–25°/o)

Fig. 7.5. The effect of various climatological perturbations upon surface temperatures as predicted by a 1-D RC climate model (see sect. 7.2.2). The model used is that described as Model 2 with reference to Table 7.2. One of the major perturbations causing cooling is seen to be increased loading of stratospheric aerosols. Column I shows that increasing cirrus cloud operates to warm rather than to cool cf. increases in middle and low level clouds (columns G and H). Warming by a 2% increase in high clouds is greater than the cooling caused by a 2% increase in middle level clouds.

are extremely stable and have residence times of decades. However, when they reach altitudes above 25 km, they undergo photochemical dissociation, releasing free chlorine. This chlorine reacts with ozone, destroying it and replacing it with oxygen. The chlorine itself will combine with other compounds and be transported down to the troposphere. The details of the various reactions, particularly the rates at which they occur, are not well known. Indeed, some observations suggest 'natural' fluctuations in ozone concentrations of magnitudes considerably greater than those postulated from the impact of freons. Thus it is by no means clear which of the reactions will dominate and whether the net result will be an increase in stratospheric ozone or a decrease, or whether the change will be climatically significant. There is, however, a significant effect due to the chlorofluoromethanes which remain in the troposphere. They have recently been shown to possess infrared absorption features which means that they will contribute increasingly to the greenhouse effect as their concentration increases.

The study of atmospheric chemistry has tended to be neglected by climatologists perhaps because the time scales of chemical reactions are so much shorter than those typically of importance in climatology. It is becoming increasingly obvious, however, that the chemistry of the atmosphere has a great bearing on radiative exchanges and on the transfer rates of atmospheric properties, and thus must be treated as an integral part of the climate system.

### Carbon dioxide

The increase in the concentration of atmospheric $CO_2$ during the last 100 years has been well documented. This increase is a worldwide phenomenon. The immediate effect of this increase in atmospheric $CO_2$ is to raise the surface temperature by the greenhouse effect (Fig. 2.20). Although $CO_2$ is itself a greenhouse gas the majority of the increased greenhouse effect induced by $CO_2$ is due to increased temperatures permitting a larger concentration of atmospheric water vapour which is a still more efficient greenhouse absorber. The consequences of increasing atmospheric $CO_2$ will be explored more fully in section 7.6.

### Surface changes

The concept of anthropogenic alteration of the climate has also been stimulated by recent investigations of bio-geophysical feedback effects at the Earth's surface. These occur through interactions between biological changes and atmospheric and/or hydrological changes. The classical example is the exacerbation of drought conditions as a result of alterations in the vegetation of a region, as described in Chapter 5 (sect. 5.8). Such changes, as yet, are local rather than global, but are persistent and have the potential to create large-scale climatic changes.

## 7.2 Global-scale climate models

Any climate model is an attempt to simulate the many processes that produce climate. The objective is to understand these processes and to predict the effects of changes of the processes. This simulation is produced by applying the basic physical laws that govern the climate. Hence a model can be thought of as a series of equations expressing these laws.

For several reasons a model must be a simplification of the real atmosphere. The laws governing the processes are not fully understood, but they are known to be complex. Furthermore, they interact with each other, producing *feedbacks*, so that any solution of the governing equations must involve a great deal of computation. The solutions that are produced start from the present conditions and investigate the effects of changes in a particular component of the climate system. However, rarely are the observational data sufficient to specify completely the starting conditions, so that there is inherent uncertainty in the results. Consequently climate models can be long and costly to use, even on the fastest computer, and the results can be only approximations.

Global-scale climate models, designed to simulate the climate of the planet, must take into account the whole climate system (Fig. 1.3). All of the interactions between components must be integrated in order to develop a climate model. This presents great problems, because the various interactions operate on different time scales. For example, the effects of deep ocean overturning may be very important when considering climate averaged over a decade, while local changes in wind direction may be unimportant on this time scale. If monthly conditions are of concern, however, the relative importance would be reversed.

Global-scale models are usually concerned with long-term average conditions. It is common, for example, to develop models for average conditions in January and July. This is not to infer that a particular January in the period for which a climate model prediction is made will have these conditions, only that the conditions apply to an average January. Indeed, it is always implied that any 'new' climate predicted will have variation about the mean, just as with the present climate. The amount of this new variation is often of concern when the results of global-scale models are used to estimate the possible impact of climatic change in a local area.

The simplifications that must be made to the laws governing climatic processes can be approached in several ways. Consequently there are numerous different global-scale climatic models available. In general, two sets of simplifications must be made. The first involves the action of the processes themselves. It is usually possible to treat in detail some of the processes, specifying their governing equations fully. However, other processes must be treated in an approximate way, because of our lack either of understanding of them or of adequate computational resources to deal with them. For example, it might be decided to treat the radiation streams in great detail, but only approximate the horizontal energy flows associated with the wind. The approximation may be approached either by using available observational data, the *empirical* approach, or through specification of the physical laws involved, the *theoretical* approach.

The second set of simplifications involves the resolution of the model in both time and space. While generally the finer the spatial resolution, the more reliable the results, constraints of both data availability and computational time may dictate that a model may have to use, for example, latitudinally averaged values as the basic input. In addition, too fine a resolution may be inappropriate because processes acting on a smaller scale than the model is designed to resolve may be inadvertently incorporated. Similar considerations are involved in the

355

choice of time resolution. Most computational procedures require a 'time step' approach to calculations. The processes are allowed to act for a certain length of time and the new conditions are calculated. The process is then repeated using these new values. This continues until the conditions at the required time have been established. Again, although accuracy potentially increases as the time step decreases, there are constraints imposed by data and computational ability and by the design of the model.

Although models are designed to aid in predicting future climates, their performance can only be tested against the past or present climate. Usually an initial objective when a model is developed is to ascertain how well it compares with the present climate. Thereafter it may be used to simulate past climates, not only to see how well it performs, but also to gain insight into the causes of these climates. Although such past climates are by no means well known, this comparison provides a very useful step in establishing the validity of the modelling approach. After such tests, the model may be used to gain insight into possible future climates.

With the variety of models available and the difficulty of testing them against both past and present climates, it is not surprising that many predictions of future climate are available. Nevertheless, some must be treated with much more caution than others, since the validation process is long and complex, usually requiring many runs of the model and much fine tuning of the parameterisations and empirical assumptions that are made.

### Types of climate models

The important components to be considered in constructing or understanding a model of the climate system are the same as those considered throughout this book:

(a) *radiation* – the way in which the input and absorption of solar radiation and the emission of infrared radiation are handled.

(b) *dynamics* – the movement of energy around the globe (specifically from low to high latitudes) and vertical movements – e.g. convection.

(c) *surface processes* – inclusion of land/ocean/ice and the resultant change in albedo, emissivity and surface–atmosphere energy interchanges.

(d) *resolution in both time and space* – the time step of the model and the horizontal and vertical scales resolved.

356

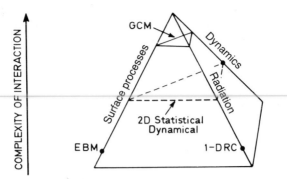

Fig 7.6. The climate modelling pyramid. The three major features to be included in climate models (surface processes, radiation and dynamics) are shown on the edges of the pyramid and resolution or complexity of interaction is shown increasing towards the apex.

The relative importance of these processes and the physical (as opposed to empirical) basis for parameterisations employed in their incorporation can be discussed using the 'Climate Modelling Pyramid' (Fig. 7.6). The edges represent three of the basic elements of models, and complexity is shown increasing upwards. At the base of the pyramid are the simpler climate models which incorporate only one primary process.

Figure 7.6 indicates that there are four basic types of model:

1. General circulation models (GCMs). The full three-dimensional nature of the atmosphere (and sometimes also the oceans and cryosphere) is included. An attempt is made to represent most physical processes believed to be important although the spatial/temporal resolution is too coarse for synoptic scale processes to be modelled correctly.
2. Two-dimensional zonally averaged statistical dynamical models (2-D SD) deal explicitly with surface processes and dynamics in a zonally averaged model and have a vertically resolved atmosphere.
3. Energy balance models (EBMs) are one-dimensional models of the variation of the surface (strictly the sea-level) temperature with latitude. Simplified relationships are used to calculate the terms contributing to the energy balance in that latitude zone.
4. One-dimensional radiative convective models (1-D RC) compute the vertical temperature profile by explicit modelling of the radiative processes and a 'convective adjustment' which re-establishes a predetermined lapse rate.

## *A simple global energy balance model*

The concepts used in each of these types of climate models can be illustrated by a discussion of the simplest type of climate model: the

357

*global energy balance model.* This is the globally averaged case of model group (3), the EBMs. In this simple model the incoming and outgoing energy for the globe are balanced and a single climatic parameter (the surface temperature, $T$) is calculated, i.e. $T$ is the dependent variable for which the 'climate equations' are solved. The rate of change of temperature, $\Delta T$, with time, $\Delta t$, is caused by a difference between the top-of-the-atmosphere (or planetary) net incoming $R\downarrow$ and net outgoing $R\uparrow$ radiant energies (per unit area).

$$\rho C \; \Delta T/\Delta t = R\downarrow - R\uparrow \qquad\qquad [7.2]$$

where $\rho C$ is the heat capacity of the system (per unit area). This is a very general equation with a variety of uses. Thus, for example, if the system we wish to model is an outdoor swimming pool we can calculate the rate of temperature change in time steps of 1 day from equation [7.2]. Suppose the pool has surface dimensions 30 m × 10 m, is well-mixed and is 2 m deep. Since 4200 J of energy are needed to raise the temperature of 1 kg of water 1 K, and 1 m³ of water weighs 1000 kg, the pool has a total thermal capacity (i.e. the amount of energy (J) needed to raise the temperature by 1 K) equal to $2.52 \times 10^9$ J K$^{-1}$. If we assume that the difference between the absorbed radiation and the emitted ratiation from the pool ($R\downarrow - R\uparrow$) is 20 W m$^{-2}$ for 24 hours, then the difference in energy content of the pool for each 24-hour time step is 20 × 30 × 10 × 24 × 60 × 60 joules. Then, from equation [7.2]

$$2.52 \times 10^9 \; \Delta T = 20 \times 30 \times 10 \times 24 \times 60 \times 60, \qquad \text{or}$$
$$\Delta T \text{ (in 1 day)} = \frac{2 \times 3 \times 24 \times 36}{2.52 \times 10^4} \approx 0.2 \text{ K} \qquad\qquad [7.3]$$

Thus at this rate it would take about a month to raise the temperature of the pool water by 6 K.

On the Earth the value of $\rho C$ is largely determined by the oceans. For instance if we assume that the energy is absorbed in the first 70 m of the ocean (the average global depth of the top or mixed layer) and that approximately 70% of the Earth's surface is covered by oceans, then the value for $\rho C$ comes from

$$\rho C = \rho_w \, c_w \, d \; 0.7 = 2.06 \times 10^8 \text{ J m}^{-2} \text{ K}^{-1}. \qquad\qquad [7.4]$$

where $\rho_w$ is the density of water, $c_w$ its specific heat at constant pressure and $d$ is the depth of the mixed layer.

For our simple energy balance model of the Earth, the energy emitted, $R\uparrow$, can be estimated using the Stefan–Boltzmann law (equation [2.13]) and the surface temperature. This value must be corrected to take into account the infrared transmissivity of the atmosphere $\tau_a$ since it is the planetary flux. Therefore we can write

$$R\uparrow \simeq \epsilon \ \sigma \ T^4 \ \tau_a \qquad [7.5]$$

The absorbed energy, $R\downarrow$, is a function of the solar flux, $S$, and the planetary albedo such that $R\downarrow = S(1-A)$.

Equation [7.2] therefore becomes

$$\Delta T/\Delta t = \{ S (1 - A) - \epsilon\tau_a\sigma T^4 \}/\rho C \qquad [7.6]$$

This equation can be used to ascertain the equilibrium climatic state by setting $\Delta T/\Delta t = 0$. Then

$$S (1 - A) = \epsilon\tau_a \ \sigma \ T^4 \qquad [7.7]$$

Using values of $S = 342.5$, $A = 0.3$, $\epsilon\tau_a = 0.62$ and $\sigma = 5.67 \times 10^{-8}$ gives rise to a surface temperature of 287 K which is in good agreement with the globally averaged surface temperature today. Note that in this calculation the solar constant ($S_F = 1370$ W m$^{-2}$) is divided by 4 to give $S$, since instantaneously the incoming radiation is incident on a smaller area than the area over which radiation is emitted. The ratio, 4 (Fig. 7.7), makes $R\downarrow$ directly comparable with $R\uparrow$.

An alternative use of equation [7.6] is similar to the calculation of the swimming pool warming rate made above. Here a time step calculation of the change in $T$ is made. This could be a response to an 'external' forcing agent, such as a change in solar flux or in the heat capacity of the oceans resulting from changes in their depth or area. Alternatively, the response could be determined by an 'interactive' climate calculation when one of the internal variables (e.g. $A$) alters.

### Equilibrium climatic states

As an example of the change in an internal variable we can consider the variation in $A$ as a function of the mean global temperature. Above a certain temperature, $T_{no\ ice}$, the planet is ice free and albedo is independent of temperature. As it becomes colder we expect the albedo to increase as a direct result of increases in ice and snow cover. Eventually the Earth becomes completely ice-covered, at temperature $T_{ice}$, and further cooling will produce no further albedo change. This could be expressed in the form

$$A(T) = A_{ice} \text{ for } T \leqslant T_{ice}$$
$$A(T) = A_{no\ ice} \text{ for } T \geqslant T_{no\ ice} \qquad [7.8]$$
and $A(T)$ is a linear function of $T$ for $T_{ice} < T < T_{no\ ice}$

**ABSORBED SOLAR RADIATION**

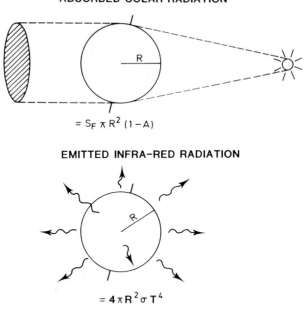

$$= S_F \, \pi \, R^2 \, (1-A)$$

**EMITTED INFRA-RED RADIATION**

$$= 4 \pi R^2 \sigma T^4$$

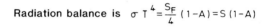

**Radiation balance is** $\quad \sigma T^4 = \dfrac{S_F}{4}(1-A) = S(1-A)$

Fig. 7.7. Diagrammatic representation of the radiation balance of the Earth. Radiation is emitted over the whole area of the sphere $(4\pi R^2)$ whilst an area $\pi R^2$ intercepts the solar radiation, $S_F$.

$T_{ice}$ is usually assumed to be 273 K but may range between 263 K and 283 K. If we are concerned with equilibrium conditions, we can calculate $R \uparrow$ for a series of temperatures and $R \downarrow$ for the series of albedos and show the results graphically. The point of intersection of the curves represents the equilibrium situation (Fig. 7.8). Any slight imbalances between the absorbed radiation, $S(1-A(T))$, and the emitted longwave flux, $\epsilon \sigma \tau_a T^4$, lead to a change in the temperature of the system at the rate $\Delta T/\Delta t$, the changes serving to return the temperature to an equilibrium state. However, there are three equilibrium solutions of equation [7.6], as shown in Fig. 7.8: a completely glaciated Earth (3), an ice-free Earth (1) and an Earth with some ice (2). All are possible.

### Equilibrium conditions and transitivity of climate systems
This simple model has some very obvious limitations. Nevertheless, it not only shows typical lines of approach, but also indicates some of the

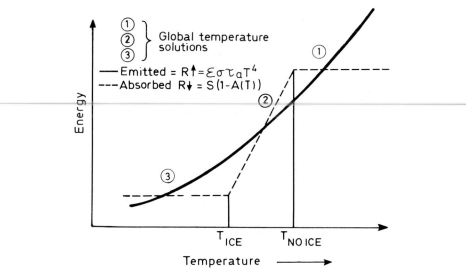

Fig. 7.8. The equilibrium temperature solutions for a zero-dimensional global climate model are shown at the intersection between the curves of emitted infrared radiation $R\uparrow$ and absorbed solar radiation $R\downarrow$.

more general problems associated with the solutions; in particular, the question of whether or not all three equilibrium states identified are 'stable' and capable of persisting for long periods of time.

Any non-linear system, even one which is far simpler than the climate system, can have a characteristic behaviour termed *almost intransitivity*. This behaviour is illustrated in Fig. 7.9. If two different One-dimensional radiative–convective models (1-D RCs) represent an the system is termed a *transitive* system. State A for this transitive system would then be considered the solution or normal state and all perturbed situations would be expected to evolve to it. At the other extreme an *intransitive* system has at least two equally acceptable solution states (A and B), depending on the initial state. Difficulty arises when a system exhibits behaviour which mimics transitivity for some time, then flips to the alternative state for another (variable) length of time and then flips back again to the initial state and so on. In such an almost intransitive system it is impossible to determine which is the 'normal state', since either of two states can continue for a long period of time, to be followed by a quite rapid and unpredictable change to the other.

At present geological and historical data are not detailed enough to determine which of these system types is typical of the Earth's climate. It is easy to see that should the climate turn out to be almost intransitive it will be extremely difficult to model.

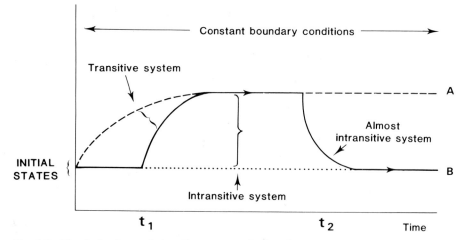

Fig. 7.9. The behaviour of the three types of climatic system: transitive, intransitive and almost intransitive with respect to an initial state. In a transitive system two different initial states evolve to the same resultant state, A. An intransitive system exhibits the 'opposite' behaviour with two alternative resultant states. The characteristic of an almost intransitive system is that it mimics transitive behaviour for an indeterminate length of time and then 'flips' to an alternative resultant state.

## *Importance of oceans*

One important reason why the system type is difficult to establish concerns the differing heat capacities of the components of the climate system. Each component responds to a change in energy at a different rate, usually expressed by the thermal relaxation time, the time required for the component's temperature to change by a fraction $1/e$ of the imposed temperature change. This feature is easy to illustrate by returning to our swimming pool analogy. In the heating example described above it was discovered that the pool's temperature would increase by ~0.2 K each day. However, suppose that the pool is enclosed in a glass building with dimensions twice that of the pool (i.e. 60 m × 20 m × 4 m). The thermal capacity of the air in the building is $C_{air} = 5.79 \times 10^6$ J K$^{-1}$. Using equation [7.2], it is easy to show for the building without the presence of the pool that $\Delta T_{air} \simeq 360$ K. Thus, by the end of one day, the temperature in the building alone would be unbearable. If however the pool is filled with water then the two thermal capacities must be combined leading to an overall temperature rise of $\Delta T_{total} \simeq 0.8$ K. Of course we have cheated a lot here in holding the net energy input constant. In all cases, as the temperatures rose the air and water would begin to re-emit increasingly larger amounts of infrared radiation, thus rapidly decreasing the assumed 20 W m$^{-2}$ net input energy. However, the message of the analogy is sound. The thermal capacity of water is so much greater

362

than that of air that the thermal inertia of a combined atmosphere–ocean system is *much* larger than that of the atmosphere alone.

This analogy emphasises the importance of modelling the complete climate system. In particular the vital role of the oceans is clear. Calculations suggest that a reasonable estimate of the thermal response time of the mixed layer (~70 m) of the global ocean is around 6–7 years. This time scale is important for two reasons. Firstly, it is most unlikely that any perturbation which operates on time scales significantly shorter than 6–7 years can cause a significant climatic response; these short period fluctuations will be damped and finally smoothed by the thermal inertia of the oceans. Secondly, model simulations must cover at least this time period so that the mixed layer of the oceans can come to equilibrium. In fact oceanic transport and deep ocean mixing may cause the necessary time periods of simulation to be increased to the order of 10–20 years.

### 7.2.1 Energy balance climate models

The principles of energy balance models (EBMs) were discussed above. A relatively simple extension of these principles allows the incorporation of latitudinal variations. The resulting EBMs have been very instructive in increasing our understanding of the climate system and in the development of new parameterisations and methods of evaluating sensitivity for more complex and realistic models. This type of model can be readily programmed and implemented on most small computers. The inherent simplicity of EBMs combined with the ease of interpreting results make them ideal instructional tools.

Energy balance models are called 'one-dimensional', the dimension in which they vary being latitude. Vertical variations are ignored and the models are used with surface temperature as the dependent variable. Since the energy balance is allowed to vary from latitude to latitude, a horizontal energy transfer term must be introduced, so that the basic equation for the energy balance at each latitude, $\theta$, is

$$\rho C \frac{\Delta T(\theta)}{\Delta t} = R\downarrow(\theta) - R\uparrow(\theta) + \text{transport into zone } \theta \qquad [7.9]$$

where $\rho C$ is the heat capacity of the system and can be thought of as the system's 'thermal inertia'.

The radiation fluxes at the Earth's surface must be parameterised with care since conditions in the vertical are not considered. To a large extent the effects of vertical temperature changes are treated implicitly. In a clear atmosphere convective effects tend to ensure that the lapse rate remains fairly constant. However, cloud amount depends only weakly on surface temperature, so that cloud albedo is only partially incorporated in the model. In particular, clouds in regions of high temperatures, such as the ITCZ, are ignored in the parameterisation of albedo.

The outgoing infrared radiation term in equation [7.9] is expressed in the form

$$R\uparrow = B_1 + B_2 T \tag{7.10}$$

where $B_1$ and $B_2$ are constants. These are obtained empirically by relating observed values of $R\uparrow$ to $T$. The relationship obtained combines the effects of surface emissivity and atmospheric transmissivity and thus provides a practical way of using equation [7.5].

When using the model for annual average calculations the surface albedo can be regarded as constant for a given latitude. This type of model, however, can also be used for seasonal calculations. In this case it is usual to allow the albedo to vary with temperature as in equation [7.8] to simulate the effects of changes in sea-ice extent.

The transport term must also be approximated since the dynamical events of the real world are not modelled in an EBM. It is assumed that a 'diffusion' approximation is adequate, relating energy flow directly to the latitudinal temperature gradient. This flow is usually expressed as being proportional to the deviation of the zonal temperature, $T$, from the global mean, $\bar{T}$. Thus the heat transport is set equal to $D\,(T - \bar{T})$ where $D$ is a constant of proportionality which is obtained from direct observations of latitudinal energy flux (e.g. Fig. 4.3).

The final energy balance equation for the annually averaged case then has the form

$$S(\theta)\,[1 - A(T,\theta)] - [\,B_1(\theta) - B_2(\theta)\,T(\theta)\,] = D\,(T(\theta) - \bar{T}) \tag{7.11}$$

where $\theta$ is the latitude. This equation can be solved for the complete set of latitude zones. In simple climatic simulations such as this, it is generally assumed that the Earth is symmetric about the equator and the solution for only one hemisphere is found. The basic constraints or boundary conditions placed on the model to ensure that it remains realistic are that there should be no temperature gradient or heat transport at either the equator or the poles.

Early energy balance models were originally found to be stable only for small perturbations of the present-day conditions. For instance, they predicted the existence of an ice-covered Earth for only slight reductions in the present solar constant, suggesting that the climate system is almost intransitive (Fig. 7.9). However, a number of stable solutions are possible, dependent upon the values chosen for the constants $B_1$, $B_2$ and $D$ in the parameterisations employed in the particular EBM.

### Stability of model results

In any climate model, great care must be taken in choosing the constants for any parameterisation scheme. The more 'accurate' they are for the present day, the more closely the model is tied to predictions of the present-day situation but the less likely it is to be able to respond realistically to perturbations.

The solution of equation [7.11] gives a climate prediction of the mean global temperature, $\bar{T}$, as a function of all the other parameters. As the equation is non-linear, there will be more than one solution. The inherent stability of these solutions to both internal and external perturbations must then be tested.

For the 'external stability' we can consider the variation of $\bar{T}$ as a function of solar variation since an alteration of the solar constant is a convenient method of exploring the characteristics of climate model structure. Figure 7.10 shows the way in which $\bar{T}$ changes as the total incident radiation, $\mu S$, changes. As the solar flux is reduced to some critical value ($\mu_c S$) the number of solutions is reduced from 2 to 1 at the critical value. Below $\mu_c S$ no solution is possible. This critical point is termed the bifurcation point. For values of incoming radiation, $\mu S$, less than $\mu_c S$ temperatures are so low that the albedo, $A(T,\theta)$, becomes very close to or equal to 1 and thus it is impossible to regain energy balance and the global temperature, $\bar{T}$, approaches minus infinity. If some limit is put on how high the albedo may get e.g. $A \leqslant 0.75$, the solution ends up as an 'ice-age Earth'.

The 'internal stability' concerns the response of each branch in Fig. 7.10 to perturbations from equilibrium which are created by

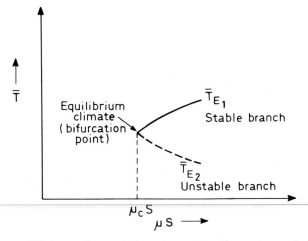

Fig. 7.10. The equilibrium climate bifurcation point. For values of the solar flux $\mu S > \mu_c S$ there are two solutions, whereas below this critical value no solutions exist. Changes in solar radiation lead to either a stable or unstable equilibrium climate illustrated here by the two equilibrium branches.

internal factors. To determine if temperatures will return to equilibrium after the perturbation we can use the time dependent formula in equation [7.6] and postulate a new value for $\bar{T}$ which is close to the equilibrium one already calculated at that level of $\mu S$. Rewriting equation [7.11] in the form of equation [7.6] allows computation of $\Delta \bar{T}/\Delta t$. This change can be computed iteratively until it is determined whether the values do regain the original $\bar{T}$ solution. If it is regained, then the solution is said to be 'internally stable'. In the case shown in Fig. 7.10 only the top branch is stable. This method of evaluation of internal stability is the way in which the transitivity or intransitivity of the climate model is established. It would not, however, permit differentiation between an almost intransitive model and an intransitive one.

The advantage of energy balance models is their simplicity of formulation. Their applicability, like that of all climate models, is restricted by the limited way in which such straightforward sets of equations can capture physical reality. We can compensate for the alteration of the solar constant by varying any or all of the parameters in the model so that a new equilibrium climate can be established by changes in the albedo, infrared emissivity or the heat transport. However, the assumptions made about the nature of these changes can affect the equilibrium climate predicted. For example, a survey of recent literature concerning EBMs shows that the change in solar constant necessary to create a totally glaciated Earth varies from 4% to 21%, depending upon the albedo and longwave parameterisations used in a particular model.

## 7.2.2 One-dimensional radiative–convective climate models

One-dimensional radiative-convective models (1-D RCs) represent an alternative approach to relatively simple modelling of the climate and they also occur at the bottom of the modelling pyramid (Fig. 7.6). In this case the 'one dimension' in the name applies to altitude. These models are basically designed to determine average global temperatures. Although surface temperatures are emphasised, temperatures at various levels in the atmosphere can be obtained. In addition, they can be extended to apply to zonally averaged conditions.

Their major characteristic is explicit and often complex calculation of the radiation streams. Both upward and downward fluxes are determined at each level and a vertical temperature profile determined. Cloud optical properties, surface albedo and cloud amount are incorporated but usually specified rather than generated by the model. From a starting condition (often an isothermal atmosphere) the heating rates for a number of layers through the atmosphere are calculated. Although the number of layers varies with the model, all models calculate the energy change in each layer which results from an imbalance between the net radiation at the top and bottom of the layer. This energy change is converted to a temperature change. At the end of each time step a new radiative temperature profile results. This temperature profile is now compared with some predetermined lapse rate. If the calculated lapse rate exceeds this 'critical' lapse rate the atmosphere is treated as being convectively unstable. Vertical mixing, equivalent to a series of convection cells, or to baroclinic instability, then takes place until the prescribed lapse rate is re-established. This 'mixing' or convective adjustment is performed instantaneously in the model and then the next radiative time step calculation is performed. This continues until the net radiative heating in every layer is no longer subject to convective readjustment and the net fluxes at the top of each layer approach zero.

All models of this type operate under the constraints, or boundary conditions, that at the top of the atmosphere the shortwave and long-wave fluxes must balance and that at the surface the net gain of energy by radiation equals the net loss by convection. However, they vary in the way they incorporate the critical lapse rate. Some use the dry adiabatic lapse rate, some the saturated one, while many use a value of 6.5 K km$^{-1}$, which, based on observational evidence, represents approximately the present-day global average lapse rate. Similarly, different humidity and cloud formulations are possible.

### *Sensitivity of a 1-D RC model*
Table 7.2 compares the predictions of $\Delta T$ (the increase in surface temperature) for differently formulated 1-D radiative–convective models for a perturbation in the form of a doubling of the atmospheric carbon dioxide from 300 ppmv to 600 ppmv. Model 1 has fixed absolute humidity. Hence the amount of water vapour in the atmosphere does not change and, in response to the external perturbation of $CO_2$, temperatures increase and relative humidity decreases. The resulting temperature increase of ~1.2 K is the simple radiative–convective result since this model does not incorporate any feedback effects. Model 2 has, by contrast, fixed relative humidity. This means that as the temperature increases the saturation vapour pressure increases and thus, because the relative humidity is the quotient of actual vapour pressure over saturated vapour pressure, the actual vapour pressure must also increase. This extra water vapour must be the result of surface evaporation. The predicted surface temperature increase is 1.94 K. The difference between the results from Models 1 and 2 illustrates the effect of evaporation on radiative exchanges and thus its importance for any climate prediction model. Model 3 uses a convective adjustment to the saturated adiabatic lapse rate rather than the value of 6.5 K km$^{-1}$ which is used in Models 1 and 2. It produces a slightly lower predicted temperature increase since the lapse rate

**Table 7.2  Equilibrium surface temperature increase due to doubled $CO_2$ (300–600 ppmv): results from a suite of 1-D RC model sensitivity experiments**

| Model | Description | $\triangle T$ (K) | Feedback factor |
|---|---|---|---|
| 1. | Fixed absolute humidity, 6.5 K km$^{-1}$ Fixed cloud altitude | 1.22 | 1 |
| 2. | Fixed relative humidity, 6.5 K km$^{-1}$ Fixed cloud altitude | 1.94 | 1.6 |
| 3. | Same as 2, except moist adiabatic lapse rate replaces 6.5 K km$^{-1}$ | 1.37 | 0.7 |
| 4. | Same as 2, except fixed cloud temperature replaces fixed cloud altitude | 2.78 | 1.4 |

*Note*: Model 1 has no feedbacks affecting the atmosphere's radiative properties. The feedback factor specifies the effect of each added process on model sensitivity to doubled $CO_2$.

decreases as additional water vapour is added to the atmosphere. The difficulty of selecting an 'appropriate' global lapse rate to which convective adjustment should be made is considerable.

Comparisons of Models 4 and 2 illustrate the importance of cloud temperature and height effects. In Model 4, the clouds, which are set at a constant, empirically determined amount, are at fixed temperatures rather than at the fixed heights used previously. The clouds therefore move to higher altitudes as the $CO_2$ perturbation increases temperatures. This is an important factor for the resultant surface temperature. For a constant amount, clouds at a higher altitude radiate less energy to space because their temperatures are lower. Hence the computed surface temperature must be raised further so that the planetary energy balance is maintained. This results in a predicted surface temperature increase $\Delta T \sim 2.8$ K – considerably larger than the $\sim 1.9$ K for fixed cloud altitude.

### 7.2.3  Two-dimensional climate models

Two-dimensional climate models represent either two horizontal dimensions or the vertical and one horizontal dimension. The latter, which are more common, combine the latitudinal dimension of the energy balance models with the vertical one of the radiative–convective models. In addition, their formulation is based on a more realistic parameterisation of the latitudinal energy transports. The general circulation is assumed to have mainly a cellular flow between latitudes. This flow is characterised using a combination of empirical and theoretical formulations. Actual observations are used to develop a set of statistics summarising the wind speeds and directions, while the laws of motion are used to obtain an energy diffusion coefficient of the type used in the energy balance models. As a consequence of this approach, these models are usually known as 'statistical-dynamical' models.

These models are obviously more complex than those considered earlier and are about half way up the modelling pyramid. Their development and use has provided insight into the operation of the present climate system. In particular, they have shown that the relatively simple diffusion coefficient approach for poleward energy transports is highly appropriate, provided the coefficient, as well as the transport itself, is allowed to vary with the temperature gradient. Similarly, advances in our understanding of baroclinic waves have resulted from studies of the results of statistical-dynamical models. Nevertheless, these formulations are still rather limited in their application as climate prediction models. Being zonally averaged, they are insensitive to changes within a latitude band. Changes such as new temperature contrasts between land and sea and new cloud regimes which might result from alterations of land surface albedo cannot be investigated. As a result, these models have largely been superseded, when considering the effects of perturbations of the present climate, by general circulation models.

### 7.2.4 General circulation climate models

Calculation of the full three-dimensional character of the climate is the aim of general circulation models. This requires the solution of a series of equations that describe the movement of energy and momentum and the conservation of mass and water vapour (Table 7.3). Physical processes such as cloud formation and heat and moisture transports within the atmosphere and between the ground and air must be included. Since our present understanding of climate dynamics dictates that temperature must be the element of prime concern, these equations are solved for the *wind field* as a function of *temperature*. The first step in obtaining a solution is to specify the atmospheric conditions at a number of 'grid points', obtained by dividing the Earth's surface into a series of rectangles, so that a regular grid results. Conditions are specified at each grid point for the surface and several layers in the atmosphere. The equations in Table 7.3 are then solved at each grid point using numerical techniques. Various techniques are available, but all use a time step approach and an interpolation scheme between grid points.

Although general circulation models formulated in this way have the potential to approach the real atmospheric situation very closely, at present there are a number of practical and theoretical limitations. The prime practical consideration is the time needed for the calculations.

## Table 7.3   Fundamental equations solved in GCMs

Conservation of momentum (Newton's second law of motion) in three
  orthogonal directions:

Force = mass × acceleration
  $\mathbf{F} = m\,\mathbf{a}$


Conservation of mass (continuity equation):
  Sum of the gradient of ($\rho\mathbf{v}$) in three orthogonal directions is zero (i.e.
  matter may not be created or destroyed)

Conservation of energy (first law of thermodynamics):
  Input energy = increase in internal energy + work done

Ideal gas law (approximate equation of state):
  (pressure × volume)/absolute temperature = gas constant
  $p\,V/T = R$

To compute each of the basic atmospheric parameters at each grid
point requires that roughly $10^5$ numbers be stored, recalculated and
re-stored at each time step. This places a strain on the resources of
even the largest and fastest computers. Since the accuracy of the model
partly depends on the spatial resolution of the grid points and the
length of the time step, a compromise must be made between the
resolution desired and the computation facilities available. At present
grid points are usually spaced approximately 5° of latitude and longitude apart and time steps of the order of 30 minutes are used. Vertical
resolution is obtained by dividing the atmosphere into about 6 to 10
levels, usually specified in terms of constant pressure surfaces.

   Very often the computational time and expense dictates that models
be run to simulate a particular season or month rather than a complete
annual cycle. Many experiments concerning the effects of carbon
dioxide on the atmosphere, for example, deal with a 'perpetual January
(or July)'. As computers improve we may expect this situation to
change. However, increased computing power also enables us to
produce models with both finer resolution and more exact solutions of
the governing equations, so there may be a tendency to continue to
produce more 'perpetual Januaries'.

   These computational constraints lead to problems of a more theoretical nature. Conditions between grid points and time steps must be
interpolated, so that some 'smoothing' always occurs. With a coarse
grid spacing small-scale atmospheric motions, such as thundercloud
formation, cannot be modelled, however important they may be for
atmospheric dynamics. Models with fine grids can, of course, be
developed, but these are mainly associated with weather, rather than

climate, prediction. On the longer time scale appropriate for climate, we must have recourse more often to parameterisation of some important processes, rather than try to deal with them explicitly.

The processes usually incorporated into general circulation models are shown in Fig. 7.11. Within the atmosphere the modellers adopt an approach similar to that used for the radiative–convective models in calculating heating rates, but also include cloud formation processes as part of the convection and consider in detail the effects of horizontal transports. The interaction between the surface and the near-surface layer of the atmosphere, however, must be parameterised. Commonly the surface fluxes of momentum, sensible heat and moisture are taken to be proportional to the product of the surface wind speed and the gradient of the property away from the surface. For land surfaces it has been found that a 'two layer' Earth is necessary to simulate both

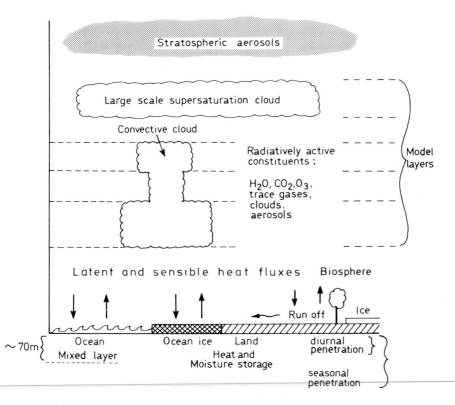

Fig. 7.11. Schematic representation of the subgrid-scale structure of a general circulation climate model. In this GCM two types of cloud are treated: grid-scale supersaturated clouds and subgrid-scale convective clouds. The surface interaction with the atmosphere is fairly complex and heat and moisture storage within the ground are treated by the use of a two layer ground model.

diurnal and seasonal heat and moisture storage and exchange with the atmosphere.

### Incorporation of oceans into GCMs

The interaction between the ocean and the atmosphere can be considered in several ways. Rather simple, but common, is the 'swamp' model where the ocean acts only as an unlimited source of moisture. The simplest models use fixed ocean temperatures based on observed average values (e.g. Fig. 4.29). However, the oceans are, in effect, the thermostat of the climate, so it is very difficult to disturb the climate away from the present day configuration whilst sea-surface temperatures are specified in an unchanging manner each month or each season, and year after year. The most sophisticated approach at present is to compute the heat storage of the mixed layer of the ocean and deduce from this the mixed layer depth and the sea-surface temperature. Usually the maximum depth of the mixed layer is prescribed to be around 70 to 100 m, allowing the lower deep ocean layer to act only as an infinite source and sink for water.

Oceans also perform the very important duty of advection of heat around the globe. Progress in simulations of the oceanic circulation has been closely related to the evolution of atmospheric models. Although the processes are similar, the transports and temperature changes in the oceans take very much longer to accomplish than in the atmosphere. Thus even if satisfactory oceanic circulation models are developed (see e.g. Fig. 4.28), there remains the problem of linking an atmospheric model with a time step (i.e. scale) of around 20–30 minutes (or less) to an oceanic model that needs between 20 and 30 years to come to equilibrium.

Many models need actual observed values for part of their input and all require observational data with which to compare the results. Some parameters, such as surface temperature, are available worldwide and pose no problem. Others, however, are sparse in either time or space. Our knowledge of sea-ice extent, for example, is dependent on satellite observations, so that there is only a short observational record. Thus it is difficult to compare observations with the long-term average values obtained from models. Similarly, global coverage for cloud observations is relatively recent (e.g. Fig. 3.16), so that it is difficult even to start those models which require the present cloud distribution as an initial input. Certainly, the improved cloud climatologies which are being developed as more satellite observations become available are needed before more complex models can be validated.

Despite the many limitations of the general circulation models, they represent the most complete, and hopefully accurate, type of climate model currently available. They certainly show tremendous advances in our understanding of the atmosphere and our ability to model it over the 25 years since the first serious climate models were produced. That they are not completely accurate is emphasised by our placing them below the peak in the climate modelling pyramid (Fig. 7.6). They can provide a tremendous amount of information about the present climate and the possible effects of future perturbations. Many future climate 'scenarios' have been generated using them. Frequently the predictions are self contradictory. This is inevitable, given our relatively poor knowledge of present conditions and understanding of the controlling processes. Provided a model is developed on sound physical principles, incorporates rational parameterisation schemes, accounts for the major processes acting in the atmosphere and has been adequately tested against the present conditions, its results should be treated with respect. The results provide, at least, an indication of the possible future climate conditions created by a perturbation in the forces which control our present climate.

### 7.2.5 Climatic feedback effects

Up to this point in our consideration of climate models we have effectively assumed that perturbation of a single parameter, whether external or internal, causes a straightforward alteration in the parameter of interest. We have failed to consider, at least explicitly, interactions internal to the climate system, the *feedback* mechanisms. These can operate either in a positive sense so as to modify the monitored parameter further in the same direction as the perturbation or in a negative sense acting to modify it in the sense opposite to the initial perturbation (Fig. 7.12(a)). Feedbacks can occur anywhere in the climate system. Some selected examples follow.

### *The ice-albedo feedback mechanism*

If some external or internal perturbation acts to decrease the surface temperature then this is likely to lead to the formation of additional snow and ice masses. Thus the surface, and probably the planetary, albedo increases. A greater amount of solar radiation is reflected away from the planet, causing temperatures to decrease further. This further decrease in temperatures leads to more snow and ice and so on. This positive feedback mechanism is known as the *ice-albedo feedback mechanism*. Of course, the ice-albedo feedback mechanism is positive in the other direction as well. With higher temperatures, a portion of the permanent cryosphere may be removed, reducing the albedo and leading to further enhancement of temperatures.

A second, but less dramatic, positive feedback mechanism occurs with the increase of atmospheric water vapour from evaporation as temperatures rise. The additional greenhouse effect of this vapour further enhances the temperature increase (Model 2 in Table 7.2). Again, in reverse this mechanism has a positive effect since as tempera-

373

## (a) FEEDBACK LOOPS

POSITIVE FEEDBACK

NEGATIVE FEEDBACK

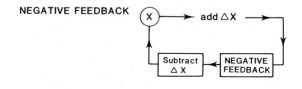

## (b) CLOUD FEEDBACK

**(i) Normal**

**(ii) More Cumulus**

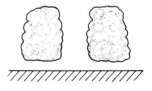

**(iii) More Stratus**

Fig. 7.12(a). Feedback mechanisms operate either in a *positive* sense to enhance the perturbation further or in a *negative* sense to damp it. (b). Cloud feedbacks are complex. A particular difficulty is whether increased cloudiness will manifest itself as (ii) increased cumuliform clouds or (iii) increased stratiform clouds. The surface area covered and hence the planetary albedo differs significantly in these two cases.

tures fall (perhaps as a result of the ice-albedo feedback) the atmosphere is able to hold less water vapour and therefore part of the greenhouse mechanism is removed from the climate system.

### Cloud feedbacks
It is difficult to establish even the direction of the feedback associated

with clouds since they are both highly reflective (thus contributing to the albedo) and composed of water and water vapour (thus contributing to the greenhouse effect). Recent work (e.g. Fig. 7.5) suggests that for low and middle level clouds the albedo effect will dominate over the greenhouse effect, providing a *negative* feedback, so that increased cloudiness will result in an overall cooling. On the other hand, cirrus clouds, which are optically thin, have a smaller impact upon the albedo so that their overall effect is to enhance the greenhouse mechanism. Thus they tend to lead to warming.

Unfortunately cloud feedback is not this straightforward. There are two additional problems. First of all, it is uncertain whether increased temperatures will lead to greater or less total cloud cover. Whilst it is generally agreed that increased temperatures will cause higher rates of evaporation and hence make more water vapour available for cloud formation, it is much less certain whether the additional clouds which are formed will be of cumuliform or stratiform type. Even for the same 'volume' of new cloud, an increased dominance of cumuliform clouds would lower the percentage of the surface covered by clouds. Stratiform clouds would increase the area covered (Fig. 7.12(b)). Thus increased temperatures, if they led to increased cumuliform cloudiness, could lead to a positive cloudiness feedback.

The second unknown factor about cloud changes as a feedback response to a climate perturbation concerns their level of formation. If it is assumed that clouds are always formed at fixed temperatures, as seems reasonable, then as the results for Model 4 in Table 7.2 illustrated, the positive feedback effect is greater than if a constant cloud height is assumed.

## Combining feedback effects

The example of cloud feedbacks indicates that more than one feedback effect is likely to operate within the climate system in response to any perturbation. These feedback effects combine in a complex way. Consider a system in which a change of surface temperature of magnitude $\Delta T$ is effected. If there are no internal feedbacks then this temperature increment is equal to the change in the equilibrium temperature, $\Delta T_{eq}$. When feedbacks occur there will be an additional temperature change due, for example, to an increase in the atmospheric water vapour content and the value of $\Delta T_{eq}$ will be

$$\Delta T_{eq} = \Delta T + \Delta T_{feedbacks} \qquad [7.12]$$

where $\Delta T_{feedbacks}$ can be either positive or negative. This equation can be rewritten, algebraically, as

$$\Delta T_{eq} = f\Delta T \qquad [7.13]$$

375

This *feedback factor, f,* can be related to the amplification or *gain, g,* of the system, which is defined, using the analogy of gain in an electronic system, by

$$f = 1/(1-g) \qquad [7.14]$$

The most important result of this analysis occurs when more than one feedback is considered. It is easily shown that the gains are additive, i.e. the total gain $g$ can be expressed as the sum of the individual gains, $g_i$, derived for each of $i$ feedback processes. Thus

$$g = \Sigma g_i \qquad [7.15]$$

The overall feedback factor, $f$, must be calculated using equation [7.14] and substituting for $g$ from equation [7.15]. This gives, for two feedbacks,

$$f = \frac{f_1 f_2}{f_1 + f_2 - f_1 f_2} \qquad [7.16]$$

In other words the feedback factors are neither simply additive nor multiplicative. Consider a feedback with $f_1 = 1.5$, operating alone. This would result in a 50% increase in the response (equation [7.13]). If a second feedback now operates with $f_2 = 2.0$, by substituting into equation [7.16] it is seen that the overall feedback factor is 6.0. Hence an additional feedback may cause a significant increase in the response.

The importance of the combination of many feedback factors can best be illustrated with an example. Suppose, as seems reasonable from published model results, that the change in surface temperature predicted by a climate model without feedbacks after a doubling of atmospheric $CO_2$ is $\Delta T = 1.2$ K. If a snow and ice feedback factor of $f_{ice} \sim 1.2$ and a water vapour feedback factor of $f_{water\ vapour} \sim 1.6$ are incorporated, then from equation [7.16], the joint value of $f \sim 2.18$. Hence

$$\Delta T_{eq} = 2.18 \times 1.2 = 2.6 \text{ K} \qquad [7.17]$$

This result is in general agreement with the evaluations of the likely effect of doubling $CO_2$ when cloudiness changes are neglected. Suppose now that two cloudiness feedback factors, both positive, are discovered, where the effect of increased temperatures leads to increased cumuliform cloud and thus to (i) decreased cloud cover and (ii) an increased cirrus greenhouse. Suppose, in the absence of any better information, both these cloudiness feedback factors have the same magnitude as $f_{ice}$. Then just one of them (i.e. snow/ice plus water vapour plus one cloud feedback) leads to a value of $\Delta T_{eq} = 4.1$ K. With the extra cloud feedback the overall feedback factor is $f_{tot} \sim 8.00$ and thus there is a total temperature change of

$$\Delta T_{eq} = 8.00 \times 1.2 = 9.6 \text{ K} \qquad [7.18]$$

Of course it is just as likely that some or all of the cloudiness feedback effects could act in a negative sense but the power of including more and more facets of the climate system is seen to be immense.

This analysis indicates that it is important to try to establish from observational data the magnitude of the feedback factors of the real climate system, as well as any changes that occur over time. It cannot be emphasised strongly enough that, at the present time, the least understood feature of the climate system is the behaviour of cloudiness as a feedback mechanism.

# 7.3   Local climate models

Climate modelling is by no means confined to the global scale. Modelling of synoptic scale features, for example, is particularly important for the development of numerical weather prediction schemes. Similarly, active development of models of smaller scale features, such as thunderstorms, is under way. At the moment these models emphasise the short-term features associated with weather and they have not yet been used in any systematic way for climate modelling. Climate models, however, have been developed for local areas, almost always with specific applications in mind.

These local application climate models display many features in common with the global models. The system under consideration is often complex, so that there is a need to identify and parameterise the important active processes, incorporate feedback mechanisms where needed, calibrate the models for particular conditions and localities and test their predictions using real data.

Since local climate models are application oriented, they can be regarded as complementary to the global ones. The ultimate objective of the latter is to predict future climates, whilst the local models are more likely to be used to estimate the impact of these future climates on human activities.

For any particular application, there can be a variety of models available, ranging from the extremely simple to the very complex. In general, the models used have been relatively simple for situations where the impact of climate, economically or socially, is only one of many possible impacts. Examples of these have already been given, including the regression model for energy demand calculations and consideration of the effects of climate on human health. In situations where climate is seen as the most important variable, however, attempts have been made to produce detailed and accurate, and therefore complex, models. This can be illustrated by examining models for agricultural production and for water resources.

## 7.3.1 Agricultural production and climate

The impact of climate on agriculture can be felt in at least two ways. Firstly, there is a direct connection between plant growth, especially crop growth, and climate, so *crop yield models* have been developed. These relate growth to weather variations in a small area and are designed for use by an agriculturalist during the growing season to optimise production. The second impact of climate on agriculture concerns conditions over a much wider area. The profit from any agricultural enterprise depends not only on the quality and quantity of the produce, but also on the selling price. This itself is affected by the overall supply which in turn depends on climatic conditions in all production regions worldwide. Hence prediction of the total production worldwide is an important consideration for an individual agriculturalist. Hence wide area, less sophisticated *general productivity models* have been developed.

For crop yield models many of the ideas developed for global models are applicable. Variations in crop yield are a response to perturbations in the whole growth system. Many of the factors causing perturbations, such as type and quality of seed and fertiliser used and the tillage practices employed, are 'external' to the system. Some factors, notably crop diseases, are partly climate dependent and in fact, as the conditions in Britain in the summer of 1976 indicated (Table 6.12), can have an effect comparable to that of climate itself. Such factors should therefore be incorporated in some way if realistic results are to be obtained. The true climatic factors are solar radiation, carbon dioxide, water availability and temperature. The first three control the rate of photosynthesis, while temperature controls the rates of chemical reactions responsible for growth and influences the amount of respiration. Strictly, water availability should refer to the amount of soil moisture available, but usually precipitation, a more commonly available parameter, is used to characterise water availability. Carbon dioxide is usually taken to be invariant on the time scale appropriate for these models.

The simplest approach to crop yield modelling is to use historical data for both yield and the climatic elements and obtain a regression relationship between them. Commonly only temperature and precipitation data are used, it being assumed that solar radiation is closely connected with temperature. Since crops respond differently to climate at different stages of growth, variations in weather within the growing season must be incorporated. In addition, crops have a 'memory' for past conditions so that at a particular time in a given year their response to a given atmospheric stimulus will depend on the antecedent conditions. Thus the resulting crop yield regression equation can be very complex. One such result, as an example, is

$$Y = AL + BT_3 + CT_3{}^2 + DT_6 + E\,T_6{}^2 + F\,p_6 + G\,t \qquad [7.19]$$

Here $p$ is precipitation and $T$ is temperature, the subscripts referring to the length of averaging period, in months from the start of the growing season, used for the parameter. This equation was derived using information from many places in the world and the variable $L$ is found to be a function of location, whilst the local constant $t$ is related to the actual starting time of the growing season. The letters $A$ to $G$ represent empirical constants which do not vary with location.

Although such regression equations allow prediction of crop yields, they permit little physical insight into the relationship between crop growth and climate. More insight can be gained by using an alternative approach which considers more closely the physiological effects of climate on plants. The weight of a crop at harvest, $W$, is the product of the mean growth rate, $\bar{C}$, and the total duration of the growth, $t_g$ (i.e. the time from planting to harvest). However, during some periods growth will be retarded and growth time lost as a result of inclement weather. At other times climatic factors will not be limiting and growth rates will be at a maximum. Thus

$$W = \bar{C}t_g = C_m\,(t_g - t_l) \qquad [7.20]$$

Here $C_m$ represents the maximum growth rate during the main growth period, the value of which is obtained from experiments for a particular crop. The time lost to growth because of inclement weather is $t_l$. The value of $t_l$ must be obtained from historical records combined with an understanding of the processes, including feedbacks, acting within the crop. For example, an increase in mean temperature will maintain a faster rate of photosynthesis, which may lead to increased yields. However, higher temperatures may also lead to a shorter growth period because of increased respiration, which may decrease the yield. A general indication of the relative importance of the two factors can be obtained using regression techniques on the historical data (e.g. Fig. 7.13). Here it is seen that increasing temperatures decrease yields, at least in the conditions common in Britain. Once such effects have been isolated, the growth time lost as a result of all of the climatic factors can be determined. The results of such an analysis for the East Midlands of England are shown in Table 7.4. These results suggest that, while rainfall variation is the most important climatic factor in yield for most crops, temperature is more important in some cases, the

379

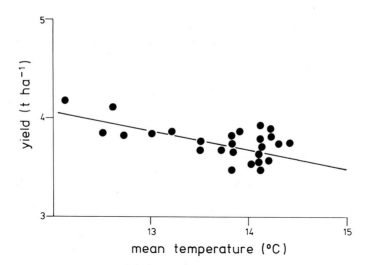

Fig. 7.13. Mean yield of wheat (tonnes ha$^{-1}$) in parts of England (1956–70) as a function of mean temperature for May, June and July for the period 1941–70.

differences depending on crop type and soil conditions. Although the variation from the three climatic parameters together rarely exceeds 25% of the total variation, much of that remaining results from effects, such as cultural practices, over which the agriculturalist has at least partial control. Hence climatic effects remain highly significant in the year-to-year variation of crop yields.

### Regional yield models and economic consequences

When general productivity models for a wide area are needed, similar approaches are used, but the models tend to have simpler formulations, largely because crop yield models are designed and calibrated for a particular crop in a particular locality. Thus wide area models give less precise results. This is rarely a great problem since they are commonly used to estimate global grain supplies and the economic consequences of scarcity or overabundance, estimates which cannot be too precise. Certainly the economic consequences flowing from climate-induced changes in productivity can be considerable. Table 7.5 indicates the income losses for the United States resulting from variations in climate about the normal. The climate effects were extracted from the total monetary variation by using a relatively simple agricultural productivity model applicable to the whole country. The table also serves to emphasise that the timing of the departure from normal has a great effect on the yield and therefore on the agricultural income.

Models relating livestock husbandry to climate are much less well developed than crop yield models. Nevertheless, those available take a similar approach, although they are considerably more complicated

**Table 7.4  Estimates of yield and variation for major arable crops in East Midlands of England**

| | Soil | Growth period $t_g$ (d) | Time lost $t_l$(d) Drought | Time lost $t_l$(d) Incomplete cover | Effective growth period $t_g-t_l$ (d) | Total dry matter (t ha$^{-1}$) | Yield[a] (t ha$^{-1}$) | Components of variation (±%) light | temp. | rain |
|---|---|---|---|---|---|---|---|---|---|---|
| Spring cereals | heavy | 90∓6 | 1±1 | 30±4 | 50±5 | 11.8±1.2 | 6.1±0.6 | 2 | 8 | 2 |
| | light | 90∓6 | 5±5 | 30±4 | 55±7 | 11.0±1.5 | 5.7±0.8 | 2 | 9 | 9 |
| Winter cereals | heavy | 90∓3 | 1±1 | 0 | 89±8 | 17.8±1.9 | 9.3±1.0 | 2 | 9 | 1 |
| | light | 90∓8 | 5±5 | 0 | 85±9 | 17.0±2.1 | 8.8±1.1 | 2 | 9 | 6 |
| Potatoes | heavy | 90∓6 | 5±5 | 30±4 | 55±7 | 11.0±1.5 | 37±5 | 2 | 9 | 9 |
| | light | 90∓6 | 8±8 | 30±4 | 52±9 | 10.4±2.0 | 35±7 | 2 | 10 | 15 |
| Sugar beet | heavy | 120∓8 | 7±7 | 30±4 | 83±6 | 16.6±1.5 | 30±3 | 2 | 4 | 8 |
| | light | 120∓8 | 10±10 | 30±4 | 80±11 | 16.0±2.5 | 29±5 | 2 | 4 | 13 |

(a) Yield estimated from total dry weight using these factors: grain 0.52; tubers 3.4; beet roots 1.8.

381

**Table 7.5  Gross spring wheat income losses in the United States due to deviations of temperature and precipitation from normal (US $ millions)**

|              | Temperature |        | Precipitation |         |        |        |
|--------------|-------------|--------|---------------|---------|--------|--------|
|              | +1 °C       | −1 °C  | +1 mm         | −1 mm   | +20%   | −20%   |
| Pre-season   | –           | –      | +7            | −8      | +21    | −30    |
| April        | +40         | −40    | +3            | −3      | +22    | −25    |
| May          | −22         | −13    | +2            | −2      | +4     | −37    |
| June         | −70         | +70    | +2            | −2      | +37    | −44    |
| July         | −78         | +92    | –             | –       | −2     | −2     |
| Whole season | −131        | +136   | +14           | −15     | +82    | −139   |

because they must take into account not only the effects of climate on the animals directly, but also its effect on the forage available and on pests. That there is a relationship between climate and income from livestock is demonstrated in Table 7.6. In this western Australian case income is sensitive to fluctuations in both temperature and precipitation. As with crops, however, the overall climatic situation, the type of livestock involved and its stocking rate will have a great influence on the way climate and its various components influence the year-to-year variations in agricultural income.

**Table 7.6  The effects of climatic changes upon income from sheep rearing in three areas in Western Australia**

|           |                                      | Average temperature |        |        | Rainfall |        |        |
|-----------|--------------------------------------|---------------------|--------|--------|----------|--------|--------|
|           |                                      | Normal              | +1 °C  | −1 °C  | +10%     | −10%   | Normal |
| Bakers Hill | Stocking rate (sheep per hectare) | 14                  | 14     | 14     | 14       | 14     | 12     |
|           | Income A$(000s)                      | 68.0                | 79.0   | 57.0   | 78.0     | 55.0   | 66.0   |
| Esperance | Stocking rate                        | 10                  | 10     | 10     | 10       | 10     | 9      |
|           | Income                               | 49.9                | 52.8   | 45.3   | 54.9     | 47.8   | 48.0   |
| Merredin  | Stocking rate                        | 7                   | 7      | 7      | 7        | 7      | 5      |
|           | Income                               | 31.5                | 30.5   | 17.6   | 27.6     | 13.8   | 20.0   |

## 7.3.2 Water resources

Most areas on Earth depend, directly or indirectly, on precipitation to provide their water resources. The annual precipitation amount, its temporal distribution within the year and its reliability from year to year, as well as its spatial distribution, all play a vital role in determining the type of water supply system which is economically feasible in a particular area.

In rural areas with abundant precipitation evenly spaced throughout the year, for example, an adequate water supply may be maintained by constructing individual storage tanks. The major climatological concern would be to ensure that these were large enough to maintain the supply through any brief dry spell that occurred, given the constant user demand and evaporative losses.

In urban areas in the same type of climate and, much more acutely, in any area where precipitation is unevenly distributed or which is subject to extended drought periods, this simple approach is not feasible. In such regions the construction of a storage reservoir is one obvious solution. However, the process of selection of a dam site for a storage reservoir is complex and involves much more than climatological considerations.

In such situations the first step is usually to develop a model to determine the size of storage reservoir needed to meet the demand. The change in storage at a particular site, $\Delta S$, is given by

$$\Delta S = I - D - E - L - O \qquad [7.21]$$

where $I$ is the inflow to the site, $D$ is the withdrawal to meet the user demand, $E$ are evaporative losses, $L$ represents other losses, such as seepage into the ground and around the dam and $O$ is any required outflow downstream. A time step calculation can be used, with historical records of $I$ and appropriate values for the other parameters, to determine the size needed to ensure that the reservoir survives the longest likely drought period without itself running dry.

Siting decisions can be made after the size is determined. These must take into account such non-climatological factors as geologically and topographically suitable sites, construction techniques and costs and the economic feasibility of the whole project.

Although the factor $I$ is likely to be composed mainly of streamflow into the reservoir, that streamflow is dependent on precipitation. Furthermore, for many areas streamflow measurements are sparse or of short duration, while precipitation records may be more abundant and cover a longer period. Consequently it is often necessary to establish a relationship between precipitation and streamflow before the storage model can be run. Frequently a correlation between precipitation and streamflow can be established using available short period streamflow records. The longer period precipitation records can then be used to estimate streamflow over the longer period required to obtain reliable results from the storage model.

A simple correlation model obviously represents a gross simplification and is analogous to models at the base of the modelling pyramid in Fig. 7.6. Although this may be appropriate for water supply planning, given possible uncertainties in the other parameters of equation [7.21], it is entirely possible to develop hydrological models which are closer to the apex of the modelling pyramid.

A hydrological model that has been developed for the central Pennine region of northern England is one such example. Figure 7.14 shows schematically the forcing factors and storage components

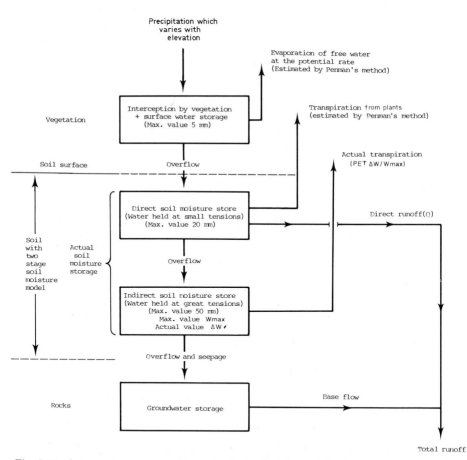

Fig. 7.14. Schematic representation of a hydrological model of the central Pennines, UK. Storage is shown by boxes and transfer by arrows. Rainfall is treated as a linear function of elevation. Total soil moisture storage, direct runoff and base flow are all computed. The results for soil moisture and direct runoff are illustrated in Fig. 7.15.

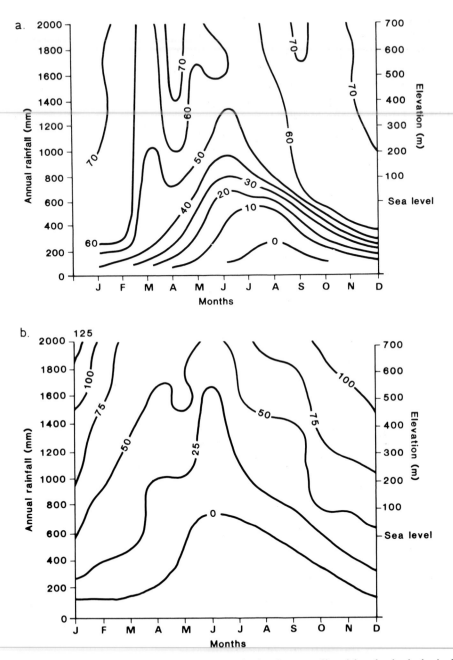

Fig. 7.15. (a) Soil moisture (mm rainfall equivalent) as predicted by the hydrological model for the last day of each month. Soil is saturated at 70 mm. (b) Monthly values of direct runoff (mm rainfall equivalent) as predicted by the hydrological model.

included. The primary inputs are precipitation and solar radiation and the predictions which are output by the model are of total evapotranspiration and the two components of runoff: direct runoff and groundwater discharge. Within the model soil moisture is modelled in a two-layer scheme. Evapotranspiration is assumed to occur from the upper layer at the potential rate and precipitation is absorbed by this upper layer. If the upper layer becomes dry evapotranspiration, $E$, can take place directly from the lower layer following the relationship

$$E = \text{PET } \Delta W/W_{max} \qquad\qquad [7.22]$$

where PET is the potential evapotranspiration, $W_{max}$ is the maximum available water content of the lower layer and $\Delta W$ is the actual available water content of the lower layer. If the upper layer becomes saturated, water which is not removed by evapotranspiration or runoff is transferred to the lower layer and stored. Saturation in the lower layer leads to transfer to the groundwater store (Fig. 7.14).

For this area, where precipitation is linearly related to elevation, the model predictions show that surface runoff and groundwater recharge are strong functions of precipitation, but vary only slightly in response to changes in evaporation resulting from solar radiation. Soil moisture values (Fig. 7.15(a)) indicate a marked summer dry period followed by a rapid recharge in the autumn for elevations below ~100 m. However, at higher levels the summer dry period is less well marked, being broken by intervals when the soil was rather wet. A similar type of relationship with rainfall and elevation is exhibited by the predicted direct runoff shown in Fig. 7.15(b).

## 7.4 Changes and cycles in the climate system

Atmospheric conditions vary on all time scales from day-to-day weather fluctuations to climatic changes which span millennia. Although past climates have an intrinsic interest, we are particularly concerned here with the aspects of changes that permit us to understand present atmospheric processes, and which can be used to test models and provide clues to future conditions. The ways in which past climatic conditions are deduced need not overly concern us. It is sufficient to say that this field, *palaeoclimatology*, requires a highly interdisciplinary approach. Geologists, oceanographers, glaciologists, botanists, archaeologists and historians are all involved with climatologists in unravelling the patterns of past climates.

Although our knowledge comes from many sources, several generalisations are possible. As we recede into the past from our present sophisticated observational network, data become increasingly sparse in time and space. From written records and botanical and archaeological evidence, we can deduce a great deal about conditions in many parts of the world over the last 1000 years. Conditions during the last 100,000 years or so are becoming increasingly clear from studies of glacial and other landscape-forming activities, pollen analysis and the analysis of lake sediments. Farther back than this we must rely, at present, mainly on geological evidence. This can be extremely sparse. Our estimates of the very earliest temperatures on Earth, for example, come from deductions from a single exposure of Precambrian rock in Greenland.

These various approaches, and the differing amounts and sophistication of information they yield, make it convenient to divide our discussion of past climates into three rough divisions: geological, covering the entire history of the Earth, intermediate, dealing primarily with the latest great Ice Ages, and historical, where more or less direct evidence of climatic conditions is available.

### 7.4.1  Geological record of climate

Over the longest possible time scale, the evolutionary history of the Earth itself, the climate system, typified by ambient surface temperature, has been remarkably stable. Geological data suggest that temperatures have varied only between 275 K and 305 K during the last 3.8 Ga*. Certainly during the last 2.3 Ga there have been a series

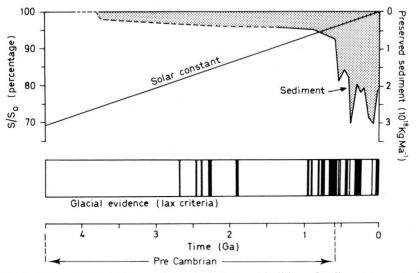

Fig. 7.16. The geological history of the glaciation and buildup of sedimentary material as a function of time. The increasing value of the solar constant (expressed as a percentage of present day) is also shown. The increasing mass of sedimentary material makes retrieval of more recent geological information easier than retrievals of data pertaining to Precambrian conditions. In order to try to show as much information as possible fairly lax criteria for glacial activity have been used.

* Ga = $10^9$ years BP

of alternations between glacial and interglacial periods suggesting relatively minor temperature excursions (Fig. 7.16). The first glaciation appears to have been initiated about 2.3 Ga, which was around the time that free oxygen began to evolve in the atmosphere. Glacial periods seem to have been relatively common in these early times, only to be replaced by a long period without glaciers prior to the onset of a new cooler phase about 1 Ga. This evidence suggests that there is a very long-term fluctuation in climate, which is of relatively small magnitude. It has been hypothesised that this long-term stability is the result of the interaction between the climate and the biosphere. This *Gaia hypothesis* suggests that biologically induced surface and atmospheric changes generate negative feedbacks which stabilise the whole system.

Over geological time there have been significant changes both in the types of rocks produced and in the biospheric composition, as preserved in the fossil record. These changes are displayed systematically in the Geological Column (Table 7.7). Climate and climate change must be a causal factor in these features. However, it is extremely difficult to isolate the climatic effect from others. Furthermore, even if the climate can be estimated, continental drift creates a

## Table 7.7  Scale of Geological Time and Biospheric Evolution

| Subdivisions of geologic time | | Apparent ages (years before present) | Notable events in evolution of organisms |
|---|---|---|---|
| **Eras** Periods | Epochs | | |
| **Cenozoic** | | | |
| Quaternary | Holocene (Recent) | | |
| | Pleistocene | | Man appears |
| | | $2 \times 10^6$ | |
| Tertiary | Pliocene | | Elephants, horses, large carnivores become dominant |
| | | $10 \times 10^6$ | |
| | Miocene | | Mammals diversify |
| | | $25 \times 10^6$ | |
| | Oligocene | | Grasses become abundant, grazing animals spread |
| | | $36 \times 10^6$ | |
| | Eocene | | Primitive horses appear |
| | | $58 \times 10^6$ | |
| | Palaeocene | | Mammals develop rapidly. Dinosaurs become extinct, flowering plants appear |
| | | $63 \times 10^6$ | |

| Subdivisions of geologic time | Apparent ages (years before present) | Notable events in evolution of organisms |
|---|---|---|
| **Eras**<br>Periods    Epochs | | |
| **Mesozoic** | | |
| Cretaceous | | |
| ——————————————— | $135 \times 10^6$ | |
| Jurassic | | Dinosaurs reach climax. Birds appear |
| ——————————————— | $180 \times 10^6$ | |
| Triassic | | Primitive mammals appear; conifers and cycads become abundant. Dinosaurs appear |
| ——————————————— | $230 \times 10^6$ | |
| **Palaeozoic** | | |
| Permian | | Reptiles spread, conifers develop |
| ——————————————— | $280 \times 10^6$ | |
| Carboniferous[a] | | Primitive reptiles appear, insects become abundant. Coal-forming forests widespread. Fishes diversify |
| ——————————————— | $340 \times 10^6$ | |
| Devonian | | Amphibians, first known land vertebrates, appear. Forests appear |
| ——————————————— | $400 \times 10^6$ | |
| Silurian | | Land plants and animals first recorded |
| ——————————————— | $440 \times 10^6$ | |
| Ordovician | | Primitive fishes, first known vertebrates, appear |
| ——————————————— | $500 \times 10^6$ | |
| Cambrian | | Marine invertebrate faunas become abundant |
| ——————————————— | $570 \times 10^6$ | |
| **Precambrian** | | |
| Proterozoic | | Life forms abundant |
| ——————————————— | $2.5 \times 10^9$ | |
| Archaean | | Primitive life forms (e.g. blue-green algae) |

[a] In North America the Carboniferous is divided into two periods (the Pennsylvanian and the Mississippian) at about $310 \times 10^6$ years.

problem of locating the geographical position of the rocks during their formative period. Nevertheless geological information permits the description of the long-term history of the climate.

For the *Precambrian* as discussed above, the data are sparse and relate mainly to glacial events. The earliest of these seems to have occurred around 2.7 Ga but evidence becomes widespread enough to be termed 'global' between 2.5 and 2.3 Ga. A second cooling seems to occur around 1.9 Ga (these data are shown in Fig. 7.16). It must be noted that all these data are sufficiently uncertain that these cold periods may have coincided.

The *Late Precambrian* glacial period (600 Ma†) includes one or more ice ages beginning around 850 Ma. It is possible that a sharp, thermal gradient over a continental land mass that includes a pole of rotation may be required to explain the evidence. This was followed by the *Ordovician ice age* ~450–430 Ma with a duration of up to 20–25 million years. The evidence for this ice age, which comes from the tropical regions, suggests that it affected most of the area of the supercontinent of Gondwanaland. In the *Permo-carboniferous glacial age* (300 Ma), Earth's land was in a single continent, which was symmetric about the equator, with a mid-latitude continental concentration.

The *Cretaceous* ice age began around 100 Ma, at a time when there was an essentially meridional configuration of the continents which, combined with shallow ocean ridges, prevented circumpolar ocean currents in either hemisphere. During the later part of the Mesozoic, the upper Cretaceous (100 Ma–65 Ma), there was a generally warmer global climate and the polar regions had no ice caps. About 55 Ma the global climate began a long cooling trend known as the *Cenozoic climatic decline* and about 35 Ma Antarctic waters underwent a significant cooling.

The *Oligocene* epoch (~35–25 Ma) was a period of global cooling. Around 30 Ma Australia moved far enough northeast to allow an Antarctic circumpolar current to develop. By 25 Ma ice reached the edge of the continent in the Ross Sea area. During the early *Miocene* (20–15 Ma) warm climates existed in low and mid-latitudes, but not in high southern latitudes. In the *Mid Miocene* (about 15–10 Ma) there was widespread incidence of further cooling, leading to substantial growth of Antarctic ice and to the development of some ice at the North Pole.

About 5 Ma the already substantial ice sheets in Antarctic underwent rapid growth and attained their present volume. At about the same time uplift of 'alpine' mountains in both hemispheres led to the growth of mountain glaciers. Not until as recently as 2 Ma did ice sheets became re-established in the Northern Hemisphere, spreading from a

† Ma = $10^6$ years BP
   ka = $10^3$ years BP

centre in the extreme north of the Atlantic Ocean. However, during at least the last 1 million years ice cover in the Arctic Ocean has never been less than it is today.

During the *Quaternary* period, covering the last 2 million years, glacial and interglacial periods alternated. Some workers suggest that there were as many as 20 glacial events; certainly there were at least seven. Interglacials, each characterised by a warm period lasting 10 ± 2 thousand years, have occurred on average every 100 thousand years during at least the past 0.5 million years. About 125 ka the *Eemian* (*Sangamon*) interglacial commenced and the warmest part persisted for about 10 thousand years. Between 115 ka and the *Pleistocene* glacial maximum at 18 ka, there were marked fluctuations superimposed on a generally declining temperature. Temperatures thereafter increased so that at present we are within the *Holocene* interglacial. To find conditions as warm and ice free as at present it is necessary to go back to the Eemian 125 ka.

Although we have emphasised glaciation and temperature conditions, the geological and geomorphological evidence also includes information about sea level and aridity. Consequently, except for the earliest periods, our inferences are based on evidence from several locations. Nevertheless, all data are highly specific for the locality from which they come and thus they may easily represent local anomalies, rather than large-scale variations in the climate.

## 7.4.2 Shorter time scale climatic changes

Data relating to conditions during the last 1 million years are relatively abundant and diverse compared to earlier periods. Hence we can say with some confidence, for example, that during the last 700,000 years only 8% of the time was the globe as warm as or warmer than today. Much of this confidence arises because several independent methods of obtaining climate related information are available. Figure 7.17 illustrates climatic records for the last $10^6$ years from four different observational techniques: oxygen isotope curves from deep sea cores; calcium carbonate percentage in ocean sediments; variation in populations of small marine animals, known as foraminifera, in tropical waters; and mid-European soil type sequences. Interpretations of these types of record have been underway for the last 50 years or so. Palaeoclimatologists have sought to synthesise and correlate the different data sets in an attempt to establish global, or at least regional, pictures of the climatic regime. The main events can be summarised as follows:

The last *Pleistocene Ice Age* reached a maximum 22,000–14,000 years ago (Fig. 7.18). During this period two large ice sheets have been identified in the Northern Hemisphere: the 'Laurentide Ice Sheet' covering parts of eastern North America and the 'Scandinavian Ice Sheet' covering parts of northern Europe. The 'Cordilleran Ice Sheet' over

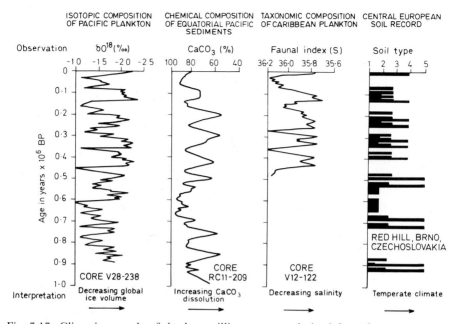

Fig. 7.17. Climatic records of the last million years as derived from four sources: the oxygen isotope curve from Pacific deep sea cores – reflecting global ice volume; calcium carbonate percentage in equatorial Pacific cores – low values are taken to indicate periods of rapid dissolution by bottom waters; faunal index – reflecting the changing composition of the Caribbean foraminiferal plankton which is calibrated as an estimate of sea-surface salinity. Glacial periods are marked by the influx of plankton preferring higher salinity waters; and the sequence of soil types accumulating at Brno, Czechoslovakia – reflecting fluctuations of temperature and, possibly, precipitation.

---

Fig. 7.18. Sea-surface temperatures, ice extent, ice elevation and continental albedo for Northern Hemisphere summer (August) as reconstructed by the CLIMAP project members for 18 ka. Contour intervals are in 1 K for isotherms and 500 m for ice elevation. Continental outlines represent a sea-level lowering of 85 m. Albedo values are given by the following key:

A   snow and ice – albedo over 0.4. Isolines show elevation (m) of ice sheet above sea level.

B   sandy deserts, patchy snow and snow-covered, dense coniferous forest. Albedos between 0.30 and 0.39.

C   loess, steppes and semi-deserts. Albedos between 0.25 and 0.29.

D   savannas and dry grass lands. Albedos between 0.20 and 0.24

E   forested and thickly vegetated land. Albedos below 0.20 (mostly 0.15–0.18)

F   ice-free oceans and lakes with isolines of sea-surface temperature (°C). Albedos below 0.10.

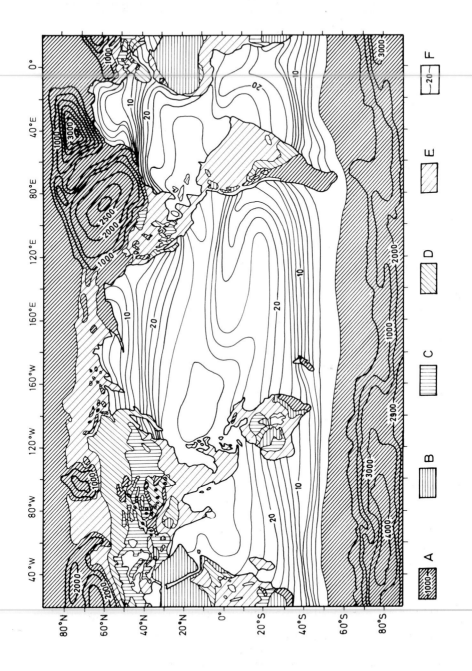

western North America achieved a maximum somewhat later than these, about 14,000 years ago. During this last ice age, the maximum area of the Northern Hemisphere ice sheets was equal to ~90% of the maximum achieved during the last million years of the Pleistocene. At the glacial maximum the sea level dropped by ~85 m and sea-surface temperature fell by as much as 10 °C in mid-latitudes of the North Atlantic and 3 °C in the Caribbean.

Widespread *deglaciation* began rather abruptly ~14,000 years ago. The Cordilleran Ice Sheet melted rapidly and was gone by ~10,000 years ago. The Scandinavian Ice Sheet lasted only slightly longer. Deglaciation was later in North America and the melting of the Laurentian Ice Sheet lagged by ~2000 years. By 8500 years ago the conditions in Europe had reached their present state while this situation was achieved in North America about 7000 years ago. Within this overall warming period were times of widespread cooling and glacial advance, seeming to occur about every 2500 years. One of these is called the *Younger Dryas event* which established itself within 100 years and lasted for 700 years (10.8–10.1 ka).

The *post-glacial climatic optimum* culminated between approximately 7000 and 5000 years ago. Summer temperatures in both Antarctica and Europe were 2°–3 °C higher than they are today. This led to a reduction of ice, mainly on land, but the sea ice in the Arctic Ocean was also reduced. Sea level rose rapidly, possibly reaching a peak of about 3 m above present levels ~4000 years ago, illustrating the lag between glacial melt and the sea-level rise.

Events of this type are sufficiently well established, and there is a sufficient worldwide network of data points, to allow attempts at reconstruction of the global climate at specific dates. Figure 7.18 shows the most, and perhaps the only, successful reconstruction to date: the 'CLIMAP' reconstruction for the time of the last glacial maximum ~18 ka. The effects of the ice sheets on continental elevation and albedo and on the lowered ocean levels are enormous. This type of reconstruction is becoming increasingly important as a source of data with which to test climate models.

Data on this time scale are also sufficiently numerous to encourage attempts to investigate the level of organisation of the fluctuations of the type seen in the curves in Fig. 7.17. For example, cyclic and quasi-cyclic activity seems to be suggested by some of the shorter period (<100,000 years) fluctuations. Such investigations are beginning to shed light on the possible mechanisms responsible for climatic change, as suggested in section 7.1.

### 7.4.3 Historical climatic change

As we approach the present time the range and diversity of climatic information increases. During the last two or three thousand years much can be inferred from archaeological evidence. Within the last thousand years documentary evidence is available, while starting in the seventeenth century direct observations become ever more numerous. Consequently for this historical period we can have more, but by no means complete, confidence in evidence for regional scale variations over relatively short time periods.

The *Iron Age*, occurring at 2.9–2.3 ka (900–300 BC), was generally wet and culminated in a cold epoch (probably around 500–300 BC). There is widespread evidence of regrowth of bogs throughout Europe.

The *Secondary optimum* in climate (AD 1000–1200) was a warming to a lesser degree and of a shorter duration than the post-glacial optimum. This led to melting of the pack ice and drift ice of the Arctic margins and was coincident in time with the founding of Norse colonies throughout the north Atlantic land areas as far as America. Summer temperatures were probably 1°C higher than today in western and central Europe, where the period lasted until AD 1300. Although Antarctica also seems to have been warmer during this secondary optimum, in China there may have been a cooling.

The *Little Ice Age* occurred between 1430 and 1850, with two intense periods around 1470 and in the late 1600s. It was most harsh in Britain during the second half of the seventeenth century. There was a preponderance over the entire period of harsher conditions than today but there were also some periods of relatively equable climate. Glaciers advanced in Europe, Asia and North America, although there were also times when summer temperatures were at or above present values. During this whole time the Arctic pack-ice expanded with important detrimental effects for Greenland and Iceland; by 1780–1820 (and probably by the late 1600s), the temperature across the north Atlantic north of 50°N was some 1–3 K less than today. The Southern Hemisphere seems to have partly escaped the cold period until 1830–1900. It may even have been warmer in Antarctica during the period 1760–1830.

The *warming trend* of the 1880s to 1940s was especially noted in the Atlantic sector of the Arctic and in northern Siberia. In the Southern Hemisphere, south of 30°S, there appears to have been a cooling.

A *cooling trend* was observed in the Northern Hemisphere following the high temperatures of the 1940s. Cooling at high latitudes (which seem to be predictive for other latitudes) stopped in the mid 1960s. Overall, sea-surface temperatures in the Northern Hemisphere cooled by 0.75 K but those in the Southern Hemisphere may have warmed during the 1935–1970 period by up to 1 K, especially at mid- and high latitudes. This has led to concern about the West Antarctic Ice Sheet, which is grounded on islands on the sea floor and so may be influenced by sea temperature or level changes. Increasing temperatures may cause a more rapid disintegration of this ice sheet and lead to a rise in sea level.

### Scale of temperature changes

The scale of temperature changes in the recent past can be put into perspective by considering temperatures over the last 100 ka (Fig. 7.19). Features such as the thermal maximum of the 1940s and the lowered temperatures of the Little Ice Age are clearly seen. However, glacial/interglacial temperature variations are of the order of 6–7 K while the temperature range over the last century is ~0.5 K.

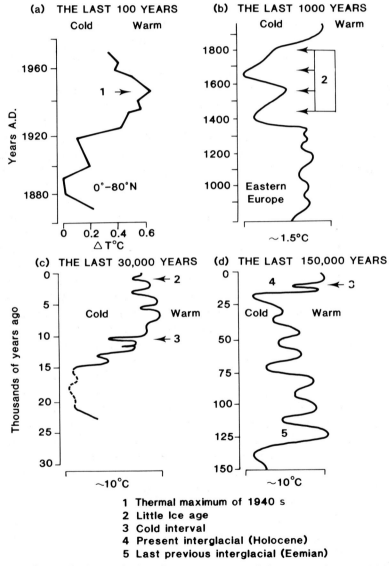

1 **Thermal maximum of 1940 s**
2 **Little Ice age**
3 **Cold interval**
4 **Present interglacial (Holocene)**
5 **Last previous interglacial (Eemian)**

Fig. 7.19. Generalised trends in climate represented by approximate mid-latitude air temperature trends over a variety of time scales for the past 150,000 years.

The present conditions are relatively warm since the temporal extent of glacial times seems to be longer than that of interglacial periods. Again the temperature fluctuations appear to be somewhat cyclic and suggest lines of approach for understanding the causes of climatic changes.

## 7.5   Climatic changes and their impact on Man

Any changes in climate, whatever the cause, must have an impact on human activity. Several examples of possible impacts have already been given implicitly in the sections dealing with applications of climate information. Since crop yield models clearly indicate that yields depend on weather, it might be postulated, for example, that a change in climate may lead to consistently low yields in a particular area, which in turn may lead to a human response in terms of a change in agricultural practice. Such simple postulates, however, can be very misleading, since they conceal several problems which are inherent in relating climate change to human impact. These concern the nature of climatic changes themselves, the strength of the relationship between climate changes and human response and the availability of past climatic data.

We can conveniently think of climatic changes as being represented by changes in the long-term mean values of a particular climatic parameter. Superimposed on this mean value will be decadal fluctuations and year-to-year variations (Fig. 7.20). These short period variations, of course, may themselves be influenced by the change in the mean. However, on the human time scale, mean value changes are likely to be sufficiently slow to be almost imperceptible. The changes in the last few decades, for example, can be detected only by careful analysis of instrument records. Much more noticeable will be changes in the variability, expressed in terms such as 'a run of exceptionally wet years', 'there seem to have been more hail storms in recent years', or even, the human memory being what it is, 'we don't get the long summers that we used to'. Any human response must depend on such a perception, whether conscious or unconscious. Furthermore, the change must be in a parameter which is conceived as significant for a particular activity. Consequently 'climate change' *per se* may not lead to any response. On the other hand, a small change in a particular parameter, expressed as a perceived change in variability, may have a profound impact on human activity.

Even if a change occurs which potentially has a significant impact on human activity, a response will not necessarily follow. Between the 'climate change' and the 'human response' is a chain of relationships (Fig. 7.21). Each of these links is influenced by non-climatic factors, which frequently will be more important than the climatic ones. In

397

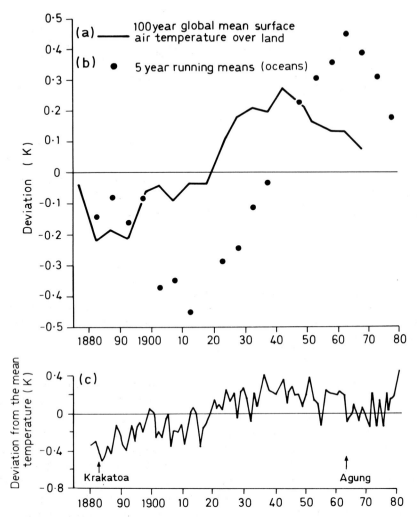

Fig. 7.20. Three reconstructions of global mean surface air temperature based on (a) land stations alone (solid line); (b) primarily on the temperature of the oceans (dots show 5-year running means) and (c) (on the lower curve) as deviations from the mean temperature over the period.

addition, connections between many of the links are weak or difficult to establish. Consequently it is impossible in most circumstances to establish readily a direct connection between the ends of the chain. All of the intervening links must be incorporated.

Any attempt to 'model' the impact of climatic changes must use historical information. For any period earlier than that of instrument records the information about climate is likely to be qualitative and selective. It is likely to emphasise information about unusual conditions which were perceived as having an impact, but is unlikely to say much

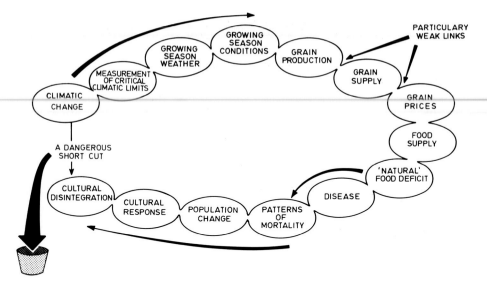

Fig. 7.21. Catenarian relationships in food supply in western Europe. This 'chain' of links between a climatic fluctuation and a cultural response can only be as strong as its weakest link. Certainly attempts to make direct correlations between climatic change and cultural response without considering all the links in the chain must be viewed with suspicion. (Courtesy: P. Laxton).

about normal conditions. Hence a great deal must be inferred about the climate and its variability before any suggestion of impacts is made.

These points can be illustrated by reference to conditions in the Little Ice Age. It has been suggested that average temperatures in mid-latitudes were about 1 K below their present levels. This suggests that the growing season was relatively shorter. This might imply that yields were low, but Fig. 7.13 indicates that this would not necessarily be the case for all crops in all areas. More significant for grain production may have been factors such as the incidence of late spring frosts, or destructive winds near harvest time, factors about which we have little or no information. It has also been suggested that cold, damp winter conditions prevailed in the Little Ice Age, making grain storage almost impossible and the population susceptible to famine. However, other factors, such as population pressure and plague, could have been equally, or more, important in creating the problems of the period. Certainly it is difficult to isolate the climate effects and demonstrate how, or if, they played a significant role.

Nevertheless, it is clear that relationships between climatic change and human response can be established in some cases. Nomadic and semi-nomadic peoples will move if a climatic variation causes an unacceptable deterioration in their environment and, of course, if an alternative more acceptable area lies within their reach. In this case the inhabitants have become adapted to an environment whose variability makes it marginal for their particular mode of existence.

In general, it is in 'marginal' agricultural areas that the links between climatic change and human response can be postulated to be strong. Such areas represent the outer limits of a particular climatic region. In the centre of the region agricultural practices will be well adapted to that particular climate and year-to-year variations will pose little threat. As the margins are approached, however, variability will become more significant. Usually overall production will be low, so that little surplus can be stored against the poor years that climate variability will inevitably bring. If a climate change occurs which alters the frequency of the poor years, some human response must follow.

The definition of 'marginality' of course depends upon the climatic regime and agricultural practices of an area. Three such areas using different climatic indices for marginality are shown in Fig. 7.22. Northern Europe is divided into agriculturally marginal and submarginal areas. The limits are given by:

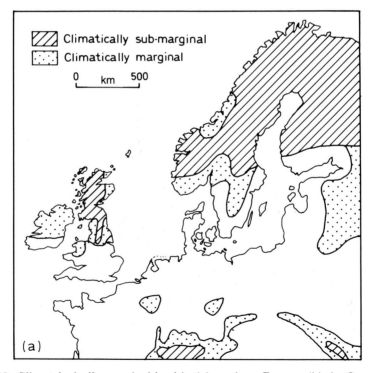

Fig. 7.22. Climatologically marginal land in (a) northern Europe, (b) the Great Plains of the USA and (c) eastern Australia. In (c) the shifts in climatic belts between 1881–1910 to 1911–40 are seen.

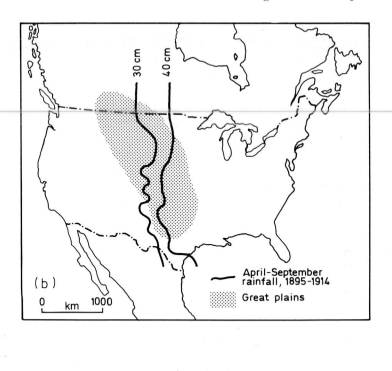

(b)

30 cm

40 cm

—— April–September
rainfall, 1895–1914

⬚ Great plains

0    km    1000

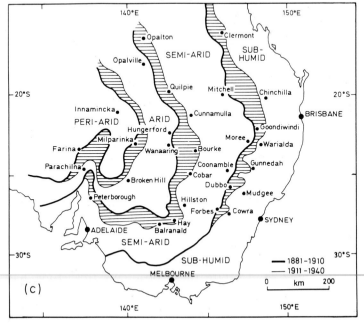

140°E      150°E

• Opalton    •Clermont

SEMI-ARID    SUB-HUMID

Opalville•

•Quilpie    Mitchell•    •Chinchilla    20°S

20°S

Innamincka•    •Cunnamulla    BRISBANE•

PERI-ARID    ARID    •Goondiwindi

Hungerford•    Moree•    •Warialda

Farina•    •Milparinka    Wanaaring•    •Bourke

Parachilna•    •Coonamble    •Gunnedah

•Cobar    Dubbo•

•Broken Hill    •Mudgee

•Peterborough    Hillston•    Forbes•    Cowra•

ADELAIDE•    •Hay    SYDNEY•

Balranald•

SEMI-ARID

30°S    SUB-HUMID    —— 1881–1910    30°S
—— 1911–1940

MELBOURNE•    0    km    200

(c)

140°E      150°E

401

(a) Marginal: either, less than 5 months $>10\,^{\circ}C$ and no increase in precipitation deficit July–September; or less than 3 months $> 10\,^{\circ}C$.

(b) Sub-marginal: either, less than 5 months $> 10\,^{\circ}C$ and 0–50 mm increase in precipitation deficit July–September; or, 5–6 months $> 10\,^{\circ}C$ and no increase in precipitation deficit July–September.

The region of marginal cultivation identified for the United States is based upon total rainfall in the period April–September rather than upon the combination of temperature and rainfall used for northern Europe. For Australia three zones of marginality, for different climatic regimes and agricultural enterprises, are shown in Fig. 7.22 (c). The limits are based on temperature and precipitation values and ranges. The changes of these limits with time indicate the eastward encroachment of aridity and the establishment of new marginal areas.

A similar effect is seen, on the local scale, for the Lammermuir Hills in Scotland (Fig. 7.23). The effect of the temperature decline during the course of the Little Ice Age and the recovery following it is clearly illustrated. The combined isopleths of 1050 degree days plus 60 mm potential water surplus (at the end of the summer) which represents the approximate 'cultivation limit' are seen to move downslope during the cooling period (Fig. 7.23(a)) and then to be restored following the Little Ice Age (Fig. 7.23(b)).

A careful study of the records of farming and settlement in this area gives credence to the suggestion that in marginal regions human response to climatic deterioration is identifiable. Figure 7.24 shows the pattern of abandonment of land together with the temporal changes in the climatic limits to cultivation as defined above. In this region of Scotland the relics of agriculture in the Middle Ages can be clearly seen in the ridged land that remains on slopes higher than those currently farmed.

This type of study has also been undertaken for parts of Norway and for Iceland. It is important to realise that the introduction of any factor affecting agricultural success such as a new crop, new methods or, more recently, financial aid will disrupt the close connection between a simplified climatic limit to cultivation and the settlement of land illustrated in Fig. 7.24. Despite these problems this study has shown the considerable importance of the climate for crop production and the possibility of direct human response to climatic change under some circumstances.

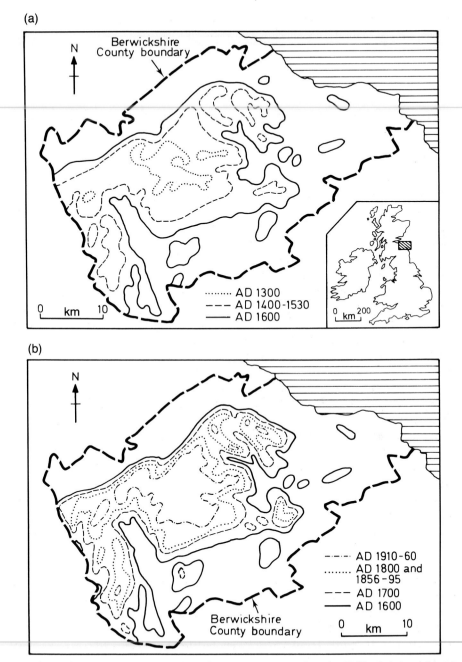

Fig. 7.23. (a) Cooling in the period AD 1300–1600 causing the 'fall' of the combined isopleths of 1050 degree days and 60 mm potential water surplus in southeast Scotland. (b) Warming in the period 1700–1910/60 resulting in the 'rise' of the same combined isopleth.

403

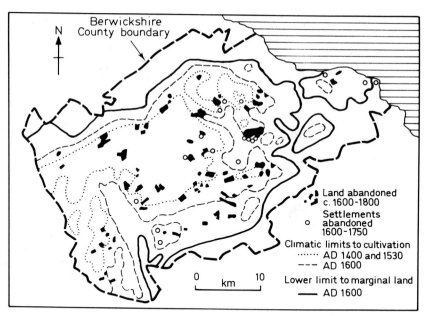

Fig. 7.24. Abandoned farmland and lowered climatic limits to cultivation in southeast Scotland, 1600–1750. The location is as shown in Fig. 7.23

## 7.6 Future climates – the probable impact of carbon dioxide

Climate changes and fluctuations which may have an impact on human society are, and have been, constantly occurring throughout the world. Until recently mankind's role in this has been restricted to the local scale. However, the increase in atmospheric carbon dioxide concentration as the result of human activity has the potential to produce worldwide climatic changes. Hence there is a possibility that, for the first time in history, Man may become a cause of global climatic change and modify the whole climate system. Such alterations are almost certain to have repercussions throughout society. Hence understanding of the processes acting and their likely climatic consequences has become a central goal of the climatologist.

Since the industrial revolution, the atmospheric concentration of $CO_2$ has been increasing as a direct consequence of fossil fuel combustion. Pre-industrial levels were probably between 288 ppmv and 295 ppmv $\pm$ 10 ppmv. The Mauna Loa record (Fig. 7.25) shows a nearly exponential rise in $CO_2$ since accurate recording began 25 years ago. The mean yearly level increased from 315 ppmv to 335 ppmv between 1958 and 1978. This trend has been calculated to be in close agreement with that expected from the fossil fuel production figures.

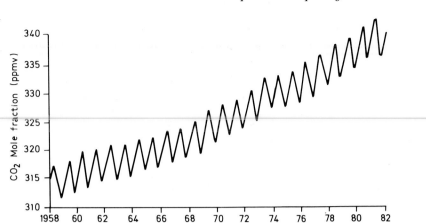

Fig. 7.25. Monthly measurements of atmospheric carbon dioxide concentration at Mauna Loa Observatory, Hawaii, USA. This is the longest continuous record of observations available. The seasonal cycle due to biospheric uptake and release is clearly seen.

Despite the concentrated nature of the major source of $CO_2$ in the industrial centres of the developed world, and despite short-term sensitivity to seasonal and other environmental changes, this rise is a worldwide phenomenon (Fig. 7.26).

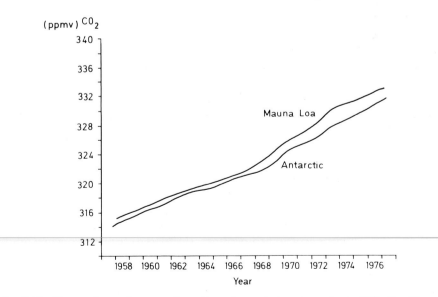

Fig. 7.26. Comparison of seasonally adjusted trends in atmospheric $CO_2$ for Mauna Loa Observatory, Hawaii, USA and the South Pole.

## Global carbon budget

Any future climate suggested as a response to carbon dioxide must depend upon the predicted atmospheric concentration. This is not easy to evaluate from the present measurements because $CO_2$ takes part in the planetary carbon cycle (Fig. 7.27), constantly being put into and removed from the atmosphere.

There is a constant cycling between the major reservoirs of carbon. $CO_2$ released by respiration and decay of vegetation is placed in the atmosphere, only to be reincorporated into vegetation as new growth develops, leading to the marked annual cycle in the atmospheric concentration (Fig. 7.25). Exchange between the atmosphere and the oceans depends on sea-surface temperatures and the oceanic circulation as well as the atmospheric and oceanic concentrations. Slow changes in the oceanic circulation, notably the Southern Oscillation, may well be the cause of the longer term variations in atmospheric concentration seen in Fig. 7.26.

All of the exchanges shown in Fig. 7.27 are attempts to create a dynamic equilibrium within the global carbon cycle. Human intervention, both through industrial production and agricultural activity, has unbalanced the system and increased the atmospheric component. As a consequence the fluxes will naturally change in an effort to restore equilibrium. Predicting these changes is extremely difficult. We have relatively little information about the pre-industrial level of the fluxes, so that little can be said about how they have already changed. In addition, the actual mechanisms of exchange are poorly understood, although it is clear that the various mechanisms act on various time scales. For example, the biotic response is relatively rapid, but oceanic uptake is very slow.

As the atmospheric concentration increases, therefore, there will be transfers of $CO_2$ from the atmosphere to the land and the oceans. Although the biotic response is rapid, vegetation represents a relatively small sink, so that most of the excess, if it is to be removed, must go into the oceans. However, their uptake is slow and a considerable time will be needed before the oceans and atmosphere can come into equilibrium. It is estimated, for example, that if anthropogenic production stopped now, about 85% of the excess $CO_2$ would be removed, most of it going into the oceans. However, the time period required to reach equilibrium would be of the order of several thousand years. A reasonable estimate of the partitioning on the decadal or century time scale is that 57% remains in the atmosphere, 34% goes into the ocean, while vegetation receives 9%.

## Models of $CO_2$ induced changes

The uncertainties associated with the carbon cycle make it extremely difficult to predict rates of increase of atmospheric $CO_2$ concentration.

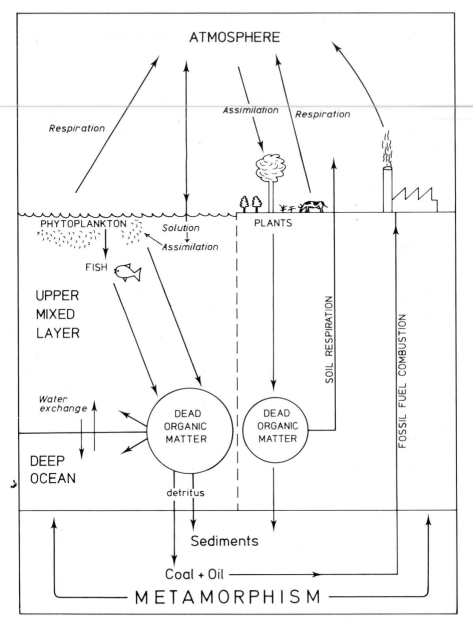

Fig. 7.27. The global carbon budget is primarily controlled by the biosphere, although there are sources and sinks for $CO_2$ other than those shown here (e.g. volcanoes). Note that the atmospheric $CO_2$ is the result of separate land and ocean surface interchanges. The carbon reserves represented by land animals and fish are very much smaller than those represented by land plants and phytoplankton. The carbon reserve incorporated into the oceanic dead organic matter is believed to be about five times larger than the terrestrial reserve. The sediments are a massive carbon sink containing 30,000 tonnes more carbon than the atmosphere.

Hence most climate modellers have concentrated their efforts on situations where the concentration has reached a specified level, usually double or quadruple the present value. The increased level of $CO_2$ must produce a greenhouse temperature increase but it is the nature, direction and magnitude of the feedback effects within the climate system that will determine the actual temperature change. These feedback effects are the subject of debate.

Climate models and analogue models all seem to give rise to similar results. The increase in global surface temperature for doubled $CO_2$ is around 1.5–4.0 K with an 'accepted average' value of 2.5 K (Table 7.8). Climate models suggest that increased near-surface tropospheric temperatures will be 'balanced' by cooling above the tropopause (Fig. 7.28). Some climate model experiments use $4 \times CO_2$ to be sure of obtaining a discernible result (signal) above the natural variability (noise) of the model's climate. Making the assumption that the temperature effect of $CO_2$ is linear, these results are halved to achieve the $2 \times CO_2$ impact.

The predicted global average temperature of $2.5 \pm 1.5$ K change smooths out a very wide range of temperature responses as a function of latitude. In high latitude regions the surface temperature increase seems to be closer to 3–4 K near the South Pole and 7–8 K near the North Pole (Fig. 7.29). This high latitude temperature increase enhancement is also a significant feature in the temperature maps

**Table 7.8   The air temperature increase predicted as a result of doubled $CO_2$ concentration by a range of methods**

| Estimation technique | | | |
|---|---|---|---|
| Simplified climatic models | General circulation models | Present-day climatic variations | Climatic variations in geological past |
| 2.4° (1967)[a] | 2.9°(1975) | 3.3°(1974) | 3.5°(1979) |
| 2.5–3.5° (1974) | 3.9°(1979) | 2.0–3.0°(1981) | 3.4°(1980) |
| 2.0–3.2°(1977) | 3.5°(1979) | | 2.9°(1981) |
| 3.3°(1979) | 2.0°(1980) | | |
| 4.4°(1980) | 2.0°(1980) | | |
| 3.2°(1980) | 3.0°(1980) | | |
| 3.5°(1981) | 2.5°(1981) | | |
| | 2.4°(1982) | | |
| | 4.0°(1984) | | |

[a] The date is the date of publication of the prediction.

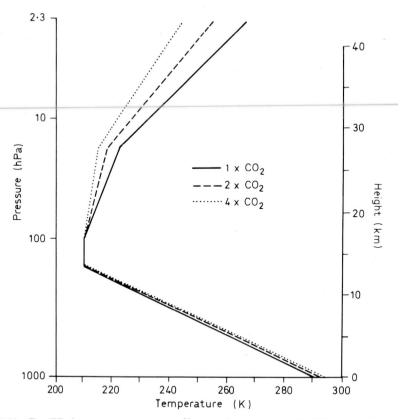

Fig. 7.28. Equilibrium temperature profiles produced by a 1-D RC global model for the normal $(1 \times CO_2)$, twice the normal $(2 \times CO_2)$ and four times the normal $(4 \times CO_2)$ amount of atmospheric carbon dioxide. The tropospheric lapse rate was adjusted to 6.5 K km$^{-1}$ in all cases.

shown in Fig. 7.30 with larger variations showing up in the two seasonal simulations than in the annual average.

Numerous other tests with various models lead to a consensus that the high latitudes will suffer a considerably greater temperature increase than elsewhere. Model results generally show that the warming in the Arctic is two to three times more than the global average and greatest in winter. The temperature response of the Southern Hemisphere is likely to be smaller. This, physical reasoning suggests, is due primarily to the greater thermal inertia provided by the proportionately larger oceanic area in the Southern Hemisphere. Although this high latitude enhancement is seen in most climate model simulations, the precise nature and location of temperature and precipitation changes which are vital for planning are still in dispute.

There is a fundamental difference between the numerical simulations

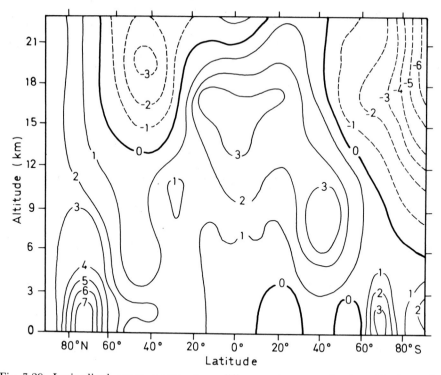

Fig. 7.29. Latitudinal mean temperature difference between the last 30 days of a $2 \times CO_2$ and a $1 \times CO_2$ atmosphere mixed layer ocean coupled model. The results show a warming (positive values) of between 1 and 3 K in the tropical troposphere and up to 7 K in polar regions near the Earth's surface. (NCAR Annual Report, 1980).

and the real-world occurrence: atmospheric $CO_2$ has been increasing *gradually* over approximately 100 years. It is not clear how representative instantaneously modelled climatic changes are of the real geophysical experiment occurring now. Furthermore, whilst the results from various models suggest average global temperature increases of similar magnitude, identification of geographically sensitive areas differs from model to model. The most important impact of any change is obviously on agricultural production, which could lead to global economic effects. The other major $CO_2$ induced impact would be that ice melting would place low lying centres of population in jeopardy from increased sea levels.

### Use of historical analogues to understand the $CO_2$ effect

An alternative approach for simulating possible $CO_2$ induced climate changes is to use relevant historical and geological analogues. Some of these results are included in Table 7.8. Table 7.9 lists some of the geological epochs in which conditions similar to those likely to be

410

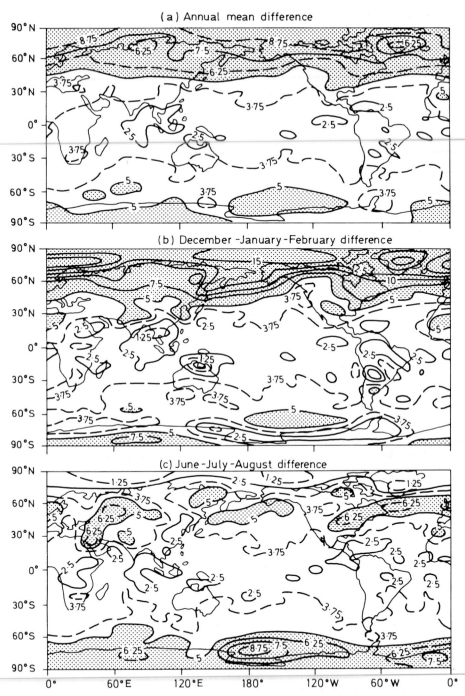

Fig. 7.30. Geographical distribution of the difference in surface air temperature (K) between a $4 \times CO_2$ and a $1 \times CO_2$ GCM simulation. Maps show (a) annual mean difference. Regions where the difference exceeds 5 K are shaded. (Manabe, S. and Stouffer, R.J. 1980, Sensitivity of a global climate model to an increase of $CO_2$ concentration in the atmosphere, *JGR*, **85**).

**Table 7.9 Comparison between warm climate phases of the past and the $CO_2$ levels currently believed to be able to produce similar changes in the average global surface temperature ($\Delta T$)**

| Period | Date | Increase $\Delta T$ (K) | $CO_2$ concentration (ppmv) (estimate) |
|---|---|---|---|
| Mediaeval | AD 1000 | +1.0 | 385–430 |
| Holocene (Altithermal) | 6000 BP | +1.5 | 420–490 |
| Eemian Interglacial | 120,000 BP | +2.0/+2.5 | 460–555/500–610 |
| ?(Ice-free Arctic but glaciated Antarctic) | 3–15 Ma | +4.0 | 630–880 |

caused by increased $CO_2$ levels may have occurred. Possible precipitation conditions in the Altithermal (8.4 ka) are shown in Fig. 7.31. The terms used are relative to the present level of rainfall in the marked areas. Regions without any markings are not necessarily areas in which no change occurred but regions for which data have not yet been discovered. Temperatures during the Altithermal were several degrees higher than now. This would, in itself, tend to elongate the growing season as long as rainfall was adequate. The major conclusion to be drawn is that the general trend is for better conditions for more people than at present. The only region of the world that suffers significantly in this reconstruction is the Prairie plains of North America.

A second analogue 'reconstruction' uses data relating to the five coldest and five warmest years in a recent 50-year period. Maps of temperature and rainfall changes between the cold' and 'warm' periods are shown in Fig. 7.32. In both parameters there are decreases and increases of similar magnitude. Although the inference is that these maps can be used as a surrogate for the effects of doubled $CO_2$ it must be pointed out that the mean temperature increase in their analogy is only 0.6 K as compared with the anticipated 2–3 K as a result of doubling $CO_2$.

There are some similarities between Figs 7.31 and 7.32. Certain areas (e.g. Europe and possibly the USSR) seem to benefit in the sense that growing conditions would be improved since the length of the

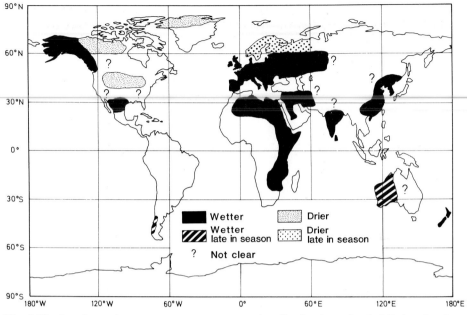

Fig. 7.31. A schematic attempt to reconstruct the distribution of rainfall (predominantly during the summer) during the Altithermal period of 4–8 ka when the world was a few degrees warmer than now. The terms 'wetter' and 'drier' are relative to the present. Blank areas are not necessarily regions of no rainfall change, just of no information.

growing season and mean temperatures are increased and rainfall is slightly enhanced. Other regions, notably the grain belt of North America, are subjected to a severe drought. However, for many areas, the two reconstructions do differ in their predictions for precipitation.

### Statistical significance of observed and predicted change

Since the increase in atmospheric $CO_2$ from anthropogenic sources has been underway for a considerable time and models predict that this should lead to a temperature increase, the observational temperature record should soon begin to show this increase. That a $CO_2$ induced temperature 'signal' has not yet been detected is not surprising since it is only in the last few decades that $CO_2$ increases should have been detectable. Furthermore, the natural year-to-year temperature variations (Fig. 7.20) introduce 'noise' which makes any signal difficult to detect. However, the temperature record is now long enough to allow statistical assessment of the annual variability of global temperatures. Thus we can predict when a $CO_2$ induced signal should be discernible above the climatic noise. Figure 7.33 suggests that this will occur by the mid 1990s.

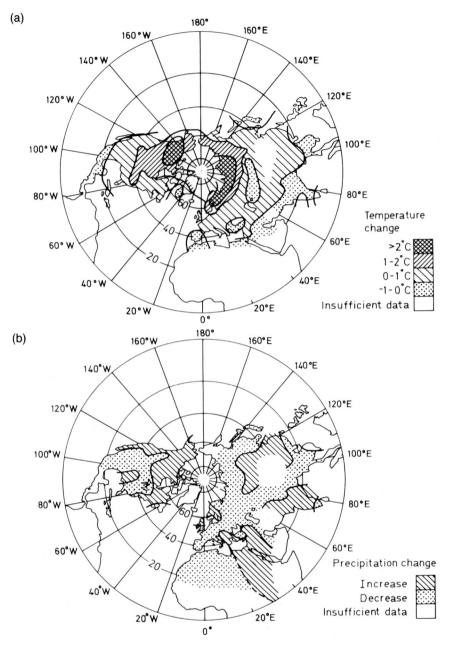

Fig. 7.32. (a) Mean annual surface temperature changes between the five coldest years and the five warmest years in the period 1925 to 1974 in the Northern Hemisphere. The corresponding change in hemispheric mean temperature is 0.6 K (cf. the expected change in global mean temperature due to a doubling of atmospheric $CO_2$ of ~2–3 K). (b) Mean annual precipitation changes from the same five coldest to the same five warmest years.

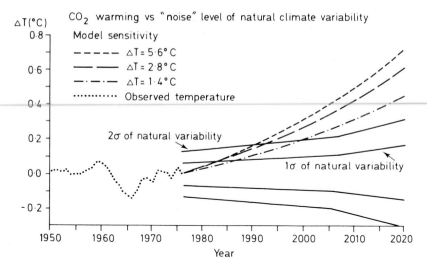

Fig. 7.33. Comparison of a series of model predicted $CO_2$ induced warmings compared with confidence intervals of $\pm$ 1σ and $\pm$ 2σ where σ is the standard deviation of the observed global temperatures computed from a 100-year temperature record. A statistically significant signal can be claimed when the temperature curve exceeds the $\pm$ 2σ level.

If, or when, such a signal occurs, it may, of course, simply be part of the natural atmospheric variability. However, it will give strong support to the contention that our model predictions of $CO_2$ influences on climate are reasonably accurate. As such it will have profound implications not only for future climates but also for many societal activities.

## 7.7  Epilogue

The realisation that human activity, through the combustion of fossil fuels, may lead to climatic change, has increased public awareness of climate, its variability and its potential impact on all aspects of human society. From the viewpoint of the climatologist the 'carbon dioxide problem' epitomises modern climatology. Society is asking for predictions of future climates and information about the potential impacts of possible climatic changes. These requests provide tremendous challenges and opportunities for the climatologist.

Climate predictions must come through climate modelling. Although there have been tremendous strides in our modelling ability over the 25 years since the first models were developed, even our most sophisticated models are far from satisfactory. Further advances are certainly possible, but they must be associated with a better description and understanding of the present climate and with more information about

415

past climates. Then the influence of possible mechanisms of change can be tested.

The present surface-based climatological observation network allows us to describe the present climate with reasonable accuracy over much of the Earth's surface. However, it is clear that more and better observations, particularly from satellites, are needed if we are to improve our specification of the processes acting to create and maintain that climate. Certainly, as much of this book indicates, the processes are known in a qualitative way. A few are known quantitatively on the global scale, a few more on the local scale. Without further knowledge, however, it will remain extremely difficult to identify realistically the feedback mechanisms that are of vital importance for the climate system.

As our understanding of the present climate increases, it becomes increasingly possible to suggest mechanisms of climatic change. External mechanisms, such as changes in solar output and the Earth's orbital geometry, are being explored. Changes associated with continental drift and the location of major topographic features, which can also be regarded as external to the climate system, may be important in producing long-term climate shifts. Internal changes, ranging from changes in atmospheric composition associated with air pollution and with volcanoes, to climatically induced biological changes at the surface, have also been proposed as possible climate change mechanisms. Feedbacks between them all make it difficult to isolate a single individual prime causal mechanism, but it seems likely that, for example, volcanoes can have an impact on a relatively short time scale, while solar luminosity affects the climate on a much longer time scale.

Such suggestions must remain as speculations unless they can be tested. Such tests can only be made using information about past climates. Hence it is vital that efforts continue to be made to unravel the intricacy and complexity of these. This is a formidable task, requiring cooperation between many disciplines if reliable conclusions are to be drawn. Geological evidence must be used when considering climate over millennia. Glaciologists, limnologists, chemists and geomorphologists contribute to our understanding of climate over thousands of years. Historians and archaeologists need to be consulted when dealing with more recent changes. A great deal of cooperation is underway and, although our knowledge is still woefully inadequate, rapid advances are being made. It is becoming increasingly possible, for example, to produce with some confidence maps of global conditions for selected times in the past.

As our knowledge increases, it must be incorporated into the climate models. Better understanding of individual atmospheric processes, for example, can be used in relatively simple models to improve our insight into the climate system. This new insight can then be incorporated into the more complex models to give more reliable predictions. Better and more extensive observations allow us to parameterise processes more fully where full physical understanding, or simply sufficient computer power, is lacking. Certainly faster computers should help increase the accuracy of our models, provided we have the physical insight to incorporate additional processes in a reasonable way.

The ultimate test of any model, of course, is whether its predictions are correct. For many local application models, the time period of the prediction is fairly short. Crop yield models are 'tested' at the end of every growing season. Hence there are frequent opportunities to incorporate new physical insights and modify and refine the models in the light of their performance. For global climate models, on the other hand, the period of prediction is much longer: decades or centuries. Hence testing in this sense is virtually impossible on a routine basis. Nevertheless, testing with historical data and careful checking of model results when small 'artificial' perturbations are introduced, can lead to predictions which can be offered with some confidence.

Predictions of the effect of carbon dioxide on climate are long-term predictions and hence not susceptible to direct testing. However, enhanced $CO_2$ levels resulting from anthropogenic activity have been established sufficiently long that we are rapidly approaching the time when, theoretically at least, we should be able to detect a $CO_2$ induced temperature 'signal' above the 'noise' of the natural year-to-year variations in temperature. However, the nature of climatic change makes such recognition difficult. Climate fluctuates on all time scales so that annual variations are superimposed on decadal ones, which in turn are superimposed on longer term ones. Essentially with regard to $CO_2$ induced changes, we are looking for a small shift in the decadal mean temperature, which is superimposed on large annual temperature variations. The current consensus from a large group of climate models is that temperatures will increase. If such an increase is detected in the 1990s it will provide strong evidence that our model predictions are reasonably accurate. There remains the possibility, of course, that any temperature increase that occurs is of 'natural' origin. Only monitoring of climatic conditions well into the next century will 'prove' whether or not the models are truly realistic. By then, of course, it will be rather late to take corrective measures should the changes prove to be generally detrimental.

Although the $CO_2$ problem has received most attention from the public, regional scale and local scale climate changes have been, and are being, induced by human activity. Theoretically, almost any activity which changes the surface of the Earth or the composition of the atmosphere will lead to a climatic change. The development of a city certainly creates an urban heat island and alters many climatic parameters in the local area. Air pollution leads to changes in the radiative transfer properties of the atmosphere, thus affecting the climate. Both of these effects are, at present, localised and do not seem to influence the global climate significantly. Nevertheless, as both urbanisation and air pollution sources and concentrations increase, the possibility of regional effects exists. Large-scale deforestation is already beginning to have a regional effect in certain tropical areas. If this deforestation continues the strong possibility exists that before long global scale changes will be initiated.

All of these actual and potential changes are not only of interest to climatologists but also a cause of concern for the wider public. Climate plays a role in so many human activities that any climatic change will have some impact on society. The range of activities where climate is important can be suggested by considering areas where climate information is applied to solve immediate and particular problems. Climate information is used in endeavours as diverse as energy generation, crop yield optimisation and water supply planning. As more is known about the climate itself and as we explore additional ways in which it can be utilised as a resource, the range of applications will also expand.

Although these current applications of climate information are primarily intended to solve immediate problems, they can also be regarded as indicators of the potential impacts of climatic change. By relating crop yield to climate, for example, some idea of potential yields under a new climatic regime can be obtained. Indeed, a similar interpretation can be placed on many of the discussions of climate applications throughout this book. Thus it is becoming possible to assess the impact of climatic change on particular sectors of the economy in particular localities.

The extension of these specific impacts to the impact on society as a whole, however, is extremely difficult. Climate is only one of many factors which are responsible for the structure of any society. In some cases climate may be dominant. This is most likely to be the case in areas where society is structured around an agricultural economy and the climatic resource is marginal for the agriculture. In such cases it appears that there are direct responses by society to any climatic change. In most areas, however, any climatic change is likely to be absorbed without major structural change. Some adaptations may have to be made, but the links between a climate change and a societal response are so many and complex that the resulting adaptation may not be a direct response to climate. Certainly the time scale of climatic change is long, in human terms, so that any adaptation may not be perceived as a response to the change, but as a response to a multitude of other factors which themselves may only be tenuously linked to climate.

The long-term, almost imperceptible nature of climatic change raises

418

another problem in societal terms. Most policymakers and planners, in both government and private industry, are concerned with rather short-term problems. Long-range planning may involve the next few decades, but rarely extends a century or more into the future. However, it is precisely in this time frame that climatic changes may become sufficiently large to have a significant impact. Hence it is extremely difficult for policymakers to consider climatic changes, even though they may acknowledge their importance in the time frame beyond their purview.

This type of problem is especially acute when global effects, such as those postulated for the greenhouse effect due to $CO_2$, are involved. We are by no means sufficiently confident in our models to state with certainty that there will be a global warming of a specific magnitude and can only vaguely estimate the changes in climatic parameters other than temperature. Furthermore, even if we could give some confident climatic statements, neither we nor specialists in other fields could state with confidence whether or not the changes would be detrimental to society as a whole. Avoiding the possible problem by stopping further placement of $CO_2$ into the atmosphere would mean a major dislocation in the present economic structure of most of the major industrialised nations. Hence any policymaker is faced with the dilemma of having to decide whether to recommend a major dislocation now or risk only vaguely specified consequences in the future.

This suggests that one role of the climatologist, given our present understanding of the climate system, is to 'warn' of possible climatic changes and their potential effects. This seemingly negative role, however, represents a tremendous leap forward over the position only 25 or so years ago. At that time a reasonably adequate description of the Earth's climate was available, together with some understanding of the processes controlling it. Furthermore, past climate changes were known to exist and there was speculation that these would continue in the future. The links between these three aspects, however, were very poorly understood. Thus there was no possibility of predicting future climates. Similarly, many applications of climate information had been developed and climate was being seen as a resource to be used. The climatic understanding necessary to refine the applications and use them to investigate the potential impacts of future conditions was not available. During the last 25 years, therefore, we have been able to build on the foundation of the traditional climatology which was mainly concerned with present conditions, to develop a modern climatological approach which is essentially forward looking.

The fact that at present we can only 'warn' in a tentative way about future conditions also indicates that there is still a tremendous amount of work for the climatologist. There is scope for increasing our understanding of the processes controlling the climate system, further develop-

419

ment of models, investigations of past climates, exploration of methods of application of climate information and analysis of climatic impact. Some of these areas are the preserve of the climatologists alone, but many require cooperative efforts between climatologists and experts in other disciplines. All contribute to the ultimate goal of modern climatology – to understand the climate system, predict its future course, utilise it fully as a resource and understand and predict its impacts on human society.

# Suggested Further Reading

(\* The starred books may be difficult for students without a maths/physics background but they are worth the struggle.)

\*Atkinson, B. W. (ed.), 1981 *Dynamical Meteorology: An Introductory Selection*. Methuen, 228 pp.

Barry, R. G. and Chorley, R. J., 1982 *Atmosphere, Weather and Climate* (4th edn). Methuen, 432 pp.

Budyko, M. I., 1982, *The Earth's Climate: Past and Future*. Academic Press, 307 pp.

Chandler, T. J. and Gregory, S., 1976, *The Climate of the British Isles*. Longman, 390 pp.

Chen, R. S., Boulding, E. and Schneider, S. H. (eds), 1983, *Social Science Research and Climate Change: An Interdisciplinary Approach*. Reidel, 255 pp.

Edholm, O. G., 1978, *Man–Hot and Cold* (*Studies in Biology* no. 97). Edward Arnold, 60 pp.

Gribbin, J. (ed.), 1978, *Climatic Change*. Cambridge University Press, 280 pp.

HMSO, 1968, *Meteorological Office Observers Handbook*. Her Majesty's Stationery Office, London, 242 pp.

HMSO, 1972, *Meteorological Glossary*, Her Majesty's Stationery Office, London, 320 pp.

Hobbs, J. E., 1980, *Applied Climatology*, in series *Studies in Physical Geography* (ed. K. J. Gregory). Butterworths, 218 pp.

\*Houghton, J. T., 1977, *The Physics of the Atmosphere*. Cambridge University Press, 203 pp.

Houghton, J. T. (ed.), 1984, *The Global Climate*. Cambridge University Press, 233 pp.

Imbrie, J. and Imbrie, K. P., 1980, *Ice Ages: Unlocking the Mystery*. Macmillan, 224 pp.

Lamb, H. H., 1982, *Climate, History and the Modern World*, Methuen, 387 pp.

Landsberg, H. E., 1981, *The Urban Climate*, Academic Press, 275 pp.

Lockwood, J. G., 1974, *World Climatology: An Environmental Approach*, Edward Arnold, 330 pp.

Lockwood, J. G., 1979, *Causes of Climate*, Edward Arnold, 260 pp.

Lovelock, J. E., 1979, *Gaia*, Oxford University Press, 157 pp.

Miller, D. H., 1977, *Water at the Surface of the Earth*, Academic Press, 576 pp.

Miller, D. H., 1981, *Energy at the Surface of the Earth*, Academic Press, 534 pp.

*Monteith, J. L., 1973, *Principles of Environmental Physics*, Edward Arnold (Elsevier in USA), 241 pp.

*Munn, R. E., 1966, *Descriptive Micrometeorology*, Academic Press, 245 pp.

National Academy of Science, 1975, *Understanding Climatic Change: A Program for Action*, U.S. Committee for GARP, National Research Council, Washington D.C., 239 pp.

National Academy of Sciences, 1982, *Carbon Dioxide and Climate: A Second Assessment*, Climate Board/Committee on Atmospheric Sciences, U.S. National Research Council, Washington D.C., 72 pp.

Neiburger, M., Edinger, J. G. and Bonner, W. D., 1982, *Understanding Our Atmosphere* (2nd edn). W. H. Freeman, 453 pp.

Oke, T. R., 1978, *Boundary Layer Climates*, Methuen, 372 pp.

Scorer, R. S., 1972, *Clouds of the World: A Complete Colour Encyclopedia*, David and Charles, 176 pp.

*Sellers, W. D., 1965, *Physical Climatology*, University of Chicago Press, 272 pp.

*Wallace, J. M. and Hobbs, P. V., 1977, *Atmospheric Science: An Introductory Survey*, Academic Press, 467 pp.

# Glossary

Terms are included in this glossary if they appear in textual locations separate from their initial definition or if a complete definition *in situ* would detract from textual continuity.

***Absorption***: The process by which incident radiation is taken into a body and retained without reflection or transmission. Absorption increases either the internal or the kinetic energy of the molecules or atoms composing the absorbing medium.

***Absorption band***: A wavelength region where the absorption of radiation by a particular substance, usually a gas, is large.

***Adiabatic process***: A process in which a system does not interact thermally with its surroundings (i.e. no heat is exchanged).

***Advection***: The process by which the property of a mass of air is transferred by movement, usually in the horizontal direction.

***Advection fog***: Fog resulting from warm air advection, i.e. the movement of warm air horizontally over a cooler surface, in which the air mass is sufficiently cooled that condensation occurs.

***Aerosol***: A collection of small liquid or solid particles suspended in the atmosphere.

***Albedo***: The ratio of reflected solar radiation to the total incoming solar radiation where both streams are measured across the complete wavelength range of solar radiation ($\sim$0.3–4.0 $\mu$m). At a single wavelength or for a narrow waveband the ratio is termed spectral reflectance.

***Anabatic wind***: A valley wind system developed in daytime in which airflow is directed upslope. The upper valley slopes are preferentially heated compared to the valley bottom, causing the warmer air to rise and an upslope flow to develop. Anabatic flow occurs only when regional pressure gradients are weak.

***Angular momentum***: The angular momentum of a particle rotating about a fixed axis is the product of the particle's linear momentum (its mass times the linear velocity) and its perpendicular distance from the axis of rotation.

***Analogue model***: Method of predicting a future climate by considering a known historical situation which had similar features to those anticipated in the future.

***Annulus experiment***: A simulation of some features of the Earth's atmospheric circulation undertaken by differentially heating a liquid contained in a hollow cylindrical dish which is rotated on a turntable.

***Atmospheric pressure***: The pressure created by the constant motion of atmospheric gas molecules. They exert a force whenever they impact upon a surface. The total force per unit area is the pressure.

***Backing***: The wind direction changing in a counterclockwise sense.

***Baroclinic***: The atmosphere has isotherms which are not parallel to the isobars, i.e. there is a temperature gradient along the isobars.

***Barotropic***: The atmosphere has horizontally uniform temperatures at all heights. Thus pressure gradients can exist but temperature gradients cannot.

***Bergeron–Findeisen process***: One possible initial stage of precipitation formation in which ice crystals grow in preference to water droplets in a mixed (i.e. water and ice) cloud.

***Black body***: A hypothetical body which is both a perfect absorber and a perfect emitter of radiation.

***Boundary layer***: The atmospheric boundary layer is usually considered as the lowest ~1 km of the atmosphere where motion is strongly influenced by surface characteristics, predominantly frictional drag.

***Cloud***: A visible suspension of water droplets or ice crystals, collected together in an identifiable unit and with its base above the ground.

***Cloud seeding***: A method by which the precipitation process is artificially induced in a cloud by, for example, the injection of artificial condensation nuclei.

***$CO_2$ fertilisation***: The theory that increased $CO_2$ in the atmosphere may lead to an increase in the efficiency of photosynthesis possibly producing lush plant growth where the water supply is adequate.

***Conditional instability***: Consider a 'parcel' of air rising in the atmosphere. The range of temperature profiles for which a wet parcel is unstable and a dry parcel stable defines the region of conditional instability.

***Conduction***: Collisions between fast moving molecules (high temperature regions) and slow molecules (low temperature regions) cause the slower moving molecules to speed up. This mode of heat transfer through matter is known as conduction.

***Conductive capacity***: A measure of the ability of a substance to conduct heat, given as the product of the heat capacity and the square root of the thermal diffusivity. The higher the conductive capacity, the better the conduction.

***Continentality***: A measure of the extent to which a location is outside the moderating influence of the oceans.

***Convection***: A type of heat transfer which occurs in a fluid by the vertical movement of large volumes of the heated material by differential heating (at the bottom for the atmosphere) thus creating, locally, a less dense, more buoyant fluid.

***Convective instability***: When warm, dry air overlies cooler, moist air, separated by an inversion layer, lifting the air *en masse* causes the lower air to reach saturation rapidly and become unstable, whilst the upper air remains stable for a longer period of lifting. Thus lifting creates instability although the original air mass was absolutely stable.

***Convergence***: If a constant volume of fluid has its horizontal dimensions

424

decreased it experiences convergence and, by conservation of mass, its vertical dimension must increase.

***Coriolis force***: A force experienced by any object moving over the surface of a rotating body such as the Earth. It serves to modify the direction of travel, causing a turning to the right (left) in the Northern (Southern) Hemisphere.

***Cryosphere***: The Earth's snow and (sea and land) ice masses.

***Detritus***: A collection of debris from the erosion of rocks, the remains of animals or of plants.

***Dewpoint temperature***: The temperature at which an air parcel would become saturated if it were cooled without a change in pressure or moisture content.

***Divergence***: If a constant volume of fluid has its horizontal dimensions increased it experiences divergence and, by conservation of mass, its vertical dimension must decrease.

***Downdraught***: A downward movement of air in the lee of an obstruction. Also, in thunderstorms a downdraught occurs when cold air descends from a thundercloud, 'hits' the ground and spreads out as a 'wedge' ahead of the storm.

***Downwash***: A downward movement of air caused by the negative pressure region behind a narrow obstacle.

***Dust veil index***: A scale which ranks volcanic dust veils in terms of mass of ejected material, duration and maximum extent of spread of the veil. The Krakatoa eruption (1883) is ranked as 1000 while the eruption of Mount Agung (1963) is 800.

***Easterly wave***: Troughs of low pressure sloping away eastward with height that move slowly westward. These occur in the region of the ITCZ.

***Eccentricity***: A measure of the deviation of an ellipse from a circle. It is the distance between the two foci divided by the length of the major axis. The eccentricity of the Earth's orbit is 0.018 at present.

***Ecliptic***: The great circle the Sun appears to describe on the celestial sphere. It is inclined at 23.5° to the equator at present.

***Eddy***: A disturbance in a flow of fluid which looks like a 'whirlpool'. Transient eddies move in space (e.g. cyclones), whilst stationary eddies remain in fairly fixed locations (e.g. anticyclones).

***Ekman spiral***: The change in the horizontal direction of fluid flow with the vertical coordinate. For the case of winds, there is a slow clockwise turning on ascent (anticlockwise in the Southern Hemisphere).

***Electrical conductivity***: The ability of a substance to conduct electricity.

***El Niño***: The reversal in direction of the Walker circulation, associated with replacement of the cool upwelling Peruvian coastal current by warm equatorial water which, in turn, leads to heavy rainfall in the normally arid desert. Fisheries off the Peruvian coast are adversely affected during these periods of suppression of cold coastal upwelling.

***Emissivity***: The degree to which a real body approaches a black body radiator.

***Emittance***: The rate at which radiation is emitted from unit area.

***Equivalent barotrophic***: The atmosphere has temperature gradients such that the isotherms are parallel to the isobars.

***Evapotranspiration***: The combined process of evaporation and transpiration.

***Foraminifera***: A family of small unicellular marine animals which secrete calcareous shells or 'tests' which are frequently preserved as fossils.

**Front**: A region of steep horizontal temperature gradients along the plane where two air masses having different origins and characteristics meet.

**Gaia hypothesis**: The hypothesis that the Earth's physical and biological systems are considered to be a complex and self-equilibriating entity.

**Gaussian distribution**: This, also called the normal distribution, is characterised by a bell-shaped, symmetrical curve, having its mean, mode and median at the point of symmetry.

**GCM**: A General Circulation Model is one in which the three-dimensional general circulation of the Earth's atmosphere, and sometimes oceans, is modelled.

**Geopotential heights**: The acceleration due to gravity, $g$, is approximately constant over the surface of the Earth since its magnitude is determined by the distance from the centre of the Earth. However, higher in the atmosphere this means that the value of $g$ decreases. Hence the use of formulae, such as the hydrostatic equation, at different heights in the atmosphere introduces a new variable: $g'(z)$. To obviate this variability with height a new height scale is introduced (the geopotential height) which is defined as the work done when lifting a body of unit mass against gravity (i.e. acceleration due to gravity multiplied by distance) divided by the value of $g$ at the Earth's surface.

**Geostrophic wind**: In the free atmosphere, when only the pressure gradient and Coriolis forces act on a parcel of air, they rapidly come into equilibrium to give the balanced or geostrophic wind, which blows parallel to the isobars.

**Greenhouse effect**: The effect whereby the Earth is warmed more than expected due to the atmospheric gases being transparent to incoming solar radiation, but opaque to outgoing terrestrial infrared radiation. The infrared radiation emitted from the surface is absorbed and re-emitted, some downwards, warming the Earth.

**Hadley cell**: A direct, thermally-driven circulation which comprises an upward motion of air at the ITCZ and downward motion in the subtropics, poleward movement of air at high levels and an equatorward movement at low levels. The ascending motion is the combined result of convergence, of radiative imbalances and of latent energy released during cloud condensation.

**Harmonic analysis**: Periodic features are represented by summation of simple sine and cosine waves.

**Heat capacity**: The heat required to change the temperature of unit volume of a body by 1 K.

**Heating (cooling) degree days**: Defined as the number of degrees by which the average daily temperature falls below (exceeds) a threshold or base temperature. The number of heating (cooling) degree days in a season is the summation of the heating (cooling) degree days for all days.

**Hurricane**: Storms with a mean surface wind speed exceeding 34 m s$^{-1}$ in the shape of an intense circular vortex. Hurricane is the regional name in the Caribbean and Gulf of Mexico, and typhoon in the western north Pacific and tropical cyclone in the Indian Ocean.

**Hydrostatic stability**: The atmosphere is hydrostatically stable if vertical accelerations are negligible and hence there exists a hydrostatic balance between the vertical pressure force and the gravitational force.

**Hygrometer**: Instrument for measuring humidity.

**Hythergraphs**: Plots of precipitation and temperature or humidity and temperature, usually by month.

**Ice age**: A period of time when the Earth's cryosphere is approximately double its present area.

**Index cycle**: The term used to describe alternation between periods of zonal and meridional flow. The 'index' is the pressure difference between two latitude zones.

**Interglacial**: The period between two ice ages.

**Intertropical convergence zone**: The ITCZ is a region close to the equator where the trade winds converge. Ascent of air causes low atmospheric pressure, deep convective clouds and heavy precipitation. The ITCZ changes position during the year following the seasonal insolation cycle.

**Inversion**: The situation when the environmental temperature increases with height. It is a highly stable condition in which convection and vertical dispersion cannot occur and pollution is trapped, often near the surface.

**Isotopic fractionation**: The separation of isotopes. Isotopes are atoms with the same atomic number (i.e. the same element) but a different number of neutrons (i.e. having a different weight from other isotopes of that element).

**Irradiance**: The rate at which radiation is incident upon a unit area.

**Jet stream**: A 'ribbonlike' belt of rapidly moving air found at or just below the tropopause. In mid-latitudes it is the 'core' of the Rossby or planetary waves.

**Katabatic wind**: A valley wind system developed on clear nights when regional pressure gradients are weak. Longwave radiant cooling of the upper valley slopes leads to downslope flow.

**Kinetic energy**: The energy of movement of a body. It is given by half the product of the mass and the square of the linear velocity of the body.

**Lapse rate**: The rate of decrease of temperature with height at a given time and place, i.e. a negative temperature gradient is a positive lapse rate.

**Lysimeter**: An instrument for measuring the amount of water lost by evapotranspiration.

**Milankovitch periodicities**: Periodic changes in the Earth's orbital parameters which are believed to control to some degree the onset of ice ages.

**Mixing ratio**: More strictly the 'water vapour mixing ratio', which is the ratio of the mass of water vapour to the mass of dry air occupying the same volume.

**Momentum**: The product of mass and velocity of a body.

**Monsoon**: A seasonal reversal of wind which in the summer season blows on-shore, bringing with it heavy rains, and in the winter season blows off-shore. The name is derived from the Arabic word 'mausin', meaning a season.

**Nephanalysis**: A cartographical representation of cloud amount and, often, cloud type. From the Greek 'nephos', meaning cloud.

**Noise**: Random fluctuations in a parameter caused by effects other than the one being studied.

**Oasis effect**: Hot dry air in equilibrium with the desert, flowing across an oasis edge, experiences rapid evaporation using sensible heat from the air as well as radiant energy. The air is cooled by this process until it reaches equilibrium with the new surface.

**Occlusion**: As a depression develops, the trailing cold front moves more rapidly than the leading warm front. The warm sector between them is narrowed until the cold front catches up with the warm front when occlusion takes place and the warm sector is effectively lifted off the ground.

***Optical thickness***: A measure of the attenuation of the solar radiation by the atmosphere due to scattering and absorption processes. The greater the optical thickness the greater the attenuation.

***Orographic***: Pertaining to the relief of mountains and hills.

***Photochemistry***: The study of chemical reactions which take place when substances are exposed to electromagnetic energy, especially ultraviolet and short wavelength visible radiation.

***Photosynthesis***: The process by which plants use light, absorbed through chlorophyll, to produce organic compounds from carbon dioxide and water, leaving oxygen as a byproduct.

***Planck curve***: The curve describing the amount of energy being radiated as a function of wavelength by a black body at a fixed temperature.

***Potential energy***: The energy a body has by virtue of the work done against a restoring force in attaining its position – i.e. being raised above the ground against the force of gravity. It is the product of the mass of the body, its height above zero (ground level) and the acceleration due to gravity.

***Potential temperature***: The temperature an air parcel would have if brought adiabatically to a standard pressure of $10^3$ hPa.

***Psychrometer***: Dry and wet bulb thermometers in a frame or sling which are ventilated so that a steady stream of air passes over the bulbs. The humidity parameters can be calculated from the two temperature readings.

***Radiation fog***: When net radiation is negative the air in contact with the ground is cooled. If there is sufficient moisture in the air or the cooling is sufficient, condensation will occur.

***Radiosonde***: A package, suspended below a balloon, consisting of instruments to sense and relay temperature, humidity and pressure as it ascends.

***Rain day***: A 24-hour period, usually starting at 0900Z during which 0.2 mm or more of precipitation falls.

***Rainout***: Pollution particles which act as condensation nuclei in the precipitation formation process are said to reach the ground by rainout.

***Rawinsonde***: A more sophisticated version of the radiosonde which also measures wind speed and direction.

***Regression***: Statistical derivation of a relationship between a dependent variable and one or more independent variables.

***Return period***: The probable time period between the repetition of two extreme events.

***Rossby wave***: When the temperature gradient across a rotating fluid (such as in an annulus experiment) reaches a critical value the previously symmetric flow breaks down into a wave-like flow; the waves being termed Rossby waves after the Swedish meteorologist who first recognised the importance of transient mid-latitude disturbances to the general circulation of the atmosphere.

***Sahel***: A region in Africa south of the Sahara Desert with a marginal climate for agriculture.

***Satellite (polar orbiting)***: A satellite whose orbit is approximately sun synchronous, low altitude (~1000 km) and intersects the equator at approximately $\pi/2$ thus passing close to the poles on each orbit.

***Satellite (geostationary)***: A satellite whose high altitude orbit (~35,000 km) is in the equatorial plane and orbital velocity matches that of the Earth so that its position remains constant with respect to the Earth.

**Scanning radiometer**: An instrument, generally on board a satellite or aircraft, which scans along a line perpendicular to the flight path detecting the radiant energy.

**Scattering**: The process by which some of a stream of radiation is dispersed to travel in all directions by particles suspended in the medium through which the radiation passes. Usually refers to solar photons – e.g. Rayleigh or Mie scattering.

**Shear instability**: In a stably stratified atmosphere a wave-like motion can develop when the vertical wind shear exceeds a critical value defining the onset of shear instability.

**Shelter belt**: An increase in surface roughness, such as that caused by a belt of trees, causes a reduction in the wind speed and thus provides protection to areas downwind.

**Solar constant**: The amount of energy passing in unit time through a unit surface perpendicular to the Sun's rays at the outer edge of the atmosphere at the mean distance between the Earth and the Sun. Currently believed to be $1370 \text{ W m}^{-2}$.

**Solar radiation**: The radiation the Earth receives from the Sun at wavelengths between 0.3 and 4.0 $\mu$m.

**Southern Oscillation**: A fluctuation in the intertropical atmospheric and hydro-dynamical circulations which manifests itself as a quasi-periodic (2–4 year) variation in sea-level pressure, surface wind, sea-surface temperature and rain-fall over a wide area of the Pacific Ocean. It is dominated by an exchange of air between the southeast Pacific subtropical high and the Indonesian equatorial low.

**Spectral analysis**: A type of Fourier (harmonic) analysis which identifies cycles in atmospheric features.

**Squall line**: A series of thundercells (cumulonimbus towers) aligned at right angles to the direction of motion. A leading cell is often 'fed' by adjacent cells so that the storm can persist for longer than individual cells.

**Standard deviation**: A measure of the scatter of observations (points) about the mean.

**Synoptic scale**: Literally 'at the same time'; usually pertaining to waves or eddies with horizontal scales of hundreds to thousands of kilometres. Often used to describe features (such as depression systems and anticyclones) which control the day-to-day variations in the weather.

**Tephigram (Tϕgram)**: A form of thermodynamic diagram in which the axes are logarithms of potential temperature (or entropy) and temperature, such that the dry adiabats (which are lines of constant potential temperature) are straight. The graph is usually rotated through 45° so that lines of constant pressure are approximately horizontal.

**Terrestrial radiation**: The radiation emitted by the Earth which is also known as longwave radiation.

**Thermal conductivity**: The rate at which a substance can conduct heat.

**Thermal diffusivity**: A quantity defining the ease with which a substance propagates temperature differences, being the thermal conductivity divided by the product of the specific heat capacity and the density of the substance.

**Thermal relaxation (or response) time**: A measure of the time taken by a system to achieve a new equilibrium temperature following an imposed perturbation.

**Thermal wind**: A thermally induced gradient in wind speed.

**Thermodynamic diagram**: A diagram in which the thermodynamic variables may be plotted as functions of two other variables; for example, pressure as a function of temperature and potential temperature.

**Thermodynamics**: The science of the relationships between different forms of energy; particularly the ways in which energy can be converted from one form to another and work be done.

**Time (of observations)**: By international agreement many meteorological and climatological observations are made at fixed times in a 24-hour system identical to Greenwich Mean Time and denoted by the hour number followed by an upper case Z (e.g. 0900Z)

**Trade winds**: The quasi-geostrophic north (south) easterly flow towards the ITCZ in the Northern (Southern) Hemisphere which is the surface 'return flow' of the Hadley cell circulations.

**Transmittance**: The ability of a substance to allow radiant energy to pass through it.

**Transpiration**: The removal of water from the interior of a plant through pores (stomata) located predominantly in the leaves.

**Urban heat island**: The effect of an urban area on the regional temperatures such that the city is a few degrees warmer than its surroundings.

**Vapour pressure**: The force per unit area created by the motions of the vapour molecules treated in isolation from all other gases in the atmosphere.

**Veering**: The wind direction changing in a clockwise sense.

**Venturi effect**: The increased velocity of a fluid flowing through a constriction, such as the funnelling of air between two converging buildings. A consequence of the requirement to conserve mass.

**Vorticity**: Twice the angular velocity of a fluid particle about a local axis through the particle. It is thus a measure of rotation of an air mass.

**Walker circulation**: A thermally driven longitudinal cellular circulation extending across the Pacific Ocean from Indonesia to close to the Peruvian coast and forming a component of the Southern Oscillation.

**Washout**: Pollution particles removed from the atmosphere by being incorporated into falling rain drops by collision and coalescence.

**Whiteout**: The multiple scattering of sunlight between the base of a low cloud layer and a snow-covered surface making it difficult to see surface features and to judge where the horizon is located.

**Wind chill**: The cooling of a body caused by the wind removing sensible and latent heat. Thus the 'perceived' temperature will be lower than the actual thermometer reading.

**Wind shear**: A condition where the wind speed (and usually direction) changes with height.

**Zenith angle**: The angular distance between the Sun's position in the sky and the local vertical (or zenith).

# Appendix: SI Units

The Système Internationale (SI), is based on seven independent fundamental units. These are the units of length, the **metre**; mass, the **kilogram**; time, the **second**; temperature, the **kelvin**; electric current, the **ampere**; amount of substance, the **mole**; and luminous intensity, the **candela**. The last three units have not been needed in material presented in this book. SI units of other quantities, such as force (the **newton**), pressure (the **pascal**), power (the **watt**), and energy (the **joule**), are defined in terms of the fundamental units. Multiples or fractions of units, in powers of ten, are designated by the following prefixes:

| | | | | |
|---|---|---|---|---|
| p | pico- | one trillionth (US) | $10^{-12}$ | 0.000 000 000 001 |
| n | nano- | one billionth (US) | $10^{-9}$ | 0.000 000 001 |
| $\mu$ | micro- | one millionth | $10^{-6}$ | 0.000 001 |
| m | milli- | one thousandth | $10^{-3}$ | 0.001 |
| c | centi- | one hundredth | $10^{-2}$ | 0.01 |
| d | deci- | one tenth | $10^{-1}$ | 0.1 |
| da | deca- | ten | 10 | 10 |
| h | hecto- | one hundred | $10^{2}$ | 100 |
| k | kilo- | one thousand | $10^{3}$ | 1 000 |
| M | mega- | one million | $10^{6}$ | 1 000 000 |
| G | giga- | one billion (US) | $10^{9}$ | 1 000 000 000 |
| T | tera- | one trillion (US) | $10^{12}$ | 1 000 000 000 000 |

## Table of SI Units and equivalents

| SI units | Conversions |
|---|---|
| | **Length** |
| metre (m) | 1 m = 3.28 ft; 1 ft = 0.304 m |
| centimetre (cm) | 1 cm = 0.3937 in; 1 in = 2.54 cm |

## Length

| | |
|---|---|
| kilometre (km) | 1 km = 0.621 mi; 1 mi = 1.61 km |
| micrometre ($\mu$m) | 1 $\mu$m = $10^{-6}$ m = $10^{-4}$ cm |
| | = $3.94 \times 10^{-5}$ in |
| | 1 Ångstrom unit (Å) = $10^{-10}$ m = $10^{-4}$ $\mu$m |
| | one degree of latitude = 111.1 km |
| | = 69.1 mi |
| | = 60 nautical miles |
| | One nautical mile = 1.15 statute miles |
| | = 1.85 km |

## Mass

| | |
|---|---|
| kilogram (kg) | 1 kg = 2.20 lb; 1 lb = 0.454 kg |
| gram (g) | 1 g = 0.0353 oz; 1 oz = 28.35 g |
| metric tonne | 1 tonne = $10^3$ kg = 1.10 short ton |
| | 1 short ton = 0.907 tonne |

## Temperature

| | |
|---|---|
| kelvin (K) | 1 K = 1 °C = 1.8 °F; |
| | 1 °F = 5/9 °C = 5/9 K |
| | $T$ °C = $(T + 273.16)$ K |

## Area

| | |
|---|---|
| square metre (m²) | 1 m² = 10.76 sq ft |
| | 1 sq ft = 0.093 m² |
| square centimetre (cm²) | 1 cm² = 0.155 sq in |
| | 1 sq in = 6.45 cm² |
| square kilometre (km²) | 1 km² = 0.386 sq mi |
| | 1 sq mi = 2.59 km² |
| hectare (ha) | 1 hectare = 10,000 m² |
| | = 2.47 acres |
| | 1 acre = 43,560 sq ft |
| | = 0.4047 hectares |

## Volume

| | |
|---|---|
| cubic metre (m³) | 1 m³ = 35.3 cu ft |
| | 1 cu ft = 0.028 m³ |
| cubic centimetre (cm³) | 1 cm³ = 0.061 cu in |
| | 1 cu in = 16.39 cm³ |
| | 1 litre (l) = $10^{-3}$ m³ = $10^3$ cm³ |
| | = 0.264 gal (US) |
| | = 0.220 gal (UK) |

## Volume

$$1 \text{ gal (US)} = 231 \text{ cu in}$$
$$= 0.0037853 \text{ m}^3$$
$$1 \text{ gal (UK)} = 277 \text{ cu in}$$
$$= 0.0045460 \text{ m}^3$$

## Speed

metre per second
(m s$^{-1}$)
centimetre per second
(cm s$^{-1}$)

$$1 \text{ m s}^{-1} = 2.24 \text{ mi h}^{-1} = 1.94 \text{ kt}$$
$$1 \text{ mi h}^{-1} = 0.447 \text{ m s}^{-1}$$
$$1 \text{ cm s}^{-1} = 0.0328 \text{ ft s}^{-1}$$
$$= 1.97 \text{ ft min}^{-1}$$
$$1 \text{ ft s}^{-1} = 30.48 \text{ cm s}^{-1}$$
$$= 0.592 \text{ kt}$$
knot (kt)
$$1 \text{ knot} = 1 \text{ nautical mile per hour}$$
$$= 0.515 \text{ m s}^{-1}$$
$$= 1.15 \text{ mi h}^{-1}$$

## Work and energy

joule (J)

$$1 \text{ joule} = 10^7 \text{ ergs}$$
$$1 \text{ calorie (cal)} = 4.186 \times 10^7 \text{ ergs}$$
$$= 4.186 \text{ joules}$$
$$1 \text{ ft lb} = 1.36 \text{ joules}$$
$$1 \text{ joule} = 0.738 \text{ ft lb}$$
$$1 \text{ BTU} = 1055 \text{ joules}$$

## Power

watt (W)

$$1 \text{ W} = 1 \text{ J s}^{-1} = 1.34 \times 10^{-3} \text{ hp}$$
$$1 \text{ horsepower (hp)} = 33{,}000 \text{ ft lb min}^{-1}$$
$$= 746 \text{ W}$$
$$1 \text{ ly min}^{-1} \quad (= 1 \text{ cal cm}^{-2} \text{ min}^{-1})$$
$$= 6.98 \times 10^2 \text{ W m}^{-2}$$

## Pressure

pascal (Pa)

$$1 \text{ Pa} = 1 \text{ N m}^{-2} = 10 \text{ dynes cm}^{-2}$$
$$= 2.45 \times 10^{-4} \text{ lb in}^{-2}$$
$$1 \text{ mbar} = 10^2 \text{ Pa} = 1 \text{ hPa}$$
$$= 10^3 \text{ dynes cm}^{-2}$$
$$= 0.750 \text{ mm Hg}$$
$$= 2.95 \times 10^{-2} \text{ in Hg}$$
$$1 \text{ mm Hg} = 133.3 \text{ Pa}$$
One standard atmosphere (atm) $= 101{,}325 \text{ Pa} = 1013.25 \text{ hPa} = 1013.25 \text{ mbar}$
$$= 760 \text{ mm Hg} = 29.92 \text{ in Hg} = 14.7 \text{ lb in}^{-2}$$

# Index

Page numbers in italics refer to pages on which cited material is to be found in Tables and/or Figures

absolute humidity 99
absorption (by gases) 59
absorptivity 60
acid rain 324–6, *325*
adiabatic lapse rate 112
adiabatic process 111
advection 172
advection fog 111
aerodynamic resistance 276
aerosols 351, *352; see also* air pollution, volcanic eruptions
agriculture, sensitivity to climate of 400–4
airflow around buildings 292–5, *292–5*
air mass climatology 243–8, *247*
air mass source regions 243, *246*
air mass thunderstorms 144
air masses 185
air pollution 318–26, *319–21, 323, 325*
albedo 45
  cloud *127*
  planetary *32, 67, 70*
    seasonal variation of *68*, 140, *156–7, 264*
  surface *32*, 46–8, *47*, 69, 73, 285
anabatic wind *300*, 303, *304*
anemometer *15*
angular momentum 159, *199–200*
annulus (dishpan) experiment 180–1, *181*
anticyclones 240–3, *244, 245*
Arctic basin (modification of drainage into) 262

atmosphere, composition of 6, *7*
atmospheric pollution *see* air pollution
atmospheric pressure 159–64
  global distribution *161*
  measurement of 161–4
  reduction to sea level 160
atmospheric stability 114–20, 173, *320, 321*
atmospheric 'window' 60
axial tilt (of the Earth) 40

backing wind *170*, 172
baroclinic atmosphere 171–3
barometer 163, *163*
barotropic (and equivalent barotropic) atmosphere 170–1
Bergen 'school' 238
Bergeron–Findeisen mechanism (for precipitation formation) 131
black body (for radiation) 35
blocking anticyclone *182*, 183
boundary layer of the atmosphere 273–8
  wind profile in 273–4
Bowen ratio 95

Campbell Stokes sunshine recorder *51*, 52
carbon budget 406–7, *407*
carbon dioxide 4
  increase of *353*, 354, 404–15, *405; see also* greenhouse effect
centrifugal force 168

434

Chinook 307
chlorofluoromethanes (from spray cans) 353
Clausius–Clapeyron equation 99
clear air turbulence (CAT) 118, 120
climate classification 204–14
climate models *see* models
climate, multidisciplinary approach 6
climate system 18–21
climate zones 8, *197*, 204–5
climatic elements 13–18
climatic normal 8, 56
climatic optimum 394
climatic record, historical 394–7, 399–404, *399*
climatic region 8–9, *196*, 210–14, *213*
climatic variations
    cultural impact of 333–40, 397–404, *399*
    human impact of 333–40, *334*, *340*
climatology, development of 7–12
cloud feedback 374–5, *374*
cloud formation 105–14, *105*, *106*
cloud seeding 131, 328–9
clouds
    classification 103–5, *106–10*
    climatic impact of 352, *353*
    condensation nuclei 128–30, *128*, *129*
    observation of 120–7, *245*
    temperatures 130–1
collision and coalescence mechanisms (for precipitation formation) 130
condensation 113, *115*
    nuclei (CCN) 128, *128*
conduction 77–9
conductive capacity 79
constant pressure surfaces 160, *162*
continentality 79, 195–8, 255–7
convection 79, 114, 116, 310–15, 367
convective instability 116, *117*
convective storms 310–14; *see also* thunderstorms
convergence 174–5
Coriolis force 164, *165*, *166*
crop growth 85–6
cryosphere 192–5; *see also* ice-albedo feedback
    effect on atmospheric circulation 192–5, 257–62

extent of *193*, *194*
    seasonal variation in 193
cyclogenesis (depression system development) 238–40
cyclone
    mid-latitude *see* depression system
    tropical *see* hurricanes

deforestation 267
depression system 183, 238–40, *241*, *242*
desert climates 222–6
    interior 252
desertification 265–7
dew 139
dew point temperature 100, *117*
diffusograph 50
divergence 174–5
Doldrums 218
dry bulb temperature *80*, 101
dust bowl, USA 10
dynamic climatology 24

Earth's orbit (around Sun) 39–40, 344–6, *345*
easterly jet 236
easterly waves 219–20, *220*
eccentricity of the Earth's orbit 39, *39*
eddy flux (of energy) *179*
effective temperature (human comfort) 330–3, *332*, *333*
Ekman spiral 169
El Niño 11, 187–9, *188*, 226
electromagnetic spectrum 31, *33*
emissivity, surface *47*, 59, 73
energy
    kinetic *20*, 21
    potential *20*, 21, 31
    radiant 29–62
energy balance (of the globe) 31, *32*, 66–73, *70–2*
energy budget (of surface) 76–7
energy cascade 20, *20*, 29, *30*
energy industry 53–8, 86–8, 147
environmental lapse rate 114–20
equatorial trough 218
evaporation 79, 93–7, 296; *see also* evapotranspiration rate
evaporation pan 96, *96*

evapotranspiration rate 94–6
  actual 96
  global distribution *98*
  measurement 96–7, *96*, *97*
  potential 95–6

fall velocity 130
fallout, of particulate material 322
feedback effects 354, 373–7, *374*; *see also*
      ice–albedo feedback, cloud
      feedback
first law of thermodynamics 111
fog 111, *217*, 259
  dissipation of 329
Föhn 307, *308*
freeze-free season (agriculture) *86*, *87*,
      *381*, 400–2
freons *see* chlorofluoromethanes
friction layer 169
frictional force 169
frontal uplift 114
frost hollow 300
frost protection 326–8, *327*

Gaia hypothesis 388
Gaussian (chimney) plume model 321
general circulation
  fluxes associated with *198–200*
  of the atmosphere 175–81
  role of 155–9
geological record (of climate) 387–94,
      *388*, *389*
geostationary (satellite) 64–5, *65*
geostrophic wind 166–8, *167*
gradient wind 168, *169*
grain imports, USSR 10
greenhouse effect 60–2, *61*, 344, 404–15

Haboob *308*
Hadley cell circulation 178–80, 214–32
hail 313–15, *314*
  formation *219*
Harmattan 310
haze 130
heating degree days *58*, 88
historical analogues *see* models
human comfort (response to climate)
      329–33

humidity
  absolute 99
  relative 99
hurricane 226–32, *227*, *228*, *230*, *231*, *232*
  effect of *232*
hydrological cycle 93, *94*, *150*
hydrometeors *133–4*
hydrostatic equation 112, 168
hygrometer
  dew point 101
  resistance 102

Ice Ages 387–91, *393*
ice–albedo feedback 195, 373
ice, on power lines 147
ice storms 313, *315*
index cycle 183–4, *182*, *184*
infrared radiation 59–62
  seasonal variation of *68*, 154, *156–7*
Intertropical Convergence Zone (ITCZ)
      178, 214, *218*, 220–2, *221*
inversion (of temperature), *63*, 116, *216*,
      259, *321*
isobars 160
isotopes *347*, *392*

jet stream 176, *177*, *178*, 181–5

Karman vortex streets *292*, *293*
katabatic wind 299, *300*
Kirchhoff's Law 35
Köppen classification 210–14, *211*, *212*,
      *213*

lapse rate 63, 111–9, *118*, *119*
level of free convection 116
lifting condensation level 113, *113*, *115*
lightning 145
Little Ice Age *347*, 395, *396*
local climates 271–317
local winds *see* winds
longwave radiation *see* infrared radiation
lysimeter 96, *97*

Man-made climate change 262–7,
      326–9; *see also* air pollution
  inadvertent 317–21

Maunder minimum 347, *348*
measurements 13–17
  accuracy 17
  contact 13
  remote sensing 13
Mediterranean climates 248–9
meridional circulation 175–7, *179*
mesosphere 62, *63*
Mie scattering 43, *43*
Milankovitch theory 344–7, *347*
Mistral *309*
mixed layer (of ocean) 185
mixing ratio 99
models
  crop yield models 378–82, *381*
  economic and regional yield models
    380–2
  energy balance models (EBMs) 357,
    363–6
  general circulation models (GCMs)
    357, 369–73, *370, 371*
  historical analogue models 410–13,
    *413, 414*
  hydrological models 384–6
  impact of doubling $CO_2$ 406–11,
    *408–10*
  local models 377–86
  oceanic circulation 189, *190*
  radiative-convective (1-D) 357, 366–8
  regression 52–3, *53*, 88
  simple global model 357–9
  statistical significance of 413–15
  two-dimensional 357, 368–9
  types of 356–9, *357*
modification of climate *see* Man-made
  climate change
moisture index 210
momentum transfer (between ocean and
  atmosphere) 187
monsoon circulation 232–7

net radiation (at surface) 73
neutral atmosphere 117
North America (climatology of) 252–5

oasis effect 279–81, *281*
occlusion *239, 242*

oceanic circulation (modelling of)
  189–90
oceans (effect on climate) 185–92,
  362–3, 372–3
optical depth 46
optical thickness (of the atmosphere) 46
orbital variations (of the Earth) *see*
  Earth's orbit
orographic uplift 113
ozone (in the stratosphere), 62–3, 352,
  *353*

palaeoclimatology 386–97
perception of climate 23, *23*
permafrost 258
peroxyacetyl nitrate (PAN) 324
photochemical smogs 324
photosynthesis *30*, 48–9
physical climatology 24
Planck's Law 33, 35
planetary waves *see* Rossby waves
Pleistocene *194*, 391, *396*
  reconstruction of Earth surface *393*
Poisson's equation 120
polar atmospheric circulation 180,
  257–62
polar cell *175*, 176, 180
polar front 181, *182*
polar orbit (satellite) 63–4, *65*
polynyi 262
potential evapotranspiration (PET)
  96
potential temperature 119
precipitable water vapour 102
  global distribution of *104*
precipitation
  formation 127–32
  global distribution *140*
  intensity *135*
  measurement of 13, 134–9, *136–7*
  pH of *see* acid rain
  types of 132–4
pressure *see* atmospheric pressure
pressure gradient force *163*
primary circulation features 153, 160
proxy climatological data 18, 386–7
pseudo-adiabatic chart 120, *121*
psychrometer 101

pyranometer 50, *50*
pyrheliometer 50

radar (for precipitation monitoring) 137, *138*
radiation fog 111
radiation instruments *50–1*
radiative equilibrium (temperature) 75
radiosonde 17, *118*
rain day 141
rain gauge 13, *136, 137*
rainout (of particulates) 322
Rayleigh scattering 42, *43*
reflectance *see* albedo
registration (of satellite images) 17
regression models *see* models
relative humidity 99, 128, *129*
remote sensing *see* satellites
residence time (of particulates) *323*
response time (of climate system) 19, 189
Rossby waves 176, *177*, 181–5, *181, 182*
rotor *305*
roughness length 274, *275*, 281–3
runoff *298*

Sahel drought 4, 10, *225*
Santa Anna 310
satellites 11, 13, 62–6, *65*, 189–92, *191*, 260–2, *260*, 262–5, *264, 303*
saturation vapour pressure 99, *100*
scales (time and space) 21, *22*
sea breeze 305–6, *306*
sea-surface temperature (observations of) 189–92, *191*
secondary circulation features 153, 160
shelter belts *282*, 283
shelter temperature 14
SI (Système Internationale) (and conversions) Appendix 431–3
smog 324
snow 46, *133–4*, 192–5, *192*; *see also* ice–albedo feedback
  clearance strategies 146–7
  depth 138–9
  line *192*
snow fences 283
solar constant 38; *see also* solar radiation

solar heating/cooling 53–8, *55, 58*
solar radiation 5, 36–41, *36, 37, 41*
  at the surface 296, *297*
  radiation measurement 50–3
  seasonal variation in *68*, 154, *156–7*
  variation in 344–9
solute effect (in clouds) 128–9
Southern Oscillation *188*, 189, 226
specific heat 75, *78*
specific humidity 103
spectral distribution of shortwave irradiance 49
spectral reflectance *48*
squall line (of thunder cells) 144, *145*
stack height 319, *320*
Stefan–Boltzmann constant 35
Stefan–Boltzmann Law 35, 59, 75
Stevenson's screen 14, *15*, 81
Stokes' Law 130
stomatal resistance 276
stratosphere 62, *63*
Sun (energy source within the) 36–8, *36, 37*
sunset colour 42
sunshine recorder *51*, 52
sunspots 347–9, *348*
surface energy budget 76–9, 272–84
surface observations, global distribution of stations for 16
surface temperatures (global) *84, 196*
surface type (effect on climatology) 278–84
synoptic scale 22, *22*

teleconnections 187–9
temperature
  atmospheric 62, *63, 64*
  diurnal variation *74*
  global distribution 81–3
  measurement of 14, *15*, 80–1, *80, 82–3*
  prediction/application of 83–9; *see also* models
  variation with depth *277*
temperature–humidity index 331, *332*
Tephigram (Tøgram) *see* thermodynamic diagram
terrestrial radiation *see* infrared radiation
thermal conductivity 77

thermal diffusivity 77
thermal properties of surfaces 77–9, *78*
thermal wind 171, *172*
thermodynamic diagram 113, *113, 117,* 120, *121, 216*
thermohydrograph *15*
thermometers 80–3
thermosphere 62, *63*
thickness (of atmospheric layer) 170
three-cell model (of general circulation), 175, *175*
thunderstorm 142–4, *143,* 206–8, *312*
trade winds 179
trade wind inversion 216
transitivity of climate system 360–2, *362*
transpiration 94
tropopause 62, *63, 64,* 116, 144, 182
troposphere 62, *63*
turbulence 79, 95, 118
typhoon *see* hurricane

ultraviolet radiation (absorption of) *33,* 62–3
urban climate 284–95
urban drain design 147–8
urban enhancement of precipitation *289*
urban heat island 287–9, *287, 288, 290*

vapour pressure 99
veering wind 172
Venturi effect *294*
volcanic eruptions 349–52, *398*
von Kármán's constant 274
vorticity (of the atmosphere) 174–5

Walker circulation 187–9
washout 322
water balance (of the surface) 148–9
water vapour (absorption by) *33,* 59–60
wavelength *see* electromagnetic spectrum
West Antarctic ice sheet 195
wet bulb temperature *80,* 101
whiteout 44
Wien's Law 35
wind chill (human comfort) 332–3
winds
    backing *170,* 172
    local 305–10
    near-surface 169, *170*
    veering 172

zenith angle 40